T0178567

D. Bump
J.W. Cogdell
E. de Shalit
D. Gaitsgory
E. Kowalski
S.S. Kudla

An Introduction to the Langlands Program

Joseph Bernstein
Stephen Gelbart
Editors

Birkhäuser
Boston • Basel • Berlin

Joseph Bernstein
Tel Aviv University
Department of Mathematics
Ramat Aviv
Tel Aviv 69978
Israel

Stephen Gelbart
The Weizmann Institute of Science
Department of Mathematics
Rehovot 76100
Israel

Library of Congress Cataloging-in-Publication Data

An introduction to the Langlands program / Joseph Bernstein, Stephen Gelbart, editors ;
[with contributions by] D. Bump ... [et al.].
p. cm.
Includes bibliographical references.
ISBN 0-8176-3211-5 (alk. paper) – ISBN 3-7643-3211-5 (alk. paper)
1. Automorphic forms. 2. L-functions. I. Bernstein, Joseph, 1945- II. Gelbart, Stephen S., 1946-

QA353.A9I59 2003
515'–dc21

2003043653
CIP

AMS Subject Classifications: 11Mxx, 11Fxx, 14Hxx, 22Exx

ISBN 0-8176-3211-5 Printed on acid-free paper.

©2004 Birkhäuser Boston

Birkhäuser 

All rights reserved. This work may not be translated or copied in whole or in part without the written permission of the publisher (Birkhäuser Boston, c/o Springer-Verlag New York, LLC, 175 Fifth Avenue, New York, NY 10010, USA), except for brief excerpts in connection with reviews or scholarly analysis. Use in connection with any form of information storage and retrieval, electronic adaptation, computer software, or by similar or dissimilar methodology now known or hereafter developed is forbidden.
The use in this publication of trade names, trademarks, service marks and similar terms, even if they are not identified as such, is not to be taken as an expression of opinion as to whether or not they are subject to property rights.

Printed in the United States of America. (MP)

9 8 7 6 5 4 3 2 SPIN 10995211

Birkhäuser is a part of *Springer Science+Business Media*

www.birkhauser.com

Contents

Preface ... vii

1. Elementary Theory of L-Functions I
E. Kowalski ... 1

2. Elementary Theory of L-Functions II
E. Kowalski ... 21

3. Classical Automorphic Forms
E. Kowalski ... 39

4. Artin L Functions
Ehud de Shalit ... 73

5. L-Functions of Elliptic Curves and Modular Forms
Ehud de Shalit ... 89

6. Tate's Thesis
Stephen S. Kudla ... 109

7. From Modular Forms to Automorphic Representations
Stephen S. Kudla ... 133

8. Spectral Theory and the Trace Formula
Daniel Bump ... 153

9. Analytic Theory of L-Functions for GL_n
J.W. Cogdell ... 197

10. Langlands Conjectures for GL_n
J.W. Cogdell ... 229

11. Dual Groups and Langlands Functoriality
J.W. Cogdell ... 251

12. Informal Introduction to Geometric Langlands
D. Gaitsgory ... 269

Preface

During the last half-century the theory of automorphic forms has become a major focal point of development in number theory and algebraic geometry, with applications in many diverse areas, including combinatorics and mathematical physics.

The 12 chapters presented in this book are based on lectures that were given in the School of Mathematics of the Institute for Advanced Studies at the Hebrew University of Jerusalem, March 12–16, 2001. The goal of these lectures was to introduce young researchers to the theory of automorphic forms, to explain its connection with the theory of L-functions, as well as to indicate connections to other fields of mathematics.

The central premise of the School was to formulate the Langlands program, which gives a very broad picture connecting automorphic forms and L-functions. To describe it, our lecturers used different technical methods to establish special cases of the program.

The Langlands program roughly states that, among other things, any L-function defined number-theoretically is the same as the one which can be defined as the automorphic L-function of some $\mathrm{GL}(n)$. In this loose way, every L-function is (conjecturally) viewed as one and the same object.

To introduce the theory of automorphic forms, we have not tried to give a complete formal introduction to the subject; this would require much more time. Instead, our lecturers concentrated on a variety of topics, and gave an "informal" personal view of number theory from the classical zeta function up to the Langlands program. We hope that the present book will be able to serve the same informal role.

The contribution of each of the lecturers and their articles may now be briefly described as follows.

The first eight chapters are devoted to the case of $\mathrm{GL}(1)$ and $\mathrm{GL}(2)$: E. Kowalski classically focuses on the basic zeta-function of Riemann and its generalizations to Dirichlet and Hecke L-functions, class field theory, and a selection of topics devoted to classical automorphic functions; E. de Shalit carefully surveys the conjectures of Artin and Shimura–Taniyama–Weil.

After discussing Hecke's L-functions, S. Kudla examines classical modular (automorphic) L-functions as $\mathrm{GL}(2)$ ones, thereby bringing into play the theory of representations. One way to study those representations which are "automorphic" is via Selberg's theory of the "trace formula"; this is introduced in D. Bump's chapter.

The last four chapters, by J. Cogdell and D. Gaitsgory, are more abstract. After starting with discussion of cuspidal automorphic representations of GL(2, (**A**)), Cogdell quickly gets to Langlands' theory for GL(n, (**A**)); then he explains why one needs the Langlands' dual group in order to formulate the general conjectures for a reductive group G different from GL(n).

Gaitsgory gives an informal introduction to the geometric Langlands program. This is a new and very active area of research which grew out of the theory of automorphic forms and is closely related to it. Roughly speaking, in this theory we everywhere replace functions—like automorphic forms—by sheaves on algebraic varieties; this allows us to use powerful methods of algebraic geometry in order to construct "automorphic sheaves."

The Editors are grateful to all six authors for their considerable skill in pulling these diverse pieces together. We also wish to thank C. J. Mozzochi for the photograph of Langlands that appears on the cover and Shlomit Davidzon for other graphic elements. Finally, we thank the Institute for Advanced Studies of the Hebrew University of Jerusalem—in particular, Dahlia Aviely, Smedar Danziger, Pnina Feldman, Shani Freiman, and Alex Levitzki, for making this Workshop possible.

Joseph Bernstein
Tel Aviv University
Ramat Aviv, Israel

Stephen Gelbart
Nicki and J. Ira Harris Professorial Chair
The Weizmann Institute of Science
Rehovot, Israel

December 2002

An Introduction to the Langlands Program

1
Elementary Theory of L-Functions I

E. Kowalski

1 Introduction

In this first chapter we will define and describe, in a roughly chronological order from the time of Euler to that of Hecke, some interesting classes of holomorphic functions with strange links to many aspects of number theory. Later chapters will explain how at least some of the mysterious aspects are understood today. But it should be emphasized that there are still many points that are not fully explained, even in a very sketchy, philosophical way.

I will particularly try to mention some of the more peculiar features of the theory of L-functions (and of automorphic forms) which arise from the point of view of analytic number theory. I will also give indications at the places where future chapters after mine will bring new perspectives.

The next chapter will develop the points presented here, and in particular will sketch proofs of some of the most important ones, especially when such a proof yields new insights into the theory.

The mathematicians whose name are most important for us now are Euler, Gauss, Dirichlet, Riemann, Dedekind, Kronecker, Hecke, Artin, and Hasse.

2 The Riemann zeta function

The first L-function has been given Riemann's name. This fact is convenient for us: it seems to call for some explanation since no one denies that other mathematicians, most notably Euler, considered this function before, and these explanations are the best entrance to our subject.

The function in question is defined by the series

$$\zeta(s) = \sum_{n \geqslant 1} \frac{1}{n^s}.$$

For integers $s \geqslant 1$, this was studied even before Euler, and even for $s \geqslant 2$, it is well known that Euler first found an exact formula (see (5.1)). However the starting point for the theory of L-functions is Euler's discovery that the existence and uniqueness of factorization of an integer as a product of prime powers implies that

$$\zeta(s) = \prod_p \frac{1}{1 - \frac{1}{p^s}}, \tag{2.1}$$

a product over all prime numbers. From the divergence of the harmonic series, Euler deduced from this a new proof of Euclid's theorem that there are infinitely many primes, and with some care be obtained the more precise formulation that

$$\sum_{p \leqslant X} \frac{1}{p} = \log\log X + O(1)$$

as $X \to +\infty$.[1] Thus (2.1) clearly contained new information about the distribution of prime numbers.

Riemann's insight [Rie] was to remark that the function $\zeta(s)$ thus defined, if s is taken to be a complex variable ($s = \sigma + it$ in his notation), is holomorphic in its region of convergence. This justifies looking for its (maximal) analytic continuation, but that too had been done before (see [We]). It is the combination of the Euler product expansion (2.1) and the analytic continuation given by the functional equation described below, which is the cause for all our rejoicing, as it reveals the strange "duality" between the complex zeros of $\zeta(s)$ and prime numbers.

To be more specific, Riemann stated that $\zeta(s)$ has a meromorphic continuation to the whole complex plane, the only singularity being a simple pole with residue 1 at $s = 1$, and that moreover this analytic continuation satisfied the following property, aptly named the *functional equation*: the function

$$\Lambda(s) = \pi^{-s/2}\Gamma(s/2)\zeta(s) \tag{2.2}$$

is meromorphic except for simples poles at $s = 0$ and $s = 1$ and satisfies for all $s \in \mathbf{C}$ the relation

$$\Lambda(1-s) = \Lambda(s). \tag{2.3}$$

From the simple poles of $\zeta(s)$ at $s = 1$ and of $\Gamma(s/2)$ at $s = 0$ one deduces in particular that $\zeta(0) = -\pi^{-1/2}\Gamma(1/2)/2 = -1/2$. Moreover, the other poles at $s = -2n$, $n \geqslant 1$ integer, of $\Gamma(s/2)$ show that (2.3) implies that $\zeta(-2n) = 0$, for $n \geqslant 1$: those zeros are called the *trivial zeros* of $\zeta(s)$.

This, and in fact all of (2.3) for integers $s \geqslant 1$, was already known to Euler in the language of divergent series!

---

[1] To avoid any controversy, here are the definitions of Landau's $O(\cdots)$ and Vinogradov's $\ll$ symbols: $f = O(g)$ as $x \to x_0$ means that there exists some (unspecified) neighborhood U of x_0, and a constant $C \geqslant 0$ such that $|f(x)| \leqslant Cg(x)$ for $x \in U$; this is equivalent to $f \ll g$ for $x \in U$ where now U is specified beforehand, and in this latter case one can speak of the "implicit constant" C in $\ll$. However we sometimes also speak of estimates involving $O(\cdots)$ being "uniform" in some parameters: this means that the U and C above can be chosen to be independent of those parameters.

The presence of the "gamma factor" in the functional equation was not well understood: indeed (2.3) was often written in completely different ways (see, e.g., [Ti, Chapter 2]). The more transparent and conceptual proofs of (2.3), and its various generalizations, by the Poisson summation formula (see the next chapter) and theta functions gave the gamma factor a clear *de facto* standing, but it is only by the time of Tate's thesis [Ta] that its role was made clear as the Euler factor at the archimedean place of **Q**. In general I should mention that an Euler product is a product over primes of the form

$$\prod_p L_p(s) \tag{2.4}$$

and that $L_p(s)$ is called the Euler factor at p.[2]

The interplay of the zeta function with primes was revealed by Riemann in the form of the so-called "explicit formula." His version (and that proved later by van Mangoldt and others) obscured somewhat the essential point in purely analytic difficulties; going around them by means of smooth test functions, one can state it in the following straightforward manner:

Proposition 2.1. *Let* $\varphi :]0, +\infty[\to \mathbf{C}$ *be a* C^∞ *function with compact support, and let*

$$\hat{\varphi}(s) = \int_0^{+\infty} \varphi(x) x^{s-1} dx,$$

be its Mellin transform, which is entire and decays rapidly in vertical strips. Let

$$\psi(x) = \frac{1}{x}\varphi(x^{-1}).$$

Then we have

$$\sum_p \sum_{k \geqslant 1} (\log p)(\varphi(p^k) + \psi(p^k)) = \int_0^{+\infty} \varphi(x) dx$$
$$- \sum_{\substack{\zeta(\rho)=0 \\ 0<\mathrm{Re}(\rho)<1}} \hat{\varphi}(\rho) + \frac{1}{2i\pi} \int_{(-1/2)} \left(\frac{1}{2}\frac{\Gamma'}{\Gamma}\left(\frac{s}{2}\right) - \frac{1}{2}\frac{\Gamma'}{\Gamma}\left(\frac{1-s}{2}\right) \right) \hat{\varphi}(s) ds. \tag{2.5}$$

A more general case of this will be sketched in the next chapter (Proposition 3.5).

In this general formula, which expresses sums over prime numbers in terms of sums over zeros of $\zeta(s)$, the first term on the right-hand side is really the contribution of the pole at $s = 1$ (very often, giving the main term in some asymptotic

[2] Be careful that when an Euler product is defined over prime ideals $\mathfrak{p}$ in a number field $\neq \mathbf{Q}$, as will be described below, the data of the Euler factors $L_\mathfrak{p}(s)$ is *stronger* than the product even if the latter defines a holomorphic function.

formula). We implicitly make use of the fact that $\zeta(s) \neq 0$ for $\mathrm{Re}(s) > 1$. It is very important to realize that this is by no means obvious from the series representation: it follows immediately from the Euler product expansion (an absolutely convergent infinite product is nonzero). This must not be underestimated: it is for instance the key analytical input in Deligne's first proof of the Riemann Hypothesis for varieties over finite fields [De, 3.8]. The functional equation then shows that $\zeta(s) \neq 0$ for $\mathrm{Re}(s) < 0$, except for the trivial zeros already identified.

Riemann immediately expresses that it is "likely" that all the (nontrivial) zeros of $\zeta(s)$ satisfy $\mathrm{Re}(s) = 1/2$. This is indeed the best possible situation if one is interested in prime numbers: if one takes for φ test functions which converge, say, in the sense of distributions to the characteristic function of $[0, X]$, one finds rather easily that one has an estimate for the error term in the prime number theorem of the form

$$\psi(X) = \sum_{p^k \leqslant X} (\log p) = X + O(X^{\beta^*}) \text{ as } X \to +\infty \tag{2.6}$$

if and only if $\beta^* > \sup\{\mathrm{Re}(\rho)\}$. Since the functional equation implies that $\sup\{\mathrm{Re}(\rho)\} \geqslant 1/2$, the Riemann Hypothesis is seen to be simply the statement that primes are distributed "in the best possible way."

Following the standard terminology, the strip $0 \leqslant \mathrm{Re}(s) \leqslant 1$, which must contain the nontrivial zeros is called the *critical strip* and the line $\mathrm{Re}(s) = 1/2$ is called the *critical line*. This will apply to all the L-functions in the first two chapters, but for automorphic L-functions, the critical strip may (depending also on normalization) be translated to the right by some amount.

3 Dirichlet L-functions

Although it is fundamental, the case of the Riemann zeta function does not exhibit some very important features of the general theory. Those, notably the notions of "conductor" and of "primitivity," and the link with class-field theory and algebraic number theory more generally, appear first in the case of Dirichlet L-functions.

Dirichlet defined those functions [Di] to prove his famous theorem:

Theorem 3.1. *Let $q \geqslant 1$ and $a \geqslant 1$ such that $(a, q) = 1$. Then there are infinity many primes $p \equiv a \pmod q$ and more precisely*

$$\sum_{\substack{p \leqslant X \\ p \equiv a \,(\mathrm{mod}\, q)}} \frac{1}{p} = \frac{1}{\varphi(q)} \log\log X + O(1)$$

as $X \to +\infty$.

He proved this result by detecting invertible congruence classes modulo q by means of harmonic analysis in $(\mathbf{Z}/q\mathbf{Z})^\times$, which lead him to Dirichlet characters:

an arithmetic function $\chi : \mathbf{Z} \to \mathbf{C}$ is a Dirichlet character modulo $q \geqslant 1$ if there exists a group homomorphism

$$\tilde{\chi} : (\mathbf{Z}/q\mathbf{Z})^\times \to \mathbf{C}^\times$$

such that $\chi(x) = \tilde{\chi}(x \pmod q)$ for $(x, q) = 1$ and $\chi(x) = 0$ if $(x, q) \neq 1$. To such a character, extending Euler's definition, one associates the L-function

$$L(\chi, s) = \sum_{n \geqslant 1} \chi(n) n^{-s} = \prod_p (1 - \chi(p) p^{-s})^{-1}$$

the last equation, the Euler product, being a consequence of unique factorization and of the complete multiplicativity of χ:

$$\chi(mn) = \chi(m)\chi(n) \text{ for all } m \geqslant 1,\ n \geqslant 1.$$

The orthogonality relations for characters of a finite group imply immediately the relation

$$\sum_{\substack{p \leqslant X \\ p \equiv a \pmod q}} \frac{1}{p} = \frac{1}{\varphi(q)} \sum_{p \leqslant X} \frac{1}{p} + \frac{1}{\varphi(q)} \sum_{\chi \neq 1} \overline{\chi}(a) \sum_{p \leqslant X} \frac{\chi(p)}{p}$$

and Dirichlet's Theorem is easily seen to be equivalent with the assertion that if $\chi \neq 1$ is a Dirichlet character, then $L(\chi, 1) \neq 0$. For Dirichlet (coming before Riemann) this is not in the sense of analytic continuation, but rather a statement about the sum of the (conditionally) convergent series which "is" $L(\chi, 1)$.[3] For the nontrivial character χ_4 modulo 4, for instance, this series is simply

$$L(\chi_4, 1) = 1 - \frac{1}{3} + \frac{1}{5} - \cdots = \frac{\pi}{4}.$$

However, since χ is a periodic function of $n \geqslant 1$, it is not difficult to extend the proof of the analytic continuation of $\zeta(s)$ and its functional equation based on the Poisson summation formula and theta functions (explained in the second chapter) to establish analogue statements for Dirichlet characters. This almost forces us to introduce the notion of primitivity: a Dirichlet character χ modulo q is called *primitive*, and q is called its *conductor* if there does not exist $\tilde{q} \mid q$, $\tilde{q} < q$ and a character $\tilde{\chi}$ of $(\mathbf{Z}/\tilde{q}\mathbf{Z})^\times$ such that

$$\chi(n) = \tilde{\chi}(n \pmod{\tilde{q}}) \text{ for } (n, \tilde{q}) = 1.$$

If q is prime, then any nontrivial character is primitive modulo q. Any Dirichlet character is *induced* in the way described above by a unique primitive character $\tilde{\chi}$, and one has the relation

$$L(\chi, s) = L(\tilde{\chi}, s) \prod_{p \mid q/\tilde{q}} (1 - \chi(p) p^{-s}) \tag{3.1}$$

[3] For $\chi \neq 1$, the partial sums are bounded so the series converges.

which shows that the analytic properties of Dirichlet L-functions can be reduced immediately to that associated to primitive characters, the second factor being a finite product (which is an entire function).

For χ primitive, the case $q = 1$ corresponds to the zeta function and is special; if $q > 1$, the character χ is nontrivial and one shows that $L(\chi, s)$ has an extension to $\mathbf{C}$ as an entire function. The functional equation requires more disctinctions: if $\chi(-1) = 1$, χ is called *even*, and one defines

$$\Lambda(\chi, s) = \pi^{-s/2}\Gamma(s/2)L(\chi, s),$$

whereas for $\chi(-1) = -1$ (χ is *odd*), one defines

$$\Lambda(\chi, s) = \pi^{-(s+1)/2}\Gamma((s+1)/2)L(\chi, s).$$

Moreover, let $\tau(\chi)$ be the Gauss sum attached to χ,

$$\tau(\chi) = \sum_{x \pmod q} \chi(x)e(x/q) \tag{3.2}$$

where $e(z) = e^{2i\pi z}$. Then the functional equation is

$$\Lambda(\chi, s) = \varepsilon(\chi)q^{1/2-s}\Lambda(\overline{\chi}, 1-s) \tag{3.3}$$

where

$$\varepsilon(\chi) = \begin{cases} \dfrac{\tau(\chi)}{\sqrt{q}} & \text{if } \chi(-1) = 1 \\ -i\dfrac{\tau(\chi)}{\sqrt{q}} & \text{if } \chi(-1) = -1. \end{cases} \tag{3.4}$$

Thus a number of features appear which are ubiquitous in the more general context:

- The functional equation (3.3) involves certain global invariants of the Dirichlet character. First, its conductor q; notice that even when starting with an imprimitive character, the functional equation will only be possible for the associated ("inducing") primitive character, and thus the a priori unknown conductor $\tilde{q}$ will appear in this way. This remark, in other contexts, is very fruitful (see Section 4).

- The second invariant is the *argument* of the functional equation $\varepsilon(\chi)$, also called the *root number*. It is a complex number of absolute value 1, as a simple calculation (or a second application of the functional equation) reveals (in the case of primitive character), and is related to the Gauss sum $\tau(\chi)$. Notice that the latter is also a kind of finite field analogue of the gamma function (multiplicative Fourier transform of an additive character). In general, the argument of the functional equation is a very delicate invariant.

- The functional equation relates $L(\chi, s)$ with the value of at $1-s$ of the *dual* character $\overline{\chi}$.

An important special case is that of real-valued characters, i.e., characters of order 2 (or quadratic characters). In such a case, we have $\overline{\chi} = \chi$ and the argument $\varepsilon(\chi)$ becomes a *sign* ± 1. Actually, Gauss had computed the exact value of $\tau(\chi)$ (Dirichlet gave a simple analytic proof) which implies that $\varepsilon(\chi) = +1$ for any primitive quadratic character χ, i.e., $\tau(\chi) = \sqrt{q}$ is $\chi(-1) = 1$ and $\tau(\chi) = i\sqrt{q}$ if $\chi(-1) = -1$.

We now come back to Dirichlet's proof of Theorem 3.1. It is enough to show that $L(\chi, 1) \neq 0$ for $\chi \neq 1$, and because of (3.1) one may assume that χ is primitive. Dirichlet easily showed that if χ is not real, then $L(\chi, 1) \neq 0$ (if not, 1 would also be zero of $L(\overline{\chi}, s) \neq L(\chi, s)$ and this is easy to exclude). The difficult case, as it has remained to this day, is that of a quadratic χ. Dirichlet's proof that $L(\chi, 1) \neq 0$ is a direct application of his analytic class number formula (see (4.5) below) and thus introduces another recurrent and all-important theme, that of linking L-functions defined by analytic means (here the Dirichlet L-function) to others coming from algebraic number theory, or algebraic geometry (here the Dedekind zeta function of a quadratic field).[4] We explain this in the next section in the more general context of number fields, and will simply conclude by saying that this proof of nonvanishing of $L(\chi, 1)$ is spectacular, but somewhat misleading. One can indeed give very natural elementary proofs of the fact that

$$L(\chi, 1) \gg \frac{1}{\sqrt{q}},$$

for $q > 1$, which is the more precise outcome of the class number formula (at least in the imaginary case.) And more importantly, all progress concerning the problem (going back to Gauss) of understanding the order of magnitude of the class number of quadratic fields has come from the opposite direction, by using the class number formula to relate it to the special value of the L-function and estimating (more often, failing to estimate) the latter quantity by analytic means (see, e.g., Landau [L], Siegel [Si], Goldfeld [Go]).

4 Dedekind zeta functions

Let K be a number field, so that $K/\mathbf{Q}$ is a finite field extension, of degree $d = [K : \mathbf{Q}]$. We denote $\mathcal{O}$, or $\mathcal{O}_K$, the ring of integers in K. Dedekind defined the zeta function of K by the series

$$\zeta_K(s) = \sum_{\mathfrak{a}} \frac{1}{(N\mathfrak{a})^s}$$

(over nonzero integral ideals $\mathfrak{a} \subset \mathcal{O}$) which is easily seen to be absolutely convergent for $\mathrm{Re}(s) > 1$. So, for $K = \mathbf{Q}$, one recovers $\zeta(s)$. Since in $\mathcal{O}$ there is unique

[4] Today one would say of "motivic" origin.

factorization of ideals into products of prime ideals, the same argument as for $\zeta(s)$ proves that there is an Euler product expansion

$$\zeta_K(s) = \prod_{\mathfrak{p}} (1 - (N\mathfrak{p})^{-s})^{-1},$$

and indeed $\zeta_K(s)$ carries much the same information about the distribution of prime ideals in K as $\zeta(s)$ does about ordinary prime numbers.

However, there is more to it: if we write $\zeta_K(s)$ as an ordinary Dirichlet series with some coefficients $r_K(n)$,

$$\zeta_K(s) = \sum_{n \geqslant 1} r_K(n) n^{-s}, \tag{4.1}$$

we immediately see that knowing $\zeta_K(s)$ implies knowing how prime numbers split in $\mathcal{O}$: indeed $r_K(p^k)$ is the number of ideals with norm $= p^k$, hence $r_K(p) = d$ if and only if p splits completely in K, $r_K(p) = 1$ means that p is totally ramified, etc.

This information is one of the most relevant to the study of the arithmetic of K, and this brings a new urgency to the exploration of the properties of $\zeta_K(s)$, which concentrates really on the zeta function itself instead of its logarithmic derivative (which is the key to the analytic distribution of prime ideals) and considers $\zeta(s)$ almost as a "trivial case."

Because of the definition by a series expansion, it is possible again to extend the method used for $\zeta(s)$ (and for $L(\chi, s)$) to derive analogous analytic properties of $\zeta_K(s)$. This is not completely straightforward however, and contains or uses most of the "basic" notions and results of algebraic number theory, which we now recall (see, e.g., [La], [CF]):

- The field K has $d = r_1 + 2r_2$ embeddings $\sigma : K \hookrightarrow \mathbf{C}$, r_1 real embeddings and r_2 pairs of complex embeddings. To each is associated an absolute value $|\cdot|_\sigma$, $|z|_\sigma = |\sigma(z)|$ if σ is real and $|z|_\sigma = |\sigma(z)|^2$ if σ is complex. Those absolute values are also called the *archimedean places* of K; note σ and $\bar{\sigma}$ give rise to the same place. We use usually v to denote a place. We let $d_v = 1$ if v is real and $d_v = 2$ otherwise. Note that the norm of an element $z \in K$ is given by

$$Nz = \prod_v |z|_v \text{ (product over the places).}$$

- The ring $\mathcal{O}$ is of rank $d = [K : \mathbf{Q}]$. Let (ω_i) be a $\mathbf{Z}$-basis, $1 \leqslant i \leqslant d$. The discriminant D of K is

$$D = \det(\omega_i^\sigma)^2_{i,\sigma}.$$

  For every prime ideal $\mathfrak{p}$ in $\mathcal{O}$, the quotient $\mathbf{F}_\mathfrak{p} = \mathcal{O}_K/\mathfrak{p}$ is a finite field and the *(absolute) norm* of $\mathfrak{p}$ is the cardinality of $\mathbf{F}_\mathfrak{p}$.

- The multiplicative group $U = \mathcal{O}^\times$ of units of $\mathcal{O}$ is a finitely generated abelian group of rank $r_1 + r_2 - 1$, the free part of which admits an injection into $\mathbf{R}^d$ through the "logarithmic" map

$$\ell : u \mapsto (\log |\sigma(u)|)_\sigma$$

the image of which is a lattice in the subspace $V = \{(x_\sigma) \mid \sum x_\sigma = 0 \text{ and } x_\sigma = x_{\overline{\sigma}}\}$. The volume $R > 0$ of a fundamental domain for $\ell(U) \subset V$ (with respect to the Lebesgue measure on V) is called the *regulator* of K.

- One denotes by $w = w_K$ the order of the torsion subgroup of U (the number of roots of unity in K).

- The group of ideal classes in K, denoted $\mathrm{Pic}(\mathcal{O})$ or $H(K)$, is a finite abelian group of order denoted $h(K) = h(\mathcal{O})$, called the *class number* of K. The ring $\mathcal{O}$ is principal (i.e., every ideal is principal) if and only if $h(K) = 1$.

If E/K is a finite Galois extension of number fields with Galois group G, for every prime ideal $\mathfrak{p}$ in K and prime ideal $\mathfrak{P} \mid \mathfrak{p}$ in E above $\mathfrak{p}$, one defines the *decomposition group* of $\mathfrak{P}$

$$G_{\mathfrak{P}} = \{\sigma \in G \mid \sigma\mathfrak{P} = \mathfrak{P}\}$$

and the *inertia subgroup*

$$I_{\mathfrak{P}} = \{\sigma \in G_{\mathfrak{P}} \mid \sigma(x) \equiv x \pmod{\mathfrak{P}} \text{ for all } x \in \mathcal{O}_E\} \lhd G_{\mathfrak{P}},$$

and $\mathfrak{P}$ is unramified if $I_{\mathfrak{P}} = 1$.

The *Frobenius conjugacy class* $\sigma_{\mathfrak{P}}$, also denoted $[\mathfrak{P}, E/K]$, is the conjugacy class in $G_{\mathfrak{P}}/I_{\mathfrak{P}}$ (i.e., in the decomposition group if $\mathfrak{P}$ is unramified) of the elements σ such that

$$\sigma(x) \equiv x^{N\mathfrak{p}} \pmod{\mathfrak{P}} \text{ for all } x \in \mathcal{O}_E,$$

i.e., the "reduction modulo $\mathfrak{P}$" of $\sigma_{\mathfrak{P}}$ is the canonical generator of the finite field extension $\mathbf{F}_{\mathfrak{P}}/\mathbf{F}_{\mathfrak{p}}$. The conjugacy class of σ *in* G only depends on $\mathfrak{p}$ and is denoted $\sigma_{\mathfrak{p}}$.

If G is an abelian group, the Frobenius conjugacy class is reduced to a single element, the Frobenius automorphism, which only depends on $\mathfrak{p}$ and is denoted $\sigma_{\mathfrak{p}}$.

The precise statement about analytic continuation and functional equation for $\zeta_K(s)$ takes the following form: let

$$\Lambda_K(s) = \pi^{-r_1 s/2}(2\pi)^{-r_2 s}\Gamma(s/2)^{r_1}\Gamma(s)^{r_2}\zeta_K(s).$$

Then $\Lambda_K(s)$ admits analytic continuation to a meromorphic function on $\mathbf{C}$ with simple poles at $s = 1$ and at $s = 0$. Thus $\zeta_K(s)$ is meromorphic with a simple pole at $s = 1$. Moreover the residue of $\zeta_K(s)$ at $s = 1$ is given by the *analytic class number formula*

$$\mathrm{Res}_{s=1}\,\zeta_K(s) = \frac{2^{r_1}(2\pi)^{r_2}h(K)R}{w\sqrt{|D|}}. \tag{4.2}$$

The functional equation satisfied by $\Lambda_K(s)$ is

$$\Lambda_K(s) = |D|^{1/2-s}\Lambda_K(1-s). \tag{4.3}$$

Compared to the previous functional equations, one notices that the discriminant appears as the conductor did for Dirichlet L-functions, and that the corresponding root number is always $+1$.

The formula (4.2) is a newcomer here, the prototypical example of a large class of formulae (proved or conjectured to hold) which express the *special values* of L-functions in terms of global invariants of the "motivic" objects they are related to. The best known generalization is the conjecture of Birch and Swinnerton–Dyer but there are many others.

The notion of a "special value" can be put in a rigorous conceptual framework (Deligne, etc.) explaining which values should have those properties for a given L-function. One should maybe mention that analytic number theorists often have reasons to consider as "special" points that escape this algebraic context: a typical example is the deformation theory of Phillips and Sarnak [PS], where the point in question is (probably) transcendental.

Let me now explain how this general formula relates to Dirichlet's proof of Theorem 3.1. Let χ be a nontrivial primitive quadratic character of conductor q. The essence of the argument is that there is a relation

$$\zeta(s)L(\chi, s) = \zeta_K(s) \tag{4.4}$$

between the Dirichlet L-function and the Dedekind zeta function of the quadratic field $K = \mathbf{Q}(\sqrt{\chi(-1)q})$ (so K is imaginary if $\chi(-1) = -1$ and real if $\chi(-1) = 1$). There is a lot of mathematics behind this seemingly simple statement, namely it is a reformulation of the quadratic reciprocity law of Gauss, and as such it has had considerable influence on the development of the whole theory of L-functions.

Before discussing this further, observe that since $\zeta(s)$ and $\zeta_K(s)$ on both sides of (4.4) have a simple pole at $s = 1$, it follows that $L(\chi, 1) \neq 0$, and more precisely that

$$L(\chi, 1) = \begin{cases} \dfrac{2\pi h(K)}{w\sqrt{q}} & \text{if } K \text{ is imaginary} \\ \dfrac{2h(K)\log\varepsilon}{\sqrt{q}} & \text{if } K \text{ is real, } \varepsilon > 1 \text{ being the fundamental unit of } K \end{cases} \tag{4.5}$$

which implies $L(\chi, 1) > 0$ (even $L(\chi, 1) \gg q^{-1/2}$ if K is imaginary); see also the discussion in Section 3.

We come back to (4.4). A prime number p is either ramified, inert, or split in the quadratic field K. Expliciting this in the Euler product expression of $\zeta_K(s)$, we have

$$\zeta_K(s) = \prod_{p \text{ split}} (1-p^{-s})^{-2} \prod_{p \text{ inert}} (1-p^{-2s})^{-1} \prod_{p \text{ ramified}} (1-p^{-s})^{-1} \tag{4.6}$$

and comparing with (4.4) it follows that the latter is equivalent with the following characterization of the splitting of primes:

$$\begin{cases} p \text{ is ramified if and only if } \chi(p) = 0, \text{ if and only if } p \mid q \\ p \text{ is split in } K \text{ if and only if } \chi(p) = 1 \\ p \text{ is inert in } K \text{ if and only if } \chi(p) = -1. \end{cases} \tag{4.7}$$

The point is that this characterization is in terms of the Dirichlet character χ which is a *finite amount of data* related to $\mathbf{Q}$ and not to K (it is after all only a particular periodic arithmetic function of period q).

Even more concretely, basic algebraic number theory shows that another characterization, involving Legendre symbols, exists:

$$\begin{cases} p \text{ is ramified if and only if } p \mid q \\ p \text{ is split in } K \text{ if and only if } \left(\frac{D}{p}\right) = 1 \\ p \text{ is inert in } K \text{ if and only if } \left(\frac{D}{p}\right) = -1 \end{cases} \tag{4.8}$$

(where D is the discriminant of K), but there is no reason a priori to expect that the map

$$p \mapsto \left(\frac{D}{p}\right)$$

(defined only on the rather chaotic set of prime numbers!) has any particularly good property. Now the primitive quadratic characters are very easy to characterize using the structure of $(\mathbf{Z}/q\mathbf{Z})^\times$. In particular, if q is squarefree and odd, there is a unique primitive character modulo q, given by

$$\chi(n) = \prod_{p|n} \left(\frac{n}{p}\right)$$

(this is clearly a quadratic character modulo q, and since for any $p \mid q$ there are quadratic residues and nonresidues modulo $p > 2$, χ cannot be induced from q/p).

In this case, one has $\chi(-1) \equiv q \pmod 4$, so the quadratic field is $K = \mathbf{Q}(\sqrt{q})$ if $q \equiv 1 \pmod 4$ and $K = \mathbf{Q}(\sqrt{-q})$ if $q \equiv 3 \pmod 4$, with discriminant $D = \chi(q)q$. Take q to be a prime > 2: then $\chi(-1) = (-1)^{(q-1)/2}$ so comparison of (4.7) and (4.8) gives (after some easy checking)

$$\left(\frac{p}{q}\right) = \left(\frac{D}{p}\right) = \left(\frac{(-1)^{(q-1)/2}}{p}\right)\left(\frac{q}{p}\right) = (-1)^{(p-1)(q-1)/4}\left(\frac{q}{p}\right),$$

the original form of the quadratic reciprocity law. Similarly, for even conductors of quadratic characters. In fact (4.4) for all quadratic fields is equivalent with quadratic reciprocity.

This explains why it may seem natural and desirable to seek generalizations of the factorization (4.4) for other L-functions, specifically at first for Dedekind zeta functions, and why one would see them as some form of "reciprocity law," if the factors appearing are (like $L(\chi, s)$) defined in terms of objects of the base field $\mathbf{Q}$, since this would amount to describing the splitting of primes in K solely in terms of data belonging to $\mathbf{Q}$.

Of course, it is also natural to expect similar situations to exist for relative extensions E/K of number fields.

The general (still very much conjectural) form of this idea belongs to the theory of Artin L-functions and to some of the conjectures of Langlands, and will be covered later. There is however one important special case where a (mostly) satisfactory theory exists and with which I will finish this first chapter. It is the case when one has an *abelian* extension L/K and the results are known as *class-field theory*.

5 Hecke L-functions, class-field theory

Let E/K be an extension of number fields and assume that it is a Galois extension with *abelian* Galois group G. It follows in particular that for every prime ideal $\mathfrak{p}$ of K unramified in E, the Frobenius element $\sigma_{\mathfrak{p}} \in G$ is well defined.

A formal computation, although nontrivial (a special case of the invariance by induction of Artin L-functions), shows that there is a factorization of $\zeta_E(s)$ as

$$\zeta_E(s) = \prod_{\rho \in \hat{G}} L(\rho, s) \tag{5.1}$$

where $\hat{G}$ is the (dual) character group of G and

$$L(\rho, s) = \prod_{\mathfrak{p} \text{ in } K} (1 - \rho(\sigma_{\mathfrak{p}})(N\mathfrak{p})^{-s})^{-1},$$

where $\rho(\sigma_{\mathfrak{p}})$ refers to the image of the Frobenius element by the Galois representation (either 0 or 1-dimensional) induced on $\mathbf{C}^{I_{\mathfrak{p}}}$, to take care of the ramification. Of course, if $\mathfrak{p}$ is unramified in E, this is literally $\rho(\sigma_{\mathfrak{p}})$. (For instance, if $\mathfrak{p}$ is totally split in E with $N\mathfrak{p} = q$, the $\mathfrak{p}$-factor on the left of (5.1) is made of $d = [E : K]$ factors $(1 - q^{-s})^{-1}$, one for each prime above $\mathfrak{p}$; on the right there are $|\hat{G}| = |G| = d$ characters ρ, and for each $\rho(\sigma_{\mathfrak{p}}) = \rho(1) = 1$, hence the $\mathfrak{p}$-factor is indeed the same).

This new Euler product is, for trivial reasons, still absolutely convergent for $\mathrm{Re}(s) > 1$ since $|\rho(\sigma)| = 1$ for any $\sigma \in G$, but its analytic continuation, for individual ρ, is by no means obvious: one can see that a simple definition as a series instead of a product does not appear immediately. There is no clear reason that the coefficients of this Dirichlet series should have any good property (compare with the discussion of the quadratic reciprocity law above).[5] In particular, at first sight,

[5]For instance, are the partial sums of the coefficients bounded, as for Dirichlet characters?

computing the coefficient of $(N\mathfrak{a})^{-s}$ seems to require knowing the factorization of $\mathfrak{a}$ in E.

Example 5.1. For a quadratic field $K/\mathbf{Q}$, there are two characters, the trivial one with L-function equal to $\zeta(s)$ and a quadratic character χ_2 with

$$L(\chi_2, s) = \prod_{p \text{ split}} (1 - p^{-s})^{-1} \prod_{p \text{ inert}} (1 + p^{-s})^{-1}$$

since $G \simeq \{\pm 1\}$ and $\chi_2(\sigma_p) = +1$ (resp., -1) if and only if p is split (resp., if p is inert).

Of course one has $L(\chi_2, s) = L(\chi, s)$ in the factorization (4.4) involving the Dirichlet character (compare with (4.8)).

The L-function part of class-field theory can be thought of as the identification of the L-functions of Galois characters in terms of generalizations of the quadratic Dirichlet character appearing in (4.4). For a general base field K, those were defined by Hecke. However, in the case of $\mathbf{Q}$ it suffices to consider the original Dirichlet characters (not necessarily quadratic), which we state, without striving for the utmost precision:

Theorem 5.2 (Kronecker–Weber theorem). *Let $K/\mathbf{Q}$ be an abelian extension with Galois group G, let $\rho : G \to \mathbf{C}^\times$ be a Galois character and $L(\rho, s)$ its L-function as above. There exists a unique primitive Dirichlet character χ modulo q for some $q \geqslant 1$ such that*

$$L(\rho, s) = L(\chi, s).$$

In particular, the analytic continuation, functional equation and other properties of $L(\chi, s)$ are thus inherited by all those L-functions. This gives as a consequence "reciprocity laws" describing the splitting of prime numbers in abelian extensions of $\mathbf{Q}$, and also the following corollary (the more usual form of the Kronecker–Weber Theorem).

Corollary 5.3. *Let $K/\mathbf{Q}$ be an abelian extension. Then K is contained in some cyclotomic field, i.e., there exists some $m \geqslant 1$ such that $K \subset \mathbf{Q}(\mu_m)$ where μ_m is the group of mth roots of unity in $\mathbf{C}$.*

Proof of the corollary. Let

$$\zeta_K(s) = \prod_{\chi} L(\chi, s) \tag{5.2}$$

be the factorization obtained from (5.1) and Theorem 5.2 in terms of (some) Dirichlet characters. Let m be the l.c.m. of the conductors q_χ of the occuring characters. Then in fact we have $K \subset \mathbf{Q}(\mu_m)$.

To prove this, we recall a simple general fact about number fields (other proofs are possible): we have $E \subset K$ for Galois extensions $E/\mathbf{Q}$ and $K/\mathbf{Q}$ if and only if all but finitely many of the prime numbers p which are totally split in K are totally split in E. Thus let p be a prime number totally split in $\mathbf{Q}(\mu_m)$. By elementary theory of cyclotomic fields [CF, Chapter 3] this means that $p \equiv 1 \pmod{m}$. Thus for any character χ modulo m, we have $\chi(p) = 1$, and in particular for all characters occuring in (5.2) we have $\chi(p) = 1$; i.e., by the Kronecker–Weber Theorem, $\rho(\sigma_p) = 1$ for all $\rho \in \hat{G}$. This means that $\sigma_p = 1 \in G$, hence that p is totally split in K. □

For more about the relationships between Dirichlet characters and cyclotomic fields, see, e.g., [Wa, Chapter 3].

Remark 5.4. The Kronecker–Weber Theorem, as stated here, bears a striking resemblance to the L-function form of the modularity conjecture for elliptic curves (explained in de Shalit's chapters). One can prove Theorem 5.2 by following the general principles of Wiles's argument [Tu] (deformation of Galois representations, and computation of numerical invariants in a commutative algebra criterion for isomorphism between two rings).

We come to the definition of Hecke L-functions of a number field K. As one may easily imagine, they are L-functions of the form

$$L(\chi, s) = \sum_{\mathfrak{a}} \chi(\mathfrak{a})(N\mathfrak{a})^{-s} = \prod_{\mathfrak{p}} (1 - \chi(\mathfrak{p})(N\mathfrak{p})^{-s})^{-1} \tag{5.3}$$

for some "arithmetic" function χ defined on integral ideals in $\mathfrak{a}$ in K. The proper definition, in the language of ideals, requires some care however; this becomes much clearer in the idèle-theoretic description. See, e.g., [Co] for further details in the "classical" language. The difficulties arise because of the various archimedean places and the class number being possibly > 1.

Let $\mathfrak{m}$ be a nonzero integral ideal of K, which will play the role of modulus. One defines the subgroups $I_{\mathfrak{m}}$ (resp., $P_{\mathfrak{m}}$) of the group I of fractional ideals in K (resp., of the subgroup of principal ideals) by

$$I_{\mathfrak{m}} = \{\mathfrak{a} \in I \mid (\mathfrak{a}, \mathfrak{m}) = 1\}$$
$$P_{\mathfrak{m}} = \{\mathfrak{a} = (\alpha) \in I_{\mathfrak{m}} \cap P \mid \alpha \equiv 1 \pmod{\mathfrak{m}}\}.$$

The finite abelian group $H_{\mathfrak{m}} = I_{\mathfrak{m}}/P_{\mathfrak{m}}$ is called the ray-class group modulo $\mathfrak{m}$. Let also $U_{\mathfrak{m}}$ be the group of units in $P_{\mathfrak{m}}$.

Let ξ_∞ be a (unitary) character

$$\xi_\infty : K^\times/\mathbf{Q}^\times \to \mathbf{C}^\times$$

(these can be easily described using the various places at infinity, see the example below) such that $U_{\mathfrak{m}} \subset \ker \xi_\infty$. Hence ξ_∞ induces a homomorphism

$$\xi_\infty : P_{\mathfrak{m}} \to \mathbf{C}^\times.$$

Definition. A Hecke character of weight ξ_∞ for the modulus $\mathfrak{m}$ is a homomorphism

$$\chi : I_{\mathfrak{m}} \to \mathbf{C}^\times$$

(unitary) such that

$$\chi((\alpha)) = \xi_\infty(\alpha)$$

if $\mathfrak{a} = (\alpha) \in P_{\mathfrak{m}}$. The character χ is extended to I by putting $\chi(\mathfrak{a}) = 0$ if $(\mathfrak{a}, \mathfrak{m}) \neq 1$.

One can define primitive Hecke characters in much the same way as for Dirichlet characters, with the same basic properties: any character is induced by a unique primitive one, and their L-functions are the same up to a finite Euler product, as in (3.1).

For a Hecke character χ the L-function $L(\chi, s)$ has the Euler product expansion (5.3) by multiplicativity and converges absolutely for $\mathrm{Re}(s) > 1$.

Example 5.5. If we take $\mathfrak{m} = 1$ and $\xi_\infty = 1$, then the corresponding Hecke characters are just the *ideal class-group characters*, i.e., the (finitely many) characters $H(K) \to \mathbf{C}^\times$ of the ideal class-group of K. More generally, if $\mathfrak{m}$ is arbitrary but $\xi_\infty = 1$, the resulting Hecke characters are called ray-class characters modulo $\mathfrak{m}$.

For the trivial Hecke character χ_0, one has $L(\chi_0, s) = \zeta_K(s)$.

Example 5.6. Let $K = \mathbf{Q}$. Then the only possibility is $\xi_\infty = 1$, $\mathfrak{m} = (m)$ for some unique $m \geqslant 1$; Hecke characters modulo $\mathfrak{m}$ are the same thing as Dirichlet characters modulo m. Primitivity also corresponds.

This extends very similarly for any field with class number 1.

Example 5.7. Let K be a quadratic field and σ the nontrivial element in the Galois group of K. Then $K^\times/\mathbf{Q}^\times \simeq \{a \in K^\times \mid Na = 1\}$ by $a \mapsto a/a^\sigma$. Assume first that K is imaginary. Then one checks easily that ξ_∞ must be of the form

$$\xi_\infty(a) = \left(\frac{a}{|a|}\right)^u$$

(i.e., $\arg(a)^u$) for some integer $u \in \mathbf{Z}$, to which the weight can be identified.

If, on the other hand, K is real, then one finds that

$$\xi_\infty(a) = \left(\frac{a}{|a|}\right)^{u_1}\left(\frac{a^\sigma}{|a^\sigma|}\right)^{u_2}$$

with $u_1, u_2 \in \{0, 1\}$.

Hecke managed to prove the fundamental analytic properties of his L-series, in complete analogy with the case of Dirichlet L-functions. The next chapter will sketch the proof, which becomes more transparent in the language of adèles: this translation is the subject of Tate's famous thesis [Ta].

Theorem 5.8 (Hecke). *Let $\chi \neq 1$ be a primitive, nontrivial Hecke character of K. Then $L(\chi, s)$ admits analytic continuation as an entire function and there is a gamma factor $\Gamma(\xi_\infty, s)$, depending only on the weight, which is a product of gamma functions, such that*

$$\Lambda(\chi, s) = \Gamma(\xi_\infty, s)L(\chi, s)$$

satisfies

$$\Lambda(\chi, s) = \varepsilon(\chi)(|D|N\mathfrak{m})^{1/2-s}\Lambda(\overline{\chi}, 1-s) \tag{5.4}$$

for some complex number $\varepsilon(\chi)$ *of absolute value* 1. *Here* D *is the (absolute) discriminant of* $K/\mathbf{Q}$ *and* $N\mathfrak{m}$ *the norm from* K *to* $\mathbf{Q}$.

We are a little vague: one can indeed write a formula for $\varepsilon(\chi)$ (in terms of a Gauss sum for χ) as well as for $\Gamma(\xi_\infty, s)$. The point is that they appear "naturally" in the course of the proof. Writing the gamma factor "as a function of the weight" is actually reminiscent of adelic arguments. Of course, one should compare this to (3.3) and (4.3).

Now one can state the analogue of Theorem 5.2 for abelian extensions of a number field K, which is extremely similar and encompasses much of class-field theory.

Theorem 5.9 (Artin). *Let* K *be a number field,* E/K *a finite abelian extension with Galois group* G*, let* $\rho : G \to \mathbf{C}^\times$ *be a Galois character and* $L(\rho, s)$ *the associated* L*-function. Then there exists a unique primitive Hecke character* χ *of* K*, of modulus* $\mathfrak{m}$ *such that*

$$L(\rho, s) = L(\chi, s).$$

Actually, this identity holds locally, i.e., the $\mathfrak{p}$*-component of the Euler products are equal for all prime ideal* $\mathfrak{p}$ *of* K*, namely*[6]

$$\rho(\sigma_{\mathfrak{p}}) = \chi(\mathfrak{p}) \textit{ for all } \mathfrak{p} \textit{ coprime to } \mathfrak{m}.$$

Remark 5.10. Actually, as in the more general case of Artin L-functions, one can define beforehand, in terms simply of ρ, a *conductor* $\mathfrak{f}(\rho)$ and a weight $\xi_\infty(\rho)$. The proof of the theorem shows that $\mathfrak{f} = \mathfrak{m}$ and the weight of χ is ξ_∞. This added precision is very important in applications, if only because it makes this statement theoretically verifiable for any given ρ by brute force search (this is similar to Weil's stipulation that the level of the modular form associated to an elliptic curve over $\mathbf{Q}$ should be its conductor).

Remark 5.11. Comparing the functional equations on both sides of (5.1) after applying Theorem 5.9, one obtains a relation between the conductors of the Galois (or Hecke) characters related to E/K and the discriminant of E. In the simple case $K = \mathbf{Q}$ this takes the form

$$|D| = \prod_{\chi} q_\chi$$

where D is the discriminant of $E/\mathbf{Q}$, χ are the Dirichlet characters in (5.2) and q_χ their conductors. For $E = \mathbf{Q}(\mu_\ell)$ for $\ell > 2$ prime, the factorization is

$$\zeta_E(s) = \prod_{\chi \pmod{\ell}} L(\chi, s).$$

[6] As often remarked by Serre, this last statement is stronger than the first one if $K \neq \mathbf{Q}$; compare the footnote after (2.4).

There are $\ell - 1$ Dirichlet characters, of which one has conductor 1 and $\ell - 2$ have conductor ℓ, hence we recover the well-known discriminant $|D| = \ell^{\ell-2}$.

Similarly, since the root number for $\zeta_E(s)$ is 1, one deduces a relation between the root numbers $\varepsilon(\chi)$. One can also fruitfully compare the residues at $s = 1$ on both sides.

Just as in the case of Dirichlet characters, the Hecke characters of a number field K have (independently of class-field theory) many applications to the equidistribution of ideals in various classes. Actually, those split naturally into two kinds: one concerns the distribution of prime ideals $\mathfrak{p}$ in K, and is based on the same methods using the logarithmic derivatives of Hecke L-functions. The other concerns the distribution (in $\mathbf{N}$) of the *norms* of integral ideals in K, i.e., the (average) properties of the arithmetic function $r_K(n)$ in (4.1). Here one uses $\zeta_K(s)$ and the Hecke L-functions themselves, and it is somewhat simpler since their singularities are completely known. One deduces for instance that

$$|\{\mathfrak{a} \subset \mathcal{O} \mid N\mathfrak{a} \leqslant X\}| = (\mathrm{Res}_{s=1}\,\zeta_K(s))X + O(X^{1-1/d})$$

as $X \to +\infty$, and that the ideals are equidistributed in ideal classes. For example, taking a quadratic field K, one gets the asymptotic formula for the number of integers $n \leqslant X$ represented by a quadratic form (with multiplicity). Standard methods of analytic number theory can also be used to deduce an asymptotic formula for the number of integers that are norm of an ideal in K (excluding multiplicity).

Example 5.12. Take $K = \mathbf{Q}(i)$. Then $\mathcal{O} = \mathbf{Z}[i]$ has class number one and the number of ideals with norm $\leqslant X$ is equal to the number of lattice points in $\mathbf{C}$ inside a disc of radius $\sqrt{R}$ (the Gauss circle problem). One gets in this case trivially

$$\sum_{n^2+m^2 \leqslant X} 1 = \pi X + O(\sqrt{X}) \text{ as } X \to +\infty$$

($r(n)$ is the number of representations of n as sum of two squares). More subtle arguments yield the same formula with $X^{1/3}$ as error term. It is conjectured that any exponent $> 1/4$ will do, which is the best possible. In suitably smoothed form, even better results are known, for instance the automorphy of theta functions (see Chapter 3) easily implies that if φ is a smooth function with compact support in $[0, +\infty[$, we have

$$\sum_{n,m} \varphi\Big(\frac{n^2+m^2}{X}\Big) = \pi X \int_0^{+\infty} \varphi(x)dx + c_0\varphi(0) + O(X^{-1/2})$$

as $X \to +\infty$, where c_0 is some constant. (Use the Mellin transform and move the line of integration to $\mathrm{Re}(s) = -1/2$.)

6 Function fields

I will only say a very few words about the "geometric" analogue (or "function field case") of the various theories described before, mostly by lack of proper competence. More detailed explanations will come in future chapters.

The analogy between the arithmetic of $\mathbf{Z}$ and that of the ring of polynomials over a field $k[X]$ is a very old one. If one takes for k a finite field $\mathbf{F}_q$ with $q = p^d$ elements, the analogy deepens. For instance, if P is an irreducible polynomial, the residue field $\mathbf{F}_q[X]/(P)$ is a finite field, as happens with number fields.

This leads to a theory that is very similar to that described previously in many respects, although geometric intuition and some other intrinsic characteristics tend to make it simpler (e.g., one can differentiate polynomials, but not integers.) It provides both a different set of problems, and a way to get evidence for conjectures which remain intractable over number fields, for instance, those concerning the zeros of L-functions (for example, the work of Katz and Sarnak [KS] on the fine distribution of zeros, or the recent proof by Lafforgue on the global Langlands correspondence for $GL(n)$ over function fields).

We will describe here briefly the case of curves. Let $C/\mathbf{F}_q$ be an algebraic curve (are there any others?) over $\mathbf{F}_q$ given as the set of zeros in $\mathbf{P}^2$ of a homogeneous polynomial $F \in \mathbf{F}_q[X, Y, Z]$. Let $D/\mathbf{F}_q$ be "the" smooth projective model of C. In the language of Hasse and Artin, this would be identified with the *function field* $K = \mathbf{F}_q(C)$ of C, which is a finite extension of the field $\mathbf{F}_q(X)$ of rational functions in one variable (geometrically, the latter corresponds to $\mathbf{P}^1$ and the extension to a ramified covering $C \to \mathbf{P}^1$).

The nonarchimedean valuations of K correspond in one-to-one fashion with the set of closed points of C (in scheme-theoretic language), in other words with the *orbits* of the action of $\mathrm{Gal}(\bar{\mathbf{F}}_q/\mathbf{F}_q)$ on the points of C defined over the algebraic closure of $\mathbf{F}_q$. If C is affine, they also correspond to the maximal ideals in the ring of functions regular on C. Define the *zeta function* of C by the Euler product

$$Z(C; s) = \prod_x (1 - (Nx)^{-s})^{-1},$$

over the set of closed points, with Nx equal to the cardinality of the residue field at x. This is clearly an analogue of the Dedekind zeta function for the field K, and it turns out to satisfy many of the same properties.

Since the norm of x is always of the form q^a for some $a \geqslant 1$, the degree $\deg(x)$ of x, it is convenient to put $T = q^{-s}$ and consider $Z(C)$ as a formal power series in T

$$Z(C) = \prod_x (1 - T^{\deg(x)})^{-1}.$$

This turns out to have another expression, peculiar to the geometric case, which gives a direct diophantine interpretation of this zeta function.

Lemma 6.1. *We have*

$$Z(C) = \exp\Big(\sum_{n\geqslant 1} \frac{|C(\mathbf{F}_{q^n})|}{n} T^n\Big),$$

as formal power series.

Sketch of proof. On applies the operator $Td\log$ to both sides. On the right the result is

$$\sum_{n\geqslant 1} |C(\mathbf{F}_{q^n})| T^n,$$

while on the right after expanding a geometric series one gets

$$\sum_x \deg(x) \sum_{k\geqslant 1} X^{k\deg(x)} = \sum_n N_n T^n,$$

with

$$N_n = \sum_{d|n} d|\{x \mid \deg(x) = d\}|.$$

The degree d is the cardinality of the Galois orbit of a closed point x, so we see that $N_n = |C(\mathbf{F}_{q^n})|$, as desired. □

The basic properties of the zeta function are as follows:

Theorem 6.2. *Suppose C is smooth and projective. Then the zeta function has analytic continuation and functional equation. More precisely (!), $Z(C)$ is a rational function of T of the form*

$$Z(C) = \frac{P_{2g}}{(1-T)(1-qT)}$$

where $P_{2g} \in \mathbf{Z}[T]$ is a monic polynomial of even degree $2g$, g being equal to the genus of C. It satisfies

$$Z(C) = \varepsilon(C) q^{-\chi/2} T^{-\chi} Z(1/(qT))$$

where $\varepsilon(C) = \pm 1$ and $\chi = 2 - 2g$ is the Euler–Poincaré characteristic of C.

The polynomial P_{2g} can be expressed over $\mathbf{C}$ as

$$P_{2g} = \prod_{i=1}^{2g} (1 - \alpha_i T)$$

with $|\alpha_i| = \sqrt{q}$ for all i.

The first part was proved by Schmidt in general; it is based on the Riemann–Roch theorem. The second follows also from this argument. The last part is the Riemann Hypothesis for $Z(C)$: it was first proved by A. Weil. Weil also made conjectures generalizing those statements to higher dimensional varieties, and they were proved in even stronger and much more powerful form through the work of Grothendieck's school and particularly, of course, by P. Deligne. It seems particularly significant that these results are based on expressions for the polynomial P_{2g} as characteristic polynomials of a certain operator (the Frobenius operator) acting on some cohomology group: a (natural) analogue of such a phenomenon is eagerly sought in the number field case.

Remark 6.3. In terms of s, the functional equation becomes

$$Z(C; s) = \varepsilon(C) q^{\chi(s-1/2)} Z(C; 1-s),$$

and the Riemann Hypothesis says that the zeros of $Z(C; s)$ satisfy

$$q^{-\sigma} = |T| = |\alpha_i|^{-1} = q^{-1/2},$$

i.e., $\text{Re}(s) = 1/2$. So the analogy is indeed very clear.

2
Elementary Theory of L-Functions II

E. Kowalski

1 Introduction

This chapter is partly a development, with some sketches of proofs, of the main points of the first chapter, and partly a survey of the important topic of the zeros of L-functions.

2 The functional equation of Hecke L-functions

We will describe the proof of the functional equation for Dedekind zeta functions and briefly mention the changes needed for the general case of Hecke L-functions. See [La, XIII-3] or Hecke's original paper for this classical approach. It is well motivated historically (based on one of Riemann's proofs), and has clear connections with modular forms via theta functions, but is not entirely satisfactory in some respects (for the appearance of the gamma factors for instance). However, its explicitness makes it still quite valuable (see the formula (2.6) below).

Let K be a number field of degree d over $\mathbf{Q}$ and

$$\zeta_K(s) = \sum_{\mathfrak{a}} (N\mathfrak{a})^{-s} \tag{2.1}$$

its Dedekind zeta function.

The fundamental tool from harmonic analysis to obtain the analytic continuation is the *Poisson summation formula*.

Proposition 2.1. *Let $f : \mathbf{R}^d \to \mathbf{C}$ be a Schwartz function on (euclidean) $\mathbf{R}^d$, and let*

$$\hat{f}(\xi) = \int_{\mathbf{R}^d} f(x)e(-<x, \xi>)dx$$

be its Fourier transform. Then we have for all $a \in \mathbf{R}^d$

$$\sum_{m\in\mathbf{Z}^d} f(m+a) = \sum_{\mu\in\mathbf{Z}^d} \hat{f}(\mu)e(<a, \mu>),$$

both series being absolutely convergent, uniformly for a in compact sets.

Sketch of proof. For the proof, one simply sees that the left-hand side is, by averaging, a function on $\mathbf{R}^d$ invariant under the translation action of $\mathbf{Z}^d$, and the right-hand side is simply its expansion into Fourier series. Compare with the discussion of Poincaré and Eisenstein series in Chapter 3. □

It is clear that such a result is relevant; one may ask why not apply it directly to $\zeta(s)$ with $f(x) = x^{-s}, x > 0$, but of course this function is not in Schwartz space.

We split the series (2.1) into ideal classes in order to obtain summation sets easily parameterized by d-tuples of integers: we have

$$\zeta_K(s) = \sum_a \zeta(s; a) \tag{2.2}$$

where

$$\zeta(s; a) = \sum_{[\mathfrak{a}]=a} (N\mathfrak{a})^{-s}$$

for any ideal class a. Using the same notation for the invariants of K as in Section 4, one has

Proposition 2.2. *Let a be any ideal class in K, let*

$$\Lambda(s; a) = \pi^{-r_1 s/2}(2\pi)^{-r_2 s}\Gamma(s/2)^{r_1}\Gamma(s)^{r_2}\zeta(s; a).$$

Then $\Lambda(s; a)$ admits analytic continuation to a meromorphic function on $\mathbf{C}$ with simple poles at $s = 1$ and at $s = 0$ and it satisfies the functional equation

$$\Lambda(s; a) = |D|^{1/2-s}\Lambda(1 - s; (a\mathfrak{d})^{-1}),$$

where $\mathfrak{d}$ is the ideal class of the different of $K/\mathbf{Q}$.

The partial zeta function $\zeta(s; a)$ is meromorphic with a simple pole at $s = 1$ with residue equal to

$$\operatorname{Res}_{s=1} \zeta(s; a) = \frac{2^{r_1}(2\pi)^{r_2} R}{w\sqrt{|D|}}.$$

Summing over $a \in H(K)$, this proposition implies the analytic continuation and functional equation of $\zeta_K(s)$ as stated in Section 4.

If one knows beforehand that $\zeta_K(s)$ has at most a simple pole at $s = 1$, Proposition 2.2 actually gives a proof of the finiteness of the class number since the series for $\zeta_K(s)$, hence the expansion (2.2), are absolutely convergent for $\operatorname{Re}(s) > 1$, and since all partial zeta functions have the same residue at $s = 1$.

Remark 2.3. The analytic continuation is crucially based on the series expression (2.1) to which the Poisson summation formula can be applied, and has nothing to do with the Euler product, as the proof through partial zeta functions (which have no Euler product if $h > 1$) shows. However it is the Euler product which has the deepest arithmetic content, in particular through its local-global interpretation and its consequences on the location of the zeros of L-functions.

Assume for simplicity that $a = 1$ is the trivial ideal class, thus represented by $\mathcal{O} = \mathcal{O}_K$. The ideals in the class thus correspond bijectively with nonzero integers $z \in \mathcal{O}$ up to multiplication by units $u \in U$.

Let $z \in \mathcal{O}$, $z \neq 0$. For each embedding σ of K, we have by definition of the gamma function

$$\pi^{-s/2}\Gamma(s/2)|z|_\sigma^{-s} = \int_0^{+\infty} \exp(-\pi y|z|_\sigma^2) y^{s/2}\frac{dy}{y},$$

$$(2\pi)^{-s}\Gamma(s)|z|_\sigma^{-s} = \int_0^{+\infty} \exp(-2\pi y|z|_\sigma^2) y^{s}\frac{dy}{y}.$$

We apply the former for real embeddings and the latter for (pairs of) complex ones, and sum over $z \in \mathcal{O}$ modulo units for $\mathrm{Re}(s) > 1$. We get a formula

$$\Lambda(s; \mathcal{O}) = \int \Theta_1(y; \mathcal{O})||y||^{s/2}\frac{dy}{y} \tag{2.3}$$

where the integral is over $(\mathbf{R}^+)^{r_1+r_2}$, the coordinates corresponding to the archimedean places of K, with

$$||y|| = \prod_v y_v^{d_v}, \quad \frac{dy}{y} = \prod_v \frac{dy_v}{y_v}$$

and the kernel is almost a theta function:

$$\Theta_1(y; \mathcal{O}) = \sum_{\substack{z\in\mathcal{O}/U \\ z\neq 0}} \exp(-\pi \sum_v y_v d_v |z|_v^2).$$

The sum is absolutely convergent, uniformly for $\mathrm{Re}(s) > 1 + \delta$ for any $\delta > 0$, since this holds for $\zeta(s, \mathcal{O})$.

We now rearrange this integral formula, integrating over $t = ||y||$ first: we let

$$G_1 = \{y \mid ||y|| = 1\};$$

observe also that U acts on G_1 (we actually let u act by the "obvious" action of u^2) and the quotient G_1/U is compact by Dirichlet's Unit Theorem. We rewrite

$$\int \sum_{\mathcal{O}/U} \cdots = \int_0^{+\infty} \int_{G_1/U} \sum_{u\in U} \sum_{\mathcal{O}/U} \cdots$$

$$= \int_0^{+\infty} \int_{G_1/U} \sum_{u\in\mathcal{O}} \cdots$$

so we get after some rearranging (and taking care of roots of unity and other details):

$$\Lambda(s; \mathcal{O}) = \frac{1}{w}\int_0^{+\infty} \int_{G_1/U} (\Theta(t^{1/d}x; \mathcal{O}) - 1)t^{s/2} d\mu(x)\frac{dt}{t}$$

($d\mu(x)$ being the appropriate measure on G_1/U) where $\Theta(x; \mathcal{O})$ is a Hecke theta function (for $a = \mathcal{O}$):

$$\Theta(x; \mathcal{O}) = \sum_{z\in\mathcal{O}} \exp(-\pi <|z|^2, x>) \tag{2.4}$$

for $x \in (\mathbf{R}^+)^{r_1+r_2}$, with the somewhat awkward shorthand notation

$$<x, y> = \sum_v d_v x_v y_v \text{ and } |z|^2 = (|z|_v^2)_v.$$

In this integral expression, there is no problem at $+\infty$ for any $s \in \mathbf{C}$ because the theta function minus its constant term 1 decays very rapidly, but there might be at 0 for s small. So we split the integral over t in two parts and for t small we need to analyze the behavior of the theta function; this is where the Poisson summation formula will be useful.

Lemma 2.4. *With notation as above, the theta function satisfies*

$$\Theta(x; \mathcal{O}) = \frac{1}{\sqrt{|D|\prod x_i}}\Theta(|D|^{-2/d}x^{-1}; \mathfrak{d}^{-1}). \tag{2.5}$$

Sketch of proof. Choosing a basis (ω_i) of $\mathcal{O}$ and letting ℓ_i be the "component of ω_i" linear form, the value of the theta function at x is of the shape

$$\sum_{n\in\mathbf{Z}^d} f(n)$$

for the Schwartz function $f(n) = \exp(-\pi Q(n))$, Q being a positive definite quadratic form on $\mathbf{R}^d$ (depending on x) defined by

$$Q(n_1, \dots, n_d) = \sum_v d_v x_v |z|_v^2 = \sum_{\sigma : K\hookrightarrow\mathbf{C}} x_\sigma \Big|\sum_{j=1}^d n_j \omega_j^\sigma\Big|^2.$$

The Fourier transform of such a function is well known (by diagonalization):

$$\hat{f}(\xi) = \frac{1}{|\det(Q)|}\exp(-\pi Q'(\xi)),$$

where $Q'(x)$ is the dual quadratic form (its matrix is the inverse of that of Q). Computing the determinant, the result follows (the discriminant arises as a determinant). □

Let

$$m = \int_{G_1/U} d\mu(x) < +\infty$$

(since G_1/U is compact). We split the integral over t at $t = \alpha$, and change t into β/t in the first integral, after separating the part involving -1 which is explicitly evaluated:

$$\begin{aligned}
w\Lambda(s;\mathcal{O}) &= \int_\alpha^{+\infty}\int_{G_1/U}(\Theta(t^{1/d}x;\mathcal{O})-1)t^{s/2}d\mu(x)\frac{dt}{t} \\
&\quad+\int_0^\alpha\int_{G_1/U}\Theta(t^{1/d}x;\mathcal{O})t^{s/2}d\mu(x)\frac{dt}{t}-\frac{2m}{s}\alpha^{s/2} \\
&= \int_\alpha^{+\infty}\int_{G_1/U}(\Theta(t^{1/d}x;\mathcal{O})-1)t^{s/2}d\mu(x)\frac{dt}{t} \\
&\quad+\beta^{s/2}\int_{\beta/\alpha}^{+\infty}\int_{G_1/U}\Theta(\beta^{1/d}t^{-1/d}x;\mathcal{O})t^{-s/2}d\mu(x)\frac{dt}{t}-\frac{2m}{s}\alpha^{s/2} \\
&= \int_\alpha^{+\infty}\int_{G_1/U}(\Theta(t^{1/d}x;\mathcal{O})-1)t^{s/2}d\mu(x)\frac{dt}{t}+|D|^{1/2}\beta^{(s-1)/2} \\
&\quad\cdot\int_{\beta/\alpha}^{+\infty}\int_{G_1/U}\Theta(|D|^{-2/d}\beta^{-1/d}t^{1/d}x;\mathfrak{d}^{-1})t^{(1-s)/2}d\mu(x)\frac{dt}{t} \\
&\quad-\frac{2m}{s}\alpha^{s/2} \\
&= \int_\alpha^{+\infty}\int_{G_1/U}(\Theta(t^{1/d}x;\mathcal{O})-1)t^{s/2}d\mu(x)\frac{dt}{t}+|D|^{1/2}\beta^{(s-1)/2} \\
&\quad\cdot\int_{\beta/\alpha}^{+\infty}\int_{G_1/U}(\Theta(|D|^{-2/d}\beta^{-1/d}t^{1/d}x;\mathfrak{d}^{-1})-1)t^{(1-s)/2}d\mu(x)\frac{dt}{t} \\
&\quad-\frac{2m}{s}\alpha^{s/2}-|D|^{-1/2}\beta^{(s-1)/2}\frac{2m}{1-s}\left(\frac{\beta}{\alpha}\right)^{(1-s)/2}.
\end{aligned}$$

One uses the fact that $d\mu$ is invariant under $x\mapsto x^{-1}$ in this computation and that $\|x\|=1$ for $x\in G_1$, when applying the transformation formula for the theta function.

In the final formula, derived under the assumption that $\mathrm{Re}(s)>1$ so that every manipulation is justified, both integrals are now entire functions of $s\in\mathbf{C}$ if $\alpha>0$, $\beta>0$. Hence it follows that $\Lambda(s;\mathcal{O})$ is meromorphic with simple poles at $s=1$ and $s=0$.

Moreover, taking $\beta=|D|^{-2}$, $\alpha=1/|D|$, one gets

$$\begin{aligned}
\Lambda(s;\mathcal{O}) = \frac{1}{w}\Bigg\{&|D|^{1/2-s}\int_{\beta/\alpha}^{+\infty}\int_{G_1/U}(\Theta(t^{1/d}x;\mathfrak{d}^{-1})-1)t^{(1-s)/2}d\mu(x)\frac{dt}{t} \\
&+\int_\alpha^{+\infty}\int_{G_1/U}(\Theta(t^{1/d}x;\mathcal{O})-1)t^{s/2}d\mu(x)\frac{dt}{t}-\frac{2m}{s}\alpha^{s/2} \\
&-|D|^{1/2-s}\frac{2m}{1-s}\left(\frac{\beta}{\alpha}\right)^{(1-s)/2}\Bigg\}.
\end{aligned}\tag{2.6}$$

and since $\beta/\alpha = \alpha$, the functional equation follows immediately.

Computing the residues explicitly requires the computation of m, which is done by describing more explicitly a fundamental domain in G_1 for the action of U.

Example 2.5. For the Riemann zeta function we get

$$\pi^{-s/2}\Gamma(s/2)\zeta(s) = \frac{1}{2}\int_0^{+\infty} (\theta(y) - 1)y^{s/2}\frac{dy}{y},$$

with

$$\theta(y) = \sum_{n\in\mathbf{Z}} e^{-\pi n^2 y}.$$

The functional equation for θ is

$$\theta(y) = y^{-1/2}\theta(y^{-1}).$$

Remark 2.6. We have used Dirichlet's Unit Theorem for K to say that G_1/U is compact: those two statements are in fact obviously equivalent, and equivalent to the statement that the measure m of G_1/U is finite. Now since the first integral expression (2.3) is a priori valid for $\mathrm{Re}(s) > 1$, it is not difficult to perform the previous computation without the information that $m < +\infty$, and *deduce* this from (2.6). (I learned this proof from Bill Duke.)

3 Zeros of L-functions, explicit formula

For all the L-functions considered in Chapter 1, the Generalized Riemann Hypothesis (abbreviated GRH) is expected to hold as for the Riemann zeta function: all nontrivial zeros of a Hecke L-function[1] $L(\chi, s)$ of a number field K, i.e., those in the critical strip $0 < \mathrm{Re}(s) < 1$, should be on the critical line $\mathrm{Re}(s) = 1/2$. Although this is still completely open, much has been proved about the zeros so that in many circumstances one can manage to prove unconditionnally results which were at first only established on the assumption of GRH.

The first basic results give a rough idea of the distribution of the zeros. Let

$$\begin{aligned} N(\chi; T_1, T_2) &= |\{\rho = \beta + i\gamma \mid L(\chi, \rho) = 0,\ 0 < \beta < 1 \text{ and } T_1 \leqslant \gamma \leqslant T_2\}| \\ N(\chi; T) &= N(\chi; -|T|, |T|). \end{aligned}$$

Also let Λ be the van Mangoldt function for K, i.e., for any nonzero integral ideal $\mathfrak{a} \subset \mathcal{O}$ we have

$$\begin{cases} \Lambda(\mathfrak{a}) = \log N\mathfrak{p} & \text{if } \mathfrak{a} = \mathfrak{p}^k \text{ for some } k \geqslant 1 \\ \Lambda(\mathfrak{a}) = 0 & \text{otherwise,} \end{cases}$$

[1] All the other L-functions of Chapter 1 are special cases of Hecke L-functions.

the point being that the Euler product (5.3) for $L(\chi, s)$ implies that the logarithmic derivative of $L(\chi, s)$ has an absolutely convergent Dirichlet series expansion for $\mathrm{Re}(s) > 1$ given by

$$-\frac{L'}{L}(\chi, s) = \sum_{\mathfrak{a}} \chi(\mathfrak{a})\Lambda(\mathfrak{a})(N\mathfrak{a})^{-s}. \tag{3.1}$$

Proposition 3.1. *Let K be a number field of degree $d = [K : \mathbf{Q}]$, χ a primitive Hecke character of K of modulus $\mathfrak{m}$. Then we have*

$$N(\chi; T) = \frac{T}{2\pi}\log\Big(\frac{T^d|D|N\mathfrak{m}}{(2\pi e)^d}\Big) + O(\log(T^d|D|N\mathfrak{m})) \tag{3.2}$$

for $T \geqslant 2$, the estimate being uniform in all parameters. For any $\varepsilon > 0$, the series

$$\sum_{\rho} \frac{1}{|\rho|^{1+\varepsilon}} \tag{3.3}$$

is absolutely convergent.

The proof of this asymptotic formula is straightforward in principle: one applies Cauchy's Theorem to the logarithmic derivative of the function $\Lambda(\chi, s)$ (which has the same zeros as $L(\chi, s)$) in the rectangle $[-1/2, 3/2] \times [-T, T]$. The major contribution comes from the gamma factor for χ. To control the part coming from $L(\chi, s)$, one uses the expansion of $\Lambda(\chi, s)$ in the Weierstrass product, due to Hadamard: if $\chi \neq 1$, then[2] there exist a and $b \in \mathbf{C}$, depending on χ, such that

$$\Lambda(\chi, s) = e^{a+bs}\prod_{\rho}\Big(1 - \frac{s}{\rho}\Big)e^{s/\rho},$$

the product being over all nontrivial zeros of $L(\chi, s)$ (this is the general theory of entire functions of finite order, of which $\Lambda(\chi, s)$ is an example; (3.3) is actually also part of this theory). This product is used to show that $-L'(\chi, s)/L(\chi, s)$ is, with good precision, the sum of the terms arising from the zeros close to s, namely those with $|s - \rho| \leqslant 1$ (we know *a fortiori* from (3.2) that there are a fair number of them, if s is in the critical strip). The constant a is not a problem, but b requires some ingenuity to be dealt with, and not too much is known about it as a function of χ.

The next important information, which is the key to the equidistribution theorems for prime ideals (the Chebotarev Density Theorem), is that there is no zero on the line $\mathrm{Re}(s) = 1$ (the edge of the critical strip). This was first proved by Hadamard and de la Vallée Poussin (independently) in 1896, and their method still remains essentially the only known to prove a zero-free region for L-functions. It is also here that the so-called Landau–Siegel zeros[3] first appear.

---

[2] One has to multiply by $s(1 - s)$ if $\chi = 1$ to get rid of the poles.

[3] Iwaniec and Sarnak have proposed this name for what was usually called Siegel zeros on the strength of Landau's contribution [L].

Theorem 3.2. *There exists an absolute constant $c_1 > 0$ with the property that for any K and χ as above, the Hecke L-function $L(\chi, s)$ has no zero in the region $s = \sigma + it$ where*

$$\sigma > 1 - \frac{c_1}{\log(T^d|D|N\mathfrak{m})},$$

except *possibly for a single simple real zero β_1. The latter can only exist if χ is a quadratic (real) character, and it is then called the* Landau–Siegel zero *for χ (and c_1).*[4]

In what follows c_1 is fixed such that the theorem holds, and we will let $\delta^*(\mathfrak{m}) = 1$ if there is a Landau–Siegel zero for a Hecke character of K modulo $\mathfrak{m}$, and $\delta^*(\mathfrak{m}) = 0$ otherwise. One can show easily that the corresponding real character modulo $\mathfrak{m}$ is unique.

Example 3.3. The classical case is when $K = \mathbf{Q}$ and then the possible existence of β_1 for a primitive odd quadratic character χ modulo q is directly related to the possibility that the class number of the corresponding imaginary quadratic field $K = \mathbf{Q}(\sqrt{-q})$ is very small. This follows easily from the class number formula (4.5). Improving the "trivial bound" $h(K) \gg q^{-1/2}$ is extremely hard, although the Generalized Riemann Hypothesis implies that, in fact,

$$L(\chi, 1) \gg \frac{1}{\log q}, \text{ hence } h(K) \gg \frac{\sqrt{q}}{\log q}.$$

Siegel [Si] has proved that for any $\varepsilon > 0$ one has

$$L(\chi, 1) \gg_\varepsilon q^{-\varepsilon} \tag{3.4}$$

for $q > 1$, but this result is *noneffective*, i.e., for given $\varepsilon > 0$ there is no (known) way of really computing a constant $C(\varepsilon)$ such that the promised inequality

$$L(\chi, 1) \geqslant C(\varepsilon) q^{-\varepsilon}$$

holds for $q > 1$. This ineffectivity is similar to that present in the Thue–Siegel theorem: the proof of Siegel's estimate is based on assuming that there exists χ with very small $L(\chi, 1)$ and using this hypothetical character to prove a lower bound for the others. As it is not expected that such a bad character exists, one understands why it seems so difficult to make Siegel's theorem effective.

Goldfeld's theorem [Go], based on known cases of the Birch and Swinnerton–Dyer conjecture for elliptic curves, is the only effective improvement on the trivial bound currently known:

$$L(\chi, 1) \gg \frac{(\log q)}{\sqrt{q}(\log\log q)}$$

for $q \geqslant 3$, with an effective implied constant (which has indeed been computed by Oesterlé). Only a logarithm factor is gained.

---

[4] This is well defined only after fixing a possible value of c_1.

Roughly speaking, the hypothetical character used by Siegel has small $L(\chi, 1)$ if and only if it has a real zero close to $s = 1$, and the ineffectivity comes here because such a zero would contradict the Riemann Hypothesis and is not believed to exist. Its good effect would be to "repulse" other zeros, thereby providing a lower bound for them. In Goldfeld's argument,[5] a *real* lever is used, namely an L-function which does have a real zero, which necessarily must be the central critical point $s = 1/2$. Being far from $s = 1$ makes its repulsing effect much weaker, with two consequences: first, that it must be a zero of order $\geqslant 3$ for the argument to go through, and secondly, that the resulting lower bound is much weaker than Siegel's.

Remark 3.4. The methods which yield Theorem 3.2 cannot be expected to yield much better results: indeed, a *lower bound* for $L(\chi, s)$ in the region described is actually produced (explaining also the example above), and one can show for instance that there is no "simple" lower-bound for $L(\chi, s)$ in (say) a strip of positive width $1 - \delta < \mathrm{Re}(s) \leqslant 1$. However the exponential sum methods of Vinogradov (see, e.g., [Ti]) can be used to enlarge a little bit the zero-free region, which is sometimes significant in applications.

From the zero-free region for the Dedekind zeta function, the original method of Hadamard and de la Vallée Poussin (or the explicit formula, see below) derives the analogue of the Prime Number Theorem. Fix a modulus $\mathfrak{m}$ and a ray-class $a \in H_{\mathfrak{m}}$ (see Section 5), and let

$$\pi_K(X; \mathfrak{m}, a) = |\{\mathfrak{p} \mid \text{the class of } \mathfrak{p} \text{ is } a \text{ and } N\mathfrak{p} \leqslant X\}|.$$

Then we have

$$\pi_K(X; \mathfrak{m}, a) = \frac{1}{|H_{\mathfrak{m}}|}\mathrm{li}(X) + \frac{1}{|H_{\mathfrak{m}}|}\delta^*(\mathfrak{m})X^{\beta_1} + O(X\exp(-c_2\sqrt{\log(X|D|N\mathfrak{m})})), \tag{3.5}$$

as $X \to +\infty$. Here the estimate is uniform in all parameters, and c_2 is absolute and effective. Since $\beta_1 < 1$, for a fixed $\mathfrak{m}$ this gives the asymptotic behavior of $\pi_K(X; \mathfrak{m}, a)$, but in most applications uniformity in $\mathfrak{m}$ is the key issue.

For example, when $K = \mathbf{Q}$ (the classical case), Siegel's estimate (3.4) immediately implies the Siegel–Walfisz theorem: for any $A > 0$, we have

$$\pi(X; q, a) \sim \frac{1}{\varphi(q)}\mathrm{li}(X)$$

uniformly for all $q < (\log X)^A$ and all a modulo q. If there are Landau–Siegel zeros, this is noneffective.

The Generalized Riemann Hypothesis implies the corresponding statement uniformly for $q < \sqrt{X}(\log X)^{-2}$. In many applications to analytic number theory,[6] it is indispensable to have such uniformity, although often only on average.

---

[5] J. Friedlander had similar ideas.

The best known results—proved using the spectral theory of automorphic forms (see Chapter 3)—for primes in arithmetic progressions go *beyond* what is immediately provable from GRH (Bombieri–Friedlander–Iwaniec); they are commonly used in applications.

The same results and difficulties occur in the Chebotarev density theorem for Artin L-functions, often exacerbated because the degree of interesting families of fields is larger than that of cyclotomic fields, making even the form of the prime ideal theorem based on GRH insufficient for applications.[7]

Considering the fact that so little progress has been made over almost a century on the Generalized Riemann Hypothesis for individual L-functions, it is hard to avoid thinking that some deep structure lies undiscovered: this is all the more tempting when compared with the case of function fields (briefly mentioned in Section 6) with the rich geometric and cohomological formalism. In recent years, a lot of work has been devoted to trying to probe evidence and clues to such a structure by analyzing (often on GRH) the finer vertical distribution of zeros of L-functions: this led to the remarkable discovery of links with the distribution of eigenvalues of random matrices of large rank. For $\zeta(s)$ this was first attempted by Montgomery (leading to the pair-correlation conjecture), and has been much generalized in particular by Rudnick and Sarnak [RS]. Some of the conjectures emerging have been proved in the function field case by Katz and Sarnak [KS]. As a spectacular vindication of the interest of such studies, Soudararajan and Conrey [CS] were able to prove, after using heuristics based on random matrices, that for a positive proportion of real Dirichlet characters, the L-function $L(\chi, s)$ has no Landau–Siegel zero!

We now come to the explicit formula. The version below obviously reduces to Proposition 2.1 when applied to $\zeta(s)$.

Proposition 3.5. *Let* $\varphi :]0, +\infty[\to \mathbf{C}$ *be a* C^∞ *function with compact support, and let*

$$\hat{\varphi}(s) = \int_0^{+\infty} \varphi(x) x^{s-1} dx,$$

be its Mellin transform, which is entire and decays rapidly in vertical strips. Let

$$\psi(x) = \frac{1}{x}\varphi(x^{-1}).$$

[6]For instance, the *Titchmarsh divisor problem* of estimating asymptotically as $X \to +\infty$ the sum

$$\sum_{p \leqslant X} d(p-1)$$

which can be rewritten as

$$2 \sum_{d \leqslant \sqrt{X-1}} (\pi(X; d, 1) - \pi(d^2+1; d, 1)) + O(\sqrt{X}).$$

[7]For instance, if $E/\mathbf{Q}$ is an elliptic curve without CM, its d-torsion $\mathbf{Q}(E[d])$ has degree roughly d^4 by Serre's Theorem, and GRH only yields an asymptotic formula uniformly for $d < X^{1/8}$.

Then we have

$$\sum_{\mathfrak{a}} \Lambda(\mathfrak{a})\big(\chi(\mathfrak{a})\varphi(N\mathfrak{a}) + \overline{\chi}(\mathfrak{a})\psi(N\mathfrak{a})\big)$$

$$= (\log|D|N\mathfrak{m})\varphi(1) + \delta(\chi)\int_0^{+\infty} \varphi(x)dx$$

$$- \sum_{\substack{\zeta(\rho)=0 \\ 0<\mathrm{Re}(\rho)<1}} \hat{\varphi}(\rho) + \frac{1}{2i\pi}\int_{(-1/2)} \left(\frac{\Gamma'}{\Gamma}(\xi_\infty, s) - \frac{\Gamma'}{\Gamma}(\xi_\infty, 1-s)\right)\hat{\varphi}(s)ds$$

$$+ \mathrm{Res}_{s=0}\left(-\frac{L'}{L}(\chi, s)\hat{\varphi}(s)\right), \tag{3.6}$$

where $\delta(\chi) = 1$ if χ is trivial and is $= 0$ otherwise, and the sum over zeros includes multiplicity.

Although seemingly complicated, the last two terms are in most circumstances very easy to deal with (using the explicit form of the gamma factor and Stirling's formula for instance).

Sketch of proof. By Mellin inversion and (3.1) one has

$$\sum_{\mathfrak{a}} \chi(\mathfrak{a})\Lambda(\mathfrak{a})\varphi(N\mathfrak{a}) = \frac{1}{2i\pi}\int_{(3/2)} -\frac{L'}{L}(\chi, s)\hat{\varphi}(s)ds.$$

One moves the line of integration to $\mathrm{Re}(s) = -1/2$ (this requires some simple estimates on the growth of the L-function in vertical strips, which is easy to get). The poles which occur are at $s = 1$ (if $\chi = 1$), possibly at $s = 0$ (depending on the shape of the gamma factor)—with contribution corresponding to the second and last term, respectively, of the right-hand side of (3.6)—and at the nontrivial zeros ρ of $L(\chi, s)$. The latter are simple with residue $-k\hat{\varphi}(\rho)$ where k is the multiplicity of ρ; thus their contribution is the third term in (3.6).

On the line $\mathrm{Re}(s) = -1/2$, one uses the functional equation (5.4) to obtain the relation

$$-\frac{L'}{L}(\chi, s) = (\log|D|N\mathfrak{m}) + \left(\frac{\Gamma'}{\Gamma}(\xi_\infty, s) - \frac{\Gamma'}{\Gamma}(\xi_\infty, 1-s)\right) - \frac{L'}{L}(\overline{\chi}, 1-s).$$

The middle term yields the fourth term on the right-hand side of (3.6). Then one writes

$$\frac{1}{2i\pi}\int_{(-1/2)} -\frac{L'}{L}(\overline{\chi}, 1-s)\hat{\varphi}(s)ds = -\frac{1}{2i\pi}\int_{(3/2)} -\frac{L'}{L}(\overline{\chi}, w)\hat{\varphi}(1-w)dw$$

and one can again apply (3.1) since w is back in the region of absolute convergence. The result follows after expanding in the Dirichlet series and observing (and using also in the term involving $\log |D|N\mathfrak{m})$ that

$$\frac{1}{2i\pi}\int_{(3/2)} \hat{\varphi}(1-w)x^{-w}dw = \psi(x)$$

by the simple fact that $\hat{\varphi}(1-w) = \hat{\psi}(w)$. □

4 The order of magnitude of L-functions in the critical strip

There are many applications where the order of magnitude of L-functions in the critical strip is very important. Of course, it is necessary to have at least some simple estimate for all the usual analytic manipulations (contour shifts, etc.) to succeed, but there are much deeper reasons. For instance, the class number formula involves interesting invariants of number fields and relates them to the value (or residue) of L-functions at $s = 1$. As for the case of imaginary quadratic fields already discussed, much of the information gained about these invariants has come from working with the L-functions. And of course, there are close links with the distribution of the zeros of the L-functions.

This first example also illustrates that "order of magnitude" does not refer only to the size as the complex argument s varies, but may refer instead to the size in terms of the conductor of the L-function: indeed, in many arithmetic applications, this is the really interesting problem and often one does not need strong bounds in terms of $|s|$.

From the proof of the analytic continuation of the L-functions, one does not get much information, but enough to bootstrap further investigations (and it is crucial to get it): namely from the formula (2.4), or its analogues, it follows that $\Lambda(\chi, s)$ is an entire[8] function of order at most 1, i.e., for $s \in \mathbf{C}$ we have

$$\Lambda(\chi, s) \ll \exp(|s|^{1+\varepsilon})$$

for any $\varepsilon > 0$, the implied constant depending on χ and ε. Since the inverse of the gamma factor is known to be an entire function of order $= 1$, it follows that $L(\chi, s)$ is also.

This is actually very far from the truth. Of course, for $\mathrm{Re}(s) > 1$, where the series converges absolutely, it follows that

$$L(\chi, s) \ll 1, \text{ uniformly for } \mathrm{Re}(s) > 1+\delta,\ \delta > 0,$$

for any χ, the implied constant depending only on K. For $\mathrm{Re}(s) < 0$, on the other hand, this implies, thanks to the functional equation, Stirling's formula and the

[8] Multiply by $s(1-s)$ if $\chi = 1$.

shape of the gamma factor, that

$$L(\chi, s) \ll |s|^{(1-\sigma)/2}, \text{ uniformly for } \sigma = \mathrm{Re}(s) < -\delta,\ \delta > 0.$$

The well-known Phragmen–Lindelöf principle of complex analysis implies that a function of order 1 polynomially bounded in terms of $|s|$ on the boundary of a vertical strip is actually polynomially bounded inside the strip, and that the "rate of growth" is a convex function of the real part. This is in general best possible but for L-functions one expects actually a very different answer:

Conjecture 4.1. *Let K be a number field, χ a Hecke character to modulus $\mathfrak{m}$ with weight ξ_∞. Then we have*

$$L(\chi, s) \ll (|D| N\mathfrak{m}(1 + |s|))^{\varepsilon}$$

for any $\varepsilon > 0$, for $1/2 \leqslant \mathrm{Re}(s) < 1$, with an implied constant depending only on ε. If $\chi = 1$, multiply on the left by $(s - 1)$.

This is called the Lindelöf Conjecture. One can quickly give two motivations: for the first, it is implied by GRH (see, e.g., [Ti, 14.2]). The second reason is more interesting and is based on a principle in analytic number theory sometimes referred to (rather incorrectly) as the *approximate functional equation*, a variant of which is the next proposition:

Proposition 4.2. *Let K and χ be as above and let $s = \sigma + it$ with $0 < \sigma < 1$. Let $G(w)$ be any function real-valued on $\mathbf{R}$ and holomorphic in the strip $-4 < \mathrm{Re}(s) < 4$, with rapid decay as $|\mathrm{Im}(w)| \to +\infty$, and such that*

$$\begin{cases} G(-w) = G(w), \\ G(0) = 1, \\ G(w)\Gamma(\xi_\infty, w) \text{ is holomorphic for } w = 0, -1, -2, -3. \end{cases}$$

Let $X, Y > 0$ be real numbers such that $XY = |D| N\mathfrak{m}$. Then we have

$$L(\chi, s) = \sum_{\mathfrak{a}} \frac{\chi(\mathfrak{a})}{(N\mathfrak{a})^s} V_s\left(\frac{N\mathfrak{a}}{X}\right) + \varepsilon(\chi)(|D| N\mathfrak{m})^{1/2-s} \sum_{\mathfrak{a}} \frac{\overline{\chi}(\mathfrak{a})}{(N\mathfrak{a})^{1-s}} W_s\left(\frac{N\mathfrak{a}}{Y}\right) \tag{4.1}$$

where V_s and W_s are the functions defined for $y > 0$ by

$$V_s(y) = \frac{1}{2i\pi} \int_{(3)} G(w) y^w \frac{dw}{w},$$

$$W_s(y) = \frac{1}{2i\pi} \int_{(3)} \frac{\Gamma(\xi_\infty, 1 - s + w)}{\Gamma(\xi_\infty, s + w)} G(w) y^w \frac{dw}{w}.$$

As the (simple) proof will show, the choice of $G(w)$ is rather arbitrary and can be modulated in many ways. It is obviously very easy to write down a concrete choice of G if need be. (Note that G may depend on s; however, very often the point s is fixed in the applications and it is the character which varies).

The point of this formula is that V_s and W_s can be easily estimated using contour shifts and the Stirling formula. This shows that they behave much as cutoff functions, i.e., they decay very rapidly for y large. More precisely, the decay rate in terms of s is such that in (4.1), one sees easily that the first sum is essentially a sum over ideals of norm $N\mathfrak{a} \leqslant X$, while the second is over ideals of norm $N\mathfrak{a} \leqslant Y(1+|s|)^{d/2}$ (this means the tails are very small). So (4.1) gives an expression of the L-function *inside the critical strip* by very rapidly convergent series. Such a formula is the basis of most (analytic) work on L-functions outside the region of absolute convergence.

In any case, choosing $X = Y = \sqrt{|D|N\mathfrak{m}}$, estimating trivially (i.e., term by term) both sums (recall that $|\varepsilon(\chi)| = 1$), one immediately gets

Corollary 4.3. *We have for* $s = \sigma + it$ *with* $0 < \mathrm{Re}(s) < 1$

$$L(\chi, s) \ll_\varepsilon ((1 + |t|)^d |D| N\mathfrak{m})^{\ell(\sigma)+\varepsilon}$$

for any $\varepsilon > 0$*, the implied constant depending only on* ε*, where* $\ell(\sigma)$ *is the affine function on* $[0, 1]$ *such that* $\ell(1) = 0$ *and* $\ell(0) = 1/2$*. If* $\chi = 1$*, multiply on the left by* $(s - 1)$*.*

In particular (the most interesting case) we have

$$L(\chi, s) \ll_\varepsilon ((1 + |t|)^d |D| N\mathfrak{m})^{1/4+\varepsilon}$$

on the critical line $\mathrm{Re}(s) = 1/2$.

The bound of the corollary is called the *convexity bound.* It can also be derived by the Phragmen–Lindelöf principle. However, from the point of view of (4.1), one sees that it amounts indeed to estimating individually each term in the sums. However those terms, for $N\mathfrak{a}$ "small," are oscillating nontrivially, this coming both from $\chi(\mathfrak{a})$ (if $\chi \neq 1$) and from $(N\mathfrak{a})^{-s} = (N\mathfrak{a})^{-\sigma-it}$. Such oscillations might be expected to yield some cancellation in the sum, hence a better estimate for $L(\chi, s)$. Indeed the well-known *square-rooting philosophy* of oscillatory sums,[9] if it applies here, would immediately yield the Lindelöf conjecture (by partial summation).

All this discussion extends quite easily to the more general automorphic L-functions discussed in the other chapters. A significant phenomemon, philosophically badly understood, is that many significant problems can be completely solved

[9] If

$$S = \sum_{1 \leqslant n \leqslant N} e(\theta_n)$$

is a truly oscillating (unbiased) exponential sum, then $|S| \ll \sqrt{N}$, "up to small amounts."

(qualitatively) by proving, for some family of L-functions, *any* estimate which improves on the convexity bound of the corollary (i.e., replacing the exponent $\ell(\sigma)+\varepsilon$ by $\ell(\sigma)-\delta$ for *some* $\delta > 0$, the size of which is irrelevant to the conclusion). Even apart from applications, the sketch above shows that any such estimate is tantamount to showing that there is some nontrivial cancellation in the two sums in (4.1) and this is obviously a deep arithmetical result.

Such *convexity breaking estimates* go back to Weyl, who proved, for $\zeta(s)$, that

$$\zeta(1/2+it) \ll_\varepsilon (1+|t|)^{1/6+\varepsilon}$$

for $t \in \mathbf{R}$, the constant depending only on $\varepsilon > 0$. Nowadays, many more cases are known for Dirichlet characters both in terms of s and in terms of the conductor, the latter being the most interesting (and most difficult) case, first proved by Burgess and only recently improved by Conrey and Iwaniec (for quadratic characters, using crucially automorphic L-functions).

Remarkable applications of these estimates for modular forms are due to Iwaniec (equidistribution of integral points on spheres), Sarnak and others. Sarnak [Sa] in particular has developed the underlying philosophy and shown how it relates to "arithmetic quantum chaos," i.e., the study of the repartition of eigenfunctions of the Laplace operator on negatively curved manifolds as the eigenvalue gets large.

We finish with the proof of the proposition.

Proof. Consider the integral over the line $\mathrm{Re}(s) = 3$:

$$I = \frac{1}{2i\pi}\int_{(3)} L(\chi, s+w)G(w)\frac{dw}{w}.$$

By expanding into Dirichlet series the L-function, it is equal to

$$\sum_{\mathfrak{a}} \frac{\chi(\mathfrak{a})}{(N\mathfrak{a})^s} V_s\left(\frac{N\mathfrak{a}}{X}\right)$$

i.e., the first sum in (4.1).

On the other hand, shifting the contour to the line $\mathrm{Re}(s) = -3$ (using the fact that $G(w)$ has been chosen to cancel any possible pole of the gamma factor), and picking up the single simple pole at $s = 0$, we obtain

$$I = G(0)L(\chi, s) + \frac{1}{2i\pi}\int_{(-3)} L(\chi, s+w)G(w)\frac{dw}{w}.$$

Now apply the functional equation of $L(\chi, s)$: the last integral (say J) becomes

$$J = \varepsilon(\chi)(|D|N\mathfrak{m})^{1/2-s} \cdot \frac{1}{2i\pi} \int_{(-3)} \frac{\Gamma(\xi_\infty, 1-s-w)}{\Gamma(\xi_\infty, s+w)} L(\overline{\chi}, 1-s-w) G(w) \left(\frac{X}{|D|N\mathfrak{m}}\right)^{w} \frac{dw}{w}$$

$$= -\varepsilon(\chi)(|D|N\mathfrak{m})^{1/2-s} \cdot \frac{1}{2i\pi} \int_{(3)} \frac{\Gamma(\xi_\infty, 1-s-w)}{\Gamma(\xi_\infty, s+w)} L(\overline{\chi}, 1-s+w) G(w) \left(\frac{X}{|D|N\mathfrak{m}}\right)^{-w} \frac{dw}{w}.$$

We are back in the region of absolute convergence, and expanding again into Dirichlet series, it follows that

$$J = -\varepsilon(\chi)(|D|N\mathfrak{m})^{1/2-s} \sum_{\mathfrak{a}} \frac{\overline{\chi}(\mathfrak{a})}{(N\mathfrak{a})^{1-s}} W_s\left(\frac{N\mathfrak{a}}{Y}\right).$$

Hence the result since $G(0) = 1$. □

5 Odds and ends, including poles

One of the features, partly known and partly conjectural, of the theory of L-functions, is that it seems that (when unitarily normalized, i.e., the critical line is translated to $\mathrm{Re}(s) = 1/2$) the point $s = 1$ is the only possible pole for an (automorphic) L-function, and further that such a pole is always accounted for by the simple pole of the Riemann zeta function, in the sense that the L-function $L(f, s)$ has a factorization

$$L(f, s) = \zeta(s)^k L(f_1, s),$$

where $L(f_1, s)$ is another L-function which is entire. This is indeed the case for the Dedekind zeta function as (5.1) shows, and is also true in a number of other cases (for the Rankin–Selberg convolution L-functions $L(f \otimes \bar{f}, s)$ on $GL(2)$, for instance, where the "quotient" f_1 is the symmetric square L-function, as will be explained in other chapters). It is known also for Artin L-functions, using Brauer's Theorem and the nonvanishing of Hecke L-functions at $s = 1$. The order of the pole at $s = 1$ of $L(\rho, s)$ is then the multiplicity of the trivial representation in ρ. This formulation is recurrent in many conjectures of arithmetic-geometry, such as the Tate conjectures about the dimension of spaces of k-cycles on varieties over finite fields.

A pole occurs typically (only?) when the coefficients of the L-function (as a Dirichlet series) are positive. Analytically, this conjecture can be paraphrased as saying that "the harmonic series is the only really divergent one."

The following cute observation (mentioned to me by Serre) gives an interesting illustration: recall Euler's formula for the values of the zeta function at negative integers

$$\zeta(0) = -1/2, \text{ and } \zeta(1-k) = -b_k/k \text{ for } k \geqslant 2 \tag{5.1}$$

where b_k is the kth Bernoulli number (see, e.g., [Wa]). Now theorems of von Staudt and J.C. Adams [IR, Chapter 15, Theorem 3 and Proposition 15.2.4] prove that the denominator of $-b_k/k$ contains all the primes p such that $p - 1 \mid 2k$, *and only them*, so that for a p in the denominator we have

$$x^{k-1} \equiv x^{-1} \pmod{p}.$$

Hence even the "poles" modulo primes of $\zeta(1-k)$ are still "explained" by the divergence of the harmonic series! In the other direction, maybe it is not so surprising that the numerator of Bernoulli numbers should remain so mysterious.

We finally only mention that the congruence properties of the values of the zeta function at negative integers were the motivation for the discovery by Leopoldt of the p-adic zeta function, later much generalized by others to p-adic L-functions of various kinds. See, for instance, [Wa, Chapter 5] for an introduction.

3
Classical Automorphic Forms

E. Kowalski

1 Introduction

With automorphic forms and their associated L-functions, one enters into new territory; the catch-phrase here is that we will describe the $GL(2)$ analogue of the $GL(1)$ theory of Dirichlet characters (i.e., over $\mathbf{Q}$). The corresponding work for more general groups and general base field K will be described in later chapters.

For more complete accounts of this beautiful theory, see for instance the books of Iwaniec ([I1], [I2]), Miyake [Mi] or Shimura [Sh], Chapter VII of Serre's book [Se1] or Chapter I of Bump's book [Bu].

2 Three motivations for automorphic forms

First we will give three reasons for introducing automorphic forms. Two are obviously related to problems mentioned in the first two chapters, and the third has a very natural algebro-geometric content. One of the amazing features of automorphic forms is that one could find many other apparently disjoint motivating examples; those included here make no mention of such important topics as the links with Galois representations, congruences, partitions, etc.

2.1 *Theta functions*

The proof of the functional equation of the Riemann zeta function (by specialization of that in Chapter 2) makes use of the basic theta function

$$\theta(z) = \sum_{n\in\mathbf{Z}} e(n^2 z/2) \tag{2.1}$$

for $z = iy$, $y > 0$. In fact, this defines a function holomorphic on the ubiquitous Poincaré upper half-plane $\mathbf{H} = \{z \in \mathbf{C} \mid y > 0\}$. The Poisson summation formula (with the fact that $x \mapsto e^{-\pi x^2}$ is self-dual for the Fourier transform) implies by analytic continuation the functional equation

$$\theta\Big(-\frac{1}{z}\Big) = (-iz)^{1/2}\theta(z),$$

where $(-iz)^{1/2}$ is given by the branch of this function on $\mathbf{H}$ which sends iy to $\sqrt{y}$.

Obviously $\theta(z)$ is also 1-periodic so there is some transformation formula for $f(\gamma z)$ where $z \mapsto \gamma z$ is any mapping in the group generated by $z \mapsto z+1$ and $z \mapsto -1/z$. This group $\bar{\Gamma}$ turns out to be isomorphic to $PSL(2, \mathbf{Z})$ acting on $\mathbf{H}$ by

$$\begin{pmatrix} a & b \\ c & d \end{pmatrix} \cdot z = \frac{az+b}{cz+d}. \tag{2.2}$$

Similarly, it follows that for any $k \geqslant 1$, $\theta(2z)^k$ satisfies similar transformation formulae. Since

$$\theta(2z)^k = \sum_{n \geqslant 0} r_k(n) e(nz)$$

with $r_k(n)$ the number of representations of n as sums of k squares of integers, it is not surprising that much classical interest was expanded around such functions. Identities such as

$$\begin{aligned} r_2(n) &= 4 \sum_{d \mid n} \chi_4(d) \\ r_4(n) &= 8(2+(-1)^n) \sum_{\substack{d \mid n \\ d \text{ odd}}} d \\ r_6(n) &= 16 \sum_{d \mid n} d^2 \chi_4\Big(\frac{n}{d}\Big) - 4 \sum_{d \mid n} d^2 \chi_4(d) \end{aligned} \tag{2.3}$$

can be derived by (more or less) elementary or ad hoc means, but the theory of modular forms can explain the general shape of such formulas, why there is none so "elementary" for k large, and yields asymptotic relations implying that the size of $r_k(n)$ is about n^{k-1} for $k \geqslant 5$. Much deeper results are due to Kloosterman for general quadratic forms in $k = 4$ variables and to Iwaniec for $k = 3$ variables (see [I1, 5-3, 11-4]), yielding for instance equidistribution results for integral points on the sphere.

2.2 *Counting solutions of determinant equations*

In studying the order of magnitude of $\zeta(s)$ or of other L-functions on the critical strip, a natural analytic approach is through the moments

$$I_k(T) = \int_{-T}^{T} |\zeta(1/2+it)|^k dt$$

since if we could prove that for all $k \geqslant 1$ we have

$$I_k(T) \ll_{k,\varepsilon} T^{1+\varepsilon} \text{ for } T \geqslant 1, \tag{2.4}$$

the Lindelöf Conjecture $\zeta(1/2+it) \ll_\varepsilon (1+|t|)^\varepsilon$ would easily follow. Note that even for a single $k \geqslant 1$, (2.4) is a confirmation of the Lindelöf Conjecture on average, stronger for larger values of k, and as such does have many applications in analytic number theory.

For $k = 2$ an asymptotic evaluation of $I_2(T)$ is rather simple and classical (see, e.g., [Ti, 7.3]), but already the case $k = 4$ is challenging and (2.4) is unknown for any value of $k > 4$. For $k = 4$, standard techniques reduce (2.4) to estimating the fourth moment of exponential sums

$$\Big(\sum_{n \leqslant X} n^{-it}\Big)^4 \quad \text{for } X \sim \sqrt{T},$$

and one expands the sum as

$$S = \sum_{a,b,c,d \leqslant X} \left(\frac{bc}{ad}\right)^{it}.$$

It is natural to split the sum according to the value of the *determinant* $h = ad - bc$ so that

$$S = \sum_{|h| \leqslant X^2} \sum_{ad-bc=h} \left(1 - \frac{h}{ad}\right)^{it}.$$

The idea is the following: if $h = 0$, we obtain the equation $ad = bc$ which is easily solved and yields a "main term." Then for $h \neq 0$, but "small," the exponential $(1 - h/ad)^{it}$ is not oscillating very much and will also yield an important contribution, whereas for h "large," the oscillations should cause a large amount of cancellation. To account for the contribution of the small values of h, one needs to be able to count with precision the number of solutions of the determinant equation

$$ad - bc = h$$

with $a, b, c, d \leqslant X$. This can be done by "usual" harmonic analysis[1] but much better results are obtained if one remarks that $SL(2, \mathbf{Z})$ acts on the solutions (when there is no size condition) and if one tries to apply a form of harmonic analysis (i.e., a form of the Poisson summation formula) adapted to this action: here we go from the classical case of $\mathbf{Z}^n$ discrete in $\mathbf{R}^n$ (and the applications to lattice-point counting in euclidean space) to that of $SL(2, \mathbf{Z})$ discrete in $SL(2, \mathbf{R})$ with corresponding lattice point problems. Those, however, are of a completely different nature because $SL(2, \mathbf{R})$ is not abelian and the whole geometry is of hyperbolic—negative curvature—type (so for instance most of the area of a domain accumulates on its boundary). For more about those aspects, see [I2].

2.3 *Invariants of elliptic curves*

Consider the standard torus $\mathbf{T} = \mathbf{R}^2/\mathbf{Z}^2$; it is a 2-dimensional smooth Lie group. What compatible structures of complex Lie group can be put on $\mathbf{T}$? The answer

[1] Actually, this quickly involves Kloosterman sums, and whichever way one estimates those, there are automorphic forms lurking in the background.

to this question naturally leads to the study of the space of lattices in $\mathbf{R}^2 = \mathbf{C}$ up to unimodular transformation: indeed if $\Lambda \subset \mathbf{C}$ is a lattice, then the quotient $\mathbf{C}/\Lambda$ is diffeomorphic to $\mathbf{T}$ and inherits from $\mathbf{C}$ a compatible complex structure and a group structure. Moreover, all such structures must arise in this way.

Since all complex automorphisms of $\mathbf{C}$ (as a group) are linear $z \mapsto \alpha z$ one sees that two such *complex tori* $\mathbf{C}/\Lambda_1$ and $\mathbf{C}/\Lambda_2$ are isomorphic as Riemann surfaces if and only if $\Lambda = \alpha\Lambda'$ for some $\alpha \in \mathbf{C}^\times$. Now if $\Lambda = \omega_1\mathbf{Z} \oplus \omega_2\mathbf{Z}$ and $\Lambda' = \omega_1'\mathbf{Z} \oplus \omega_2'\mathbf{Z}$, then $\Lambda = \Lambda'$ if and only if the elements of the basis are related by a unimodular linear transformation

$$\omega_1' = a\omega_1 + b\omega_2$$
$$\omega_2' = c\omega_1 + d\omega_2$$

with

$$\gamma = \begin{pmatrix} a & b \\ c & d \end{pmatrix} \in SL(2,\mathbf{Z}), \text{ ie } ad - bc = 1.$$

Up to homothety, $\Lambda \simeq \mathbf{Z} \oplus \tau\mathbf{Z}$ with $\tau = \omega_1/\omega_2$ (resp., $\Lambda' \simeq \mathbf{Z} \oplus \tau'\mathbf{Z}$ with $\tau' = \omega_1'/\omega_2'$) so the equation becomes

$$\tau' = \gamma \cdot \tau = \frac{a\tau + b}{c\tau + d},$$

and finally we can also still switch ω_1 and ω_2, i.e., replace τ by $1/\tau$: one and one only of the two choices is in $\mathbf{H}$.

Thus the set of complex tori can be identified with the quotient of $\mathbf{H}$ by this action of $SL(2,\mathbf{Z})$, and various invariants of such complex tori appear as functions $f : \mathbf{H} \to \mathbf{C}$ invariant under this action. Traditionally, meromorphic functions on $\mathbf{C}$ with two $\mathbf{R}$-linearly independent periods ω_1 and ω_2 are called *elliptic functions* and many naturally arise in such a way that they give rise to functions $f(z;\omega_1,\omega_2)$ homogeneous of some weight k so that

$$f(\lambda z; \lambda\omega_1, \lambda\omega_2) = \lambda^{-k} f(z;\omega_1,\omega_2),$$

and thus

$$f(u;\omega_1,\omega_2) = \omega_2^{-k} g(u/\omega_2, \omega_1/\omega_2),$$

for some function g defined on $\mathbf{C} \times (\mathbf{H} \cup \overline{\mathbf{H}})$, which restricted to $\mathbf{C} \times \mathbf{H}$ is such that

$$g(z;\gamma\tau) = (c\tau + d)^k g(z;\tau) \text{ for } \gamma \in SL(2,\mathbf{Z}).$$

Example 2.1. The Weierstrass $\wp$-function of the lattice Λ is

$$\wp(z;\omega_1,\omega_2) = \frac{1}{z^2} + \sum_{\substack{\omega \in \Lambda \\ \omega \neq 0}} \left(\frac{1}{(z-\omega)^2} - \frac{1}{\omega^2} \right),$$

amd it is Λ-invariant (this is not so obvious: differentiate to see that $\wp'$ is definitely Λ-invariant, then integrate). Clearly,

$$\wp(\lambda z; \lambda\omega_1, \lambda_2\omega_2) = \lambda^{-2}\wp(z; \omega_1, \omega_2).$$

Moreover

$$\wp(z; \Lambda) = z^{-2} + \sum_{m\geqslant 1} (m+1)z^m G_{m+2}(\Lambda)$$

where

$$G_k(\Lambda) = \sum_{\omega\neq 0} \frac{1}{\omega^k},$$

and $G_k(\lambda\Lambda) = \lambda^{-k}G_k(\Lambda)$ furnishes another example of those *modular functions*. This theme of elliptic functions is again very classical and much beloved of mathematicians of the 18th and 19th century.

3 Definitions and examples

We will now briefly survey the main definitions and results of what can be called the classical theory of automorphic forms and L-functions. This belongs historically to the early 20th century and the most important names for us are Eisenstein, Poincaré, Hecke, Ramanujan, Petersson and (later) Maass and Selberg. We will speak simultaneously of holomorphic modular forms and of nonholomorphic (Maass) forms, although the latter were introduced quite a bit later: this will strongly motivate the development of more group-theoretic methods to unify the work being done.

For simplicity we will mostly restrict our attention to congruence subgroups of $SL(2, \mathbf{Z})$, which is arithmetically the most important.

Recall that $\mathbf{H}$ is the Poincaré upper half-plane. It is a simply connected Riemann surface and a model of the hyperbolic plane (of constant negative curvature -1) when equipped with the Riemannian metric

$$ds^2 = \frac{dx^2 + dy^2}{y^2}.$$

The group $GL(2, \mathbf{R})^+$ of invertible matrices with positive determinant acts on $\mathbf{H}$ by fractional linear transformation as in (2.2). The scalar matrices act trivially so often one restricts the attention to $SL(2, \mathbf{R})$ or to the quotient $PGL(2, \mathbf{R})^+ = PSL(2, \mathbf{R})$, which is the automorphism group of $\mathbf{H}$ as a Riemann surface, and the group of orientation-preserving isometries of $\mathbf{H}$ as a Riemannian manifold; the full group of isometries is the semidirect product

$$PSL(2, \mathbf{R}) \rtimes \mathbf{Z}/2\mathbf{Z},$$

where a representative of the orientation-reversing coset is the symmetry $z \mapsto -\bar{z}$ with respect to the imaginary axis.

Associated with the Riemann metric, there is the volume element $d\mu(z) = y^{-2}dxdy$, and the Laplace operator $\Delta = -y^2(\partial_x^2+\partial_y^2)$, both of which are invariant by isometries.

In analogy with the study of periodic functions on $\mathbf{R}$, or of elliptic functions, one considers subgroups[2] $\Gamma < SL(2,\mathbf{R})$ which are discrete, and then functions on $\mathbf{H}$ which are "periodic" with respect to Γ, i.e., functions on the quotient space $\Gamma\backslash\mathbf{H}$, or differential forms, and so on.

A considerable difference is that those quotients are very diverse in shape and form. Topologically, almost all (orientable) surfaces arise as such quotients, and all Riemann surfaces of genus > 1 (it is known since Riemann that the set of all such surfaces, up to complex isomorphism, depends on $3g-3$ complex "parameters" or "moduli").

For arithmetic purposes, the subgroup $SL(2,\mathbf{Z})$ springs to attention: it is indeed discrete in $SL(2,\mathbf{R})$ and therefore so is every subgroup of $SL(2,\mathbf{Z})$. In general a *congruence subgroup* $\Gamma < SL(2,\mathbf{Z})$ is one such that $\Gamma \supset \Gamma(q) = \{\gamma \in SL(2,\mathbf{Z}) \mid \gamma \equiv 1 \,(\mathrm{mod}\, q)\}$ for some q. Among congruence groups, the *Hecke congruence groups*

$$\Gamma_0(q) = \{\gamma = \begin{pmatrix} a & b \\ c & d \end{pmatrix} \in SL(2,\mathbf{Z}) \mid q \mid c\}$$

for any integer $q \geqslant 1$ will be of particular interest. The number q is called the *level* of the group.

The index of $\Gamma_0(q)$ in $\Gamma_0(1) = SL(2,\mathbf{Z})$ is easily computed

$$[\Gamma_0(1) : \Gamma_0(q)] = q\prod_{p\mid q}\Big(1+\frac{1}{q}\Big).$$

In the "abelian" case, $\mathbf{R}/\mathbf{Z}$ is compact; however this is not so of the quotient $SL(2,\mathbf{Z})\backslash\mathbf{H}$ and this introduces the important notion of *cusps*.

We will describe geometrically the cusps as follows: one can visualize the quotient space $\Gamma\backslash\mathbf{H}$, for any $\Gamma < SL(2,\mathbf{R})$ discrete, by a *fundamental domain*, i.e., an open subset F of $\mathbf{H}$ such that

- No two points in F are Γ-equivalent.
- Any Γ-orbit Γz intersects the closure $\overline{F}$ of F.

In the case of $SL(2,\mathbf{Z})$, one can take the familiar triangle with one vertex at infinity

$$F_1 = \{z = x+iy \in \mathbf{H} \mid -1/2 < x < 1/2, \text{ and } |z| > 1\}$$

(see, e.g., [Se1, VII-1.2]).

[2] The context will always clearly distinguish between a subgroup and the gamma function.

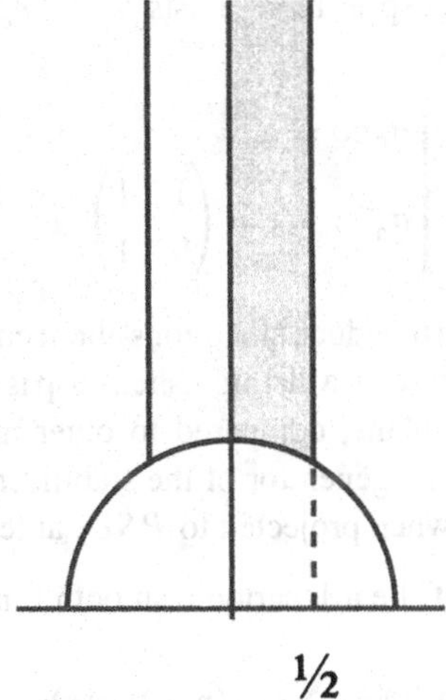

Then for $\Gamma < SL(2, \mathbf{Z})$ (the case which will occupy us), one can take

$$F_\Gamma = \bigcup_{g \in \Gamma\backslash\mathbf{H}} g \cdot F_1.$$

The Gauss–Bonnet theorem or direct integration shows that the hyperbolic volume of F_1 is finite (it is equal to $\pi/3$), but since $in \in F_1$ for $n > 1$, it is not compact. The situation is the same for $\Gamma_0(q)$, with volume multiplied by the index of $\Gamma_0(q)$ in $SL(2, \mathbf{Z}) = \Gamma_0(1)$.

The lack of compactness lies of course in the points where the fundamental domain reaches the "boundary" $\overline{\mathbf{R}} = \mathbf{R} \cup \infty$. Note that $GL(2, \mathbf{R})^+$ also acts on the boundary by the same formula (2.2).

If the covolume $\mathrm{Vol}(\Gamma\backslash\mathbf{H}) = \mathrm{Vol}(F)$ of Γ is finite, there are only finitely many cusps because around each of them small disjoint neighborhoods with constant positive volume can be constructed. There are no cusps if and only if $\Gamma\backslash\mathbf{H}$ is compact. (We will not really discuss such groups here; however, some very important arithmetic examples exist, based on unit groups of quaternion algebras, and they occur in the Jacquet–Langlands correspondance that will be discussed in other chapters). In terms of group theory, any cusp $\mathfrak{a}$ for a group Γ is the unique fixed point of some $\gamma \in \Gamma$ (such elements are called *parabolic*). The geometric action of such a γ on $\mathbf{H}$ is by translation along the horocycles "around the cusp": for $\mathfrak{a} = \infty$,

$$\gamma = \begin{pmatrix} 1 & 1 \\ 0 & 1 \end{pmatrix} \text{ acting by } z \mapsto z + 1, \tag{3.1}$$

and the horocycles are the horizontal lines $\mathrm{Im}(z) = y_0$.

By transitivity, for any cusp $\mathfrak{a}$, there exists a *scaling matrix* $\sigma_{\mathfrak{a}} \in SL(2,\mathbf{R})$ such that

$$\begin{cases} \sigma_{\mathfrak{a}}\infty = \mathfrak{a}, \\ \sigma_{\mathfrak{a}}^{-1}\gamma_{\mathfrak{a}}\sigma_{\mathfrak{a}} = \begin{pmatrix} 1 & 1 \\ 0 & 1 \end{pmatrix} \end{cases}$$

which, by conjugacy, is used to reduce all notions about cusps to the case of $\mathfrak{a} = \infty$. This is simpler to visualize (having a distinguished cusp is a very convenient feature of the Poincaré upper half-plane, compared to other models of the hyperbolic plane). Here $\gamma \in \Gamma_{\mathfrak{a}} < \Gamma$ is a generator of the stabilizer of the cusp in Γ, which is an infinite cyclic group (when projected to PSL_2 at least).

Definition. Let $f : \mathbf{H} \to \mathbf{C}$ be a 1-periodic smooth function on $\mathbf{H}$ ($f(z+1) = f(z)$),

$$f(z) = \sum_{n\in\mathbf{Z}} a_f(n; y)e(nx),$$

its Fourier expansion. One says that

(i) f is of *moderate growth* at ∞ if there exists $A > 0$ such that

$$f(z) \ll |y|^A \text{ for } y \geqslant 1;$$

(ii) f is *cuspidal* at ∞ if $a_0(y) = 0$.

Besides the natural right action of $SL(2,\mathbf{R})$ on functions on $\mathbf{C}$ by $(f \mid \gamma)(z) = f(\gamma z)$, one defines for any integer k an action

$$f \mapsto f\,|_k\,\gamma \text{ where } (f\,|_k\,\gamma)(z) = (cz+d)^{-k} f(\gamma z).$$

One checks that indeed $f\,|_k\,\gamma\,|_k\,\gamma' = f\,|_k\,\gamma\gamma'$. Also, the matrix (3.1) still acts by $(f\,|_k\,\gamma)(z) = f(z+1)$.

We can finally define modular forms.

Definition. Let $k \in \mathbf{Z}$ be an integer, $\lambda \in \mathbf{C}$, $q \geqslant 1$ an integer and χ a Dirichlet character modulo q. A smooth function $f : \mathbf{H} \to \mathbf{C}$ on $\mathbf{H}$ is called a *modular form* of weight k, level q, with nebentypus χ, resp., an automorphic function with eigenvalue λ, level q and nebentypus χ, if f is holomorphic on $\mathbf{H}$ and satisfies

$$f\,|_k\,\gamma = \chi(a)f, \text{ for } \gamma = \begin{pmatrix} a & b \\ c & d \end{pmatrix} \in \Gamma_0(q),$$

resp., f satisfies $\Delta f = \lambda f$ and

$$f \mid \gamma = \chi(a)f, \text{ for } \gamma \in \Gamma_0(q).$$

Remark 3.1. Taking $\gamma = -1$, on gets the relation $f = f\,|_k\,\gamma = \chi(-1)(-1)^k f$, so there can exist nonzero holomorphic modular forms only if the character satisfies the consistency condition $\chi(-1) = (-1)^k$, which is tacitly assumed to be the case in what follows. Similarly for automorphic functions, we must have $\chi(-1) = 1$.

One can also define nonholomorphic forms of weight $k \neq 0$, using a modified differential operator. Both holomorphic and Maass forms can be most convincingly put into a single framework through the study of the representation theory of $GL(2, \mathbf{R})$ (or of the adèle group $GL(2, \mathbf{A})$ in the arithmetic case).

Using the definition above, one can impose more regularity conditions at the cusps.

Definition. Let f be a modular form (resp., an automorphic function). One says that f is holomorphic at the cusps (resp., is a Maass form) if and only if it is of moderate growth at all cusps, i.e., for any $\mathfrak{a}$, the 1-periodic function $f_{\mathfrak{a}} = f|_k \sigma_{\mathfrak{a}}$ is of moderate growth at infinity. We let $M_k(q, \chi)$ (resp., $M^{\lambda}(q, \chi)$) denote the vector space of modular forms of weight k, level q and character χ (resp., Maass forms).

One says that f is a cusp form (resp., a Maass cusp form) if it is cuspidal at all cusps, and we denote $S_k(q, \chi) \subset M_k(q, \chi)$ (resp., $S^{\lambda}(q, \chi)$) the subspace of cusp forms (resp., Maass cusp forms).

Remark 3.2. Other equivalent formulations can be given. In particular, for $\Gamma_0(q)$, an automorphic function f is a Maass cusp form if and only if f is nonconstant (i.e., $\lambda \neq 0$) and in $L^2(\Gamma_0(q)\backslash\mathbf{H})$ (with respect to the hyperbolic measure). This amounts to saying that 0 is the only residual eigenvalue for $\Gamma_0(q)$ (see, e.g., [I2, 11.2]).

For holomorphic forms, the Fourier expansion at infinity is of the form

$$f(z) = \sum_{n \in \mathbf{Z}} a_f(n) e(nz) \text{ with } a_f(n) \in \mathbf{C}, \tag{3.2}$$

and f being holomorphic means $a_n = 0$ for $n < 0$, since $|e(nz)| = \exp(-2\pi n y)$ is not polynomially bounded for $n < 0$.

Remark 3.3. Let $Y_0(q) = \Gamma_0(q)\backslash\mathbf{H}$. This is a noncompact Riemann surface (there are some fixed points in $\mathbf{H}$ under the action, all $SL(2, \mathbf{Z})$-equivalent to either i, fixed by $z \mapsto -1/z$, or $\exp(i\pi/3)$, fixed by $z \mapsto (z-1)/z$. However, suitable coordinates at those points still provide a complex structure; see, e.g., [Sh, Chapter 1] for the details).

Moreover, by adding the finitely many cusps, one can compactify $Y_0(q)$, getting a compact Riemann surface denoted $X_0(q)$. For instance, at ∞ one uses $q = e(z)$ as a coordinate chart (compare (3.2)). For $q = 1$, one gets $X_0(1) \simeq \mathbf{P}(1)$, through the j-invariant $j(z)$ (see, e.g., [Se1, VII-3.3]).

By the general theory of Riemann surfaces, it follows that $X_0(q)$ is actually an algebraic curve. However much more is true: this algebraic curve is actually defined over $\mathbf{Q}$, and very good models over $\mathbf{Z}$ exist (so the "bad fibers" are quite well understood). All this is largely based on the interpretation of $X_0(q)$ as a *moduli space*: it "classifies" pairs (E, H), where E is an elliptic curve and H is a cyclic subgroup of order q on E. The image in $X_0(q)$ of the point $z \in \mathbf{H}$ then corresponds to the pair $(\mathbf{C}/(\mathbf{Z} \oplus z\mathbf{Z}), <1/q>)$. These are very important aspects that will be discussed in further chapters, and it is the key to many of the deep arithmetical properties of modular forms. This is related to our third motivating example for modular forms.

An easy example of those links is the important isomorphism

$$\begin{cases} S_2(q) \to \Gamma(X_0(q), \Omega_1) \\ f \mapsto f(z)dz \end{cases}$$

between the space of weight 2 cusp forms of level q (with trivial nebentypus) and the space of holomorphic 1-forms on the modular curve $X_0(q)$ (not $Y_0(q)$). This is fairly easy to prove: the weight 2 action is just what is needed to show that $f(z)dz$ is invariant under $\Gamma_0(q)$, hence descends to a 1-form on $X_0(q)$. In addition, $dq/q = 2i\pi dz$ for the local coordinate $q = e(z)$ at ∞, so the condition of vanishing at the cusp ∞ is equivalent with $f(z)dz$ being holomorphic at ∞, and similarly at the other cusps.

The last paragraph of this remark shows (since $X_0(q)$ is compact so the space of global differentials $\Gamma(X_0(q), \Omega_1)$ is finite dimensional) that $\dim_{\mathbf{C}} S_2(q) < +\infty$. Indeed, it is a general fact that $\dim M_k(q, \chi)$ (resp., $\dim M^\lambda(q, \chi)$) is finite.[3] For holomorphic forms, one can argue in an elementary way, using Cauchy's Theorem (see, e.g., [Se1, VII-3]), but deeper results, including exact formulae for $\dim M_k(q, \chi)$, when $k > 1$, follow from the Riemann–Roch theorem as in [Sh, 2.6]: for instance $\dim S_2(q)$ is the genus of $X_0(q)$, by the above, which can be computed explicitly, using the natural (ramified) covering $X_0(q) \to X_0(1) = \mathbf{P}(1)$ induced by the inclusion $\Gamma_0(q) < \Gamma_0(1)$.

With small fluctuations, the dimension increases as the level and the weight do: $\dim M_k(q)$ is of size (roughly) $qk/12$, so there are many modular forms as soon as the weight or the level is not too small (for instance, the genus of $X_0(q)$ is > 1 when $q > 49$, or if q is prime > 19).

The excluded case[4] $k = 1$ is indeed quite different: the problem of computing its dimension is of great arithmetic significance, and remains largely open (see [Se2] and [Du]). Spectrally, the weight $k = 1$ corresponds to an eigenvalue which is in the continuous spectrum so it is very hard to pick it up using the tools of harmonic analysis like the trace formula; any reasonable test function also detects the (large) contribution of eigenvalues close to it.

For the case of Maass forms, $\dim M^\lambda(q, \chi)$ is proved to be finite using the spectral theory of operators in Hilbert space. In contrast to the holomorphic case, this is just a qualitative statement and no formula for the dimension of this space is known. This is understandable, as the spectral theory also implies that $S^\lambda(q, \chi) = 0$ for all but (at most!) countably many eigenvalues $\lambda > 0$, tending to $+\infty$ (and M^λ/S^λ is easy to describe). The eigenvalues are completely mysterious and only a few special examples are known ($\lambda = 1/4$ is a special case, also of deep arithmetic significance, a recurring sentence here, and a few others can be shown to have associated Maass forms using Hecke characters of real quadratic fields, as shown by Maass; see the example in [Ge]).

---

[3]This holds in much greater generality.

[4]For $k \leqslant 0$, there are no modular forms.

Selberg developed his celebrated trace formula at least in part to address this problem of the existence of Maass cusp forms. Using it, he was able to prove (see, e.g., [I2, Chapter 11]) that *for congruence subgroups*, in particular for $\Gamma_0(q)$, the Weyl law holds:

$$|\{\lambda \mid 0 \leqslant \lambda \leqslant X \text{ and } S^\lambda(q) \neq 0\}| \sim \frac{\mathrm{Vol}(X_0(q))}{4\pi} X$$

as $X \to +\infty$, the eigenvalues being counted with multiplicity.[5]

On the other hand, it is now considered likely (contrary to the expectation when Selberg proved the result above) that for "generic" (nonarithmetic) groups Γ, this statement fails so badly that there are only finitely many (if any) eigenvalues λ for Maass cusp forms for Γ. This shift in common belief is due to the beautiful theory of Phillips and Sarnak of the disappearance of cusp forms under deformation of the group [PS].

We conclude this section by giving some concrete examples of modular forms of various types.

Example 3.4. The most natural way of constructing a function that is "periodic" under some action of a group G is by averaging (see the proof of the Poisson summation formula, Proposition 2.1) a function over the group. In fact, if the function is already invariant under a subgroup $G_1 < G$, one can average only over the cosets of G_1 in G. This leads to Poincaré and Eisenstein series.

Let $m \geqslant 0$ be an integer. Define the mth holomorphic Poincaré series of weight k by

$$P_m(z) = \sum_{\gamma \in \Gamma_\infty \backslash \Gamma_0(q)} \overline{\chi}(a)(cz+d)^{-k} e(m\gamma z).$$

For $k > 2$, this series converges absolutely uniformly on compact subsets to a holomorphic function on $\mathbf{H}$ which, for the reason just mentioned, is a modular form of level q. One can compute the expansion of P_m at the various cusps and see that $P_m \in M_k(q, \chi)$, and in fact that if $m \neq 0$, then $P_m \in S_k(q, \chi)$ (see below). For $m = 0$, this is called an *Eisensten series*. In the case of $q = 1$, we get the classical Eisenstein series

$$E_k(z) = \sum_{(a,b)=1} \frac{1}{(az+b)^k}.$$

Eisenstein series and Poincaré series can be defined in the nonholomorphic setting also with a complex variable s instead of k:

$$P_m(z; s) = \sum_{\gamma \in \Gamma_\infty \backslash \Gamma_0(q)} \overline{\chi}(a) \operatorname{Im}(\gamma z)^s e(m\gamma z), \text{ and } E(z; s) = P_0(z; s) \qquad (3.3)$$

---

[5] The problem of bounding from above in a precise way the multiplicity of Maass cusp forms is one of the most inscrutable open problems in analytic number theory.

(notice that $\mathrm{Im}(\gamma z) = |cz+d|^{-2}$ for $\gamma \in SL(2, \mathbf{R})$). This converges for $\mathrm{Re}(s) > 1$, and defines then a real-analytic function on $\mathbf{H}$ which satisfies

$$P_m(\cdot; s) \mid \gamma = \overline{\chi}(d) P_m(\cdot; s) \text{ for } \gamma \in \Gamma_0(q),$$

but it is not an eigenvalue of the Laplacian if $m \neq 0$; in fact,

$$\Delta P_m(\cdot; s) = (s(1-s) - 4\pi^2 m^2) P_m(\cdot; s) + 4\pi m s P_m(\cdot; s+1).$$

The Eisenstein series $E(z; s)$ for $\mathrm{Re}(s) > 1$, on the other hand, is indeed an eigenfunction of Δ with eigenvalue $\lambda = s(1-s)$ and it is quite easy to check that it has polynomial growth at the cusps. As will be discussed in further chapters, nonholomorphic Eisenstein series turn out to play a very important role even for studying holomorphic forms through their appearance in the Rankin–Selberg method for instance (see Section 6.2 for a very short introduction).

In general, it is often convenient to write *any* eigenvalue in this way (so two values of s exist for a given λ), and to further put $s = 1/2 + it$ with $t \in \mathbf{C}$. For cusp forms, $\lambda > 0$ translates into $\lambda \in]0, 1[$ (if $\lambda \leqslant 1/4$) or $\lambda = 1/2 + it$ with $t \in \mathbf{R}$. In the first case, the eigenvalue is called *exceptional*. Selberg made the following conjecture:

Conjecture 3.5. *For any congruence subgroup Γ, in particular $\Gamma_0(q)$, there is no exceptional eigenvalue, i.e., the first nonzero eigenvalue for Δ acting on $L^2(\Gamma\backslash\mathbf{H})$ satisfies $\lambda_1 \geqslant 1/4$.*

Moreover, Selberg proved $\lambda_1 \geqslant 3/16$. This is an arithmetic statement because it is easy to construct noncongruence subgroups for which there are arbitrarily many exceptional eigenvalues. In problems such as those mentioned in Section 2.2, exceptional eigenvalues for $\Gamma_0(q)$ have an effect such as the Landau–Siegel zero for Dirichlet characters: the uniformity in q (say in counting solutions to $ad - bc = h$ with $c \equiv 0 \pmod{q}$ with $a^2 + b^2 + c^2 + d^2 \leqslant X$) is affected by the presence of "many" exceptional eigenvalues (the closer to 0, the worse the effect). Hence Selberg's theorem indicates that the situation is a little bit better controlled. This conjecture is also clearly understood today as the archimedean analogue of the Ramanujan–Petersson conjecture (see Section 4.2 below), and indeed the significant improvements to the $3/16$ bound proved by Luo, Rudnick and Sarnak [LRS] is based on this analogy.

The spectral analysis of the general Poincaré series is the essence of the Kuznetsov–Bruggeman formula which is of great importance in the applications of Maass forms to analytic number theory (see [I2, Chapter 9], [CP]).

Properly speaking, we have defined the Poincaré and Eisenstein series relative to the cusp ∞. One can define analogous functions for any cusp, and the computation of the Fourier expansions of the Eisenstein series shows that $M_k(q, \chi) = S_k(q, \chi) \oplus E_k(q, \chi)$, where $E_k(q, \chi)$ is the vector space spanned by the Eisenstein series (at all cusps).

Example 3.6. We now come to theta functions, with notation as in [I1, Chapter 10], in a fairly general situation. Let A be a symmetric positive definite $(r \times r)$-matrix with integral coefficients, with all diagonal coefficients *even*, and $N \geqslant 1$ any integer such that NA^{-1} has integer coefficients. We let

$$A[x] = {}^t x A x, \text{ for } x \in \mathbf{R}^r$$

be the associated quadratic form. *Assume the number of variables r is even.*[6] Let

$$\Theta(z; A) = \sum_{m \in \mathbf{Z}^r} e(A[m]z/2).$$

Proposition 3.7. *We have* $\Theta(z; a) \in M_{r/2}(2N)$.

This is proved using the Poisson summation formula and generalizes, of course, the formula (2.5).

For instance, let $A_4 = \text{diag}(2, 2, 2, 2)$. We can take $N = 2$, and we have $A_4[x]/2 = x_1^2 + x_2^2 + x_3^2 + x_4^2$, so $\Theta(z; A_4) = \theta(2z)^4$ where θ is the basic theta function (2.1). Hence $\Theta(\cdot; A_4) \in M_2(4)$. Now $X_0(4)$ has genus zero hence $S_2(4) = 0$, so as explained in the previous example $\Theta \in E_2(4)$. There are 3 cusps for $\Gamma_0(4)$, namely 0, 1/2 and ∞, and one can explicitly compute the Fourier expansions of the three corresponding Eisenstein series of weight 2, E_0, $E_{1/2}$ and E_∞. Checking the first few coefficients, one identifies the linear combination equal to $\Theta(z; A_4)$:

$$\Theta(z; A_4) = \alpha E_\infty(z) + \beta E_0(z),$$

for some explicit α, β, and (2.3) follows from this.

Example 3.8. The Ramanujan Delta function is

$$\Delta(z) = e(z) \prod_{n \geqslant 1} (1 - e(nz))^{24}.$$

One has $\Delta \in S_{12}(1)$, and in fact the latter space is one dimensional. Using the Eisenstein series $E_k \in M_k(1)$, there is another expression:

$$\Delta = (60E_4)^3 - 27(140E_6)^2.$$

In terms of elliptic curves, if $E \simeq \mathbf{C}/(\mathbf{Z} \oplus z\mathbf{Z})$, $\Delta(z)$ is (up to a nonzero constant) the discriminant of the curve E.

4 Fourier expansions and L-series

4.1 *Definition*

We have already mentioned that modular forms are in particular 1-periodic on $\mathbf{H}$ so that a Fourier expansion of the type

---

[6]For r odd, the theory leads to half-integral modular forms, which we will not describe for lack of space and time.

$$f(z) = \sum_{n \geqslant 0} a_f(n) e(nz) \text{ (for a holomorphic form } f) \tag{4.1}$$

$$f(x+iy) = \sum_{n \in \mathbf{Z}} a_f(n; y) e(nx) \text{ (for a nonholomorphic one)} \tag{4.2}$$

exists. Similar expansions, after conjugating by the scaling matrices $\sigma_{\mathfrak{a}}$, hold at every cusp of the group considered.

In the case of (4.2), if f also satisfies $\Delta f = \lambda f$, applying Δ on both sides yields (separation of variable!) an ordinary differential equation of order 2 satisfied by $a_f(n; y)$, namely

$$y^2 w'' + (\lambda - 4\pi^2 n^2 y^2) w = 0,$$

which is of Bessel type. If $n \neq 0$, two linearly independent solutions are

$$w_1(y) = \sqrt{y} K_{s-1/2}(2\pi |n| y)$$
$$w_2(y) = \sqrt{y} I_{s-1/2}(2\pi |n| y)$$

(writing $\lambda = s(1-s)$ with $s \in \mathbf{C}$), where K_s and I_s denote standard Bessel functions (see, e.g., [I2, Appendix B-4]). For $n = 0$, two solutions are y^s and y^{1-s} (or $y^{1/2}$ and $y^{1/2} \log y$ for $s = 1/2$).

The important fact is that the asymptotics of I_s and K_s at infinity are different: $w_1(y) \sim \frac{\pi}{2} e^{-2\pi |n| y}$ and $w_2(y) \sim \frac{1}{2\pi} e^{2\pi |n| y}$ (there is no typo, but the legacy of wildly inconsistent normalizations). For a Maass form, of moderate growth, the second solution can therefore not appear for $n \neq 0$, and we write the Fourier expansion (at ∞) of f in the following normalized way:

$$f(z) = a_f(0) y^s + b_f(0) y^{1-s} + \sqrt{y} \sum_{n \neq 0} a_f(n) K_{s-1/2}(2\pi |n| y) e(nx), \tag{4.3}$$

with $a_f(n) \in \mathbf{C}$, $b_f(0) \in \mathbf{C}$. This represents a form cuspidal at ∞ if and only if $a_f(0) = b_f(0) = 0$.

In the case of theta functions, the Fourier coefficients at infinity are related to representations of integers by quadratic forms and are clearly of arithmetic significance. It is however not obvious at all that for more general forms there should be some interest in the Fourier coefficients, or that they should have special properties. It turns out however that this is the case, and this can be at least partly revealed through L-functions in a way closely connected to the use of theta functions in the study of Hecke L-functions.

4.2 *Examples; order of magnitude*

But before going in this direction, we mention some explicit computations of Fourier coefficients.

Example 4.1. Assume $k \geqslant 3$. Let $m \geqslant 0$ be an integer. The nth Fourier coefficient of the Poincaré series $P_m(z)$ (at ∞) is given for $n \geqslant 1$ by

$$p(m,n) = \left(\frac{m}{n}\right)^{(k-1)/2} \left\{\delta(m,n) + 2\pi i^{-k} \sum_{q|c} c^{-1} S_{\overline{\chi}}(m,n;c) J_{k-1}\left(\frac{4\pi\sqrt{mn}}{c}\right)\right\} \tag{4.4}$$

(see, e.g., [I1, 3.2] for a proof), where J_{k-1} is a Bessel function (again) and $S_{\overline{\chi}}(m,n;c)$ a *twisted Kloosterman sum*

$$S_{\overline{\chi}}(m,n;c) = \sum_{x \pmod c}^{*} \overline{\chi}(x) e\left(\frac{mx + n\bar{x}}{c}\right),$$

the sum being over invertible elements and $\bar{x}$ the inverse of x modulo c. The 0th Fourier coefficient turns out to vanish, as well as those at the other cusps, showing that P_m is a cusp form.

The Bessel function may look strange and unfamiliar, but one has

$$J_{k-1}(y) = O(y^{k-1}) \text{ as } y \to 0,$$

(from its power series expansion) showing the convergence of the series. Also, various integral expressions would reveal the fact that $J_{k-1}(y)$ is really an archimedean analogue of the Kloosterman sums (or the other way around). This is best explained, once more, in the adèlic language (see [CP]).

Using the Petersson inner product on $S_k(q,\chi)$

$$<f,g> = \int_{\Gamma_0(q)\backslash \mathbf{H}} f(z)\overline{g(z)} y^k d\mu(z) \tag{4.5}$$

(the integrand is $\Gamma_0(q)$-invariant), one shows that

$$<f, P_h> = c_k h^{1-k} a_f(h)$$

for some (explicit) constant $c_k > 0$. It follows that the Poincaré series span $S_k(q,\chi)$. However, the relations they satisfy are quite mysterious.

Example 4.2. The Fourier expansion at ∞ of the Eisenstein series of weight k (even) for $SL(2,\mathbf{Z})$ is given by

$$E_k(z) = 2\zeta(k) + \frac{2(2i\pi)^k}{(k-1)!} \sum_{n \geqslant 1} \sigma_{k-1}(n) e(nz) \tag{4.6}$$

with

$$\sigma_k(n) = \sum_{d|n} d^k.$$

For the nonholomorphic Eisenstein series for $SL(2,\mathbf{Z})$ we have

$$E(z,s) = y^s + \varphi(s) y^{1-s} + 4\sqrt{y} \sum_{n \geqslant 1} \eta_{s-1/2}(n) K_{s-1/2}(2\pi n y) \cos(2\pi n x),$$

where

$$\eta_u(n) = \sum_{ab=n} \left(\frac{a}{b}\right)^u,$$

and $\varphi(s)$ is the *scattering matrix* for $SL(2,\mathbf{Z})$, reduced to a single function:

$$\varphi(s) = \sqrt{\pi}\frac{\Gamma(s-1/2)}{\Gamma(s)}\frac{\zeta(2s-1)}{\zeta(2s)}.$$

This expansion shows that (thanks to the properties of the zeta function) $E(z,s)$ can be analytically continued to all of $\mathbf{C}$ and satisfies

$$E(z,s) = \varphi(s)E(z,1-s)$$

or symmetrically, by the functional equation of $\zeta(s)$

$$\tilde{\varphi}(s)E(z,s) = \tilde{\varphi}(1-s)E(z,1-s) \text{ with } \tilde{\varphi}(s) = \pi^{-s}\Gamma(s)\zeta(2s). \tag{4.7}$$

Analogues of this are true in much greater generality; this was proved by Selberg for finite covolume subgroups of $SL(2,\mathbf{R})$, and then extended to higher rank groups by Langlands. The analytic continuation of Eisenstein series is one of the keys to the spectral theory of automorphic forms and to their applications in analytic number theory (see, e.g., [I2, Chapter 6]).

Example 4.3. The nth Fourier coefficient of Δ at the only cusp ∞ of $SL(2,\mathbf{Z})$ is denoted $\tau(n)$; the arithmetic function τ is usually called the Ramanujan function.

Because the modular forms considered are of moderate growth, it is easily shown that the Fourier coefficients $a_f(n)$ must be of (at most) polynomial size as functions of n: for we have

$$a_f(n) = \int_x^{x+1} f(x+t+iy)e(-nt)dt$$

for any $x \in \mathbf{R}$, $y > 0$ if f is holomorphic.

For Eisenstein series, the exact Fourier expansions reveals that (up to arithmetic fluctuations, see (4.6)), $a_f(n)$ is of size about n^{k-1}.

For the same reason, the coefficients of cusp forms should be smaller since f then vanishes at infinity. Such is indeed the case and the Parseval formula applied to the bounded function $y^{k/2}f(z)$ (resp., $f(z)$ in the nonholomorphic case) proves that

$$\sum_{n\leqslant N} |a_f(n)|^2 \ll N^k \text{ for } N \geqslant 1 \tag{4.8}$$

(taking $y = N^{-1}$), resp., $\ll N$ if f is nonholomorphic.

This turns out to be close to the truth (as an average result), as the Rankin–Selberg method shows, and it also indicates that $a_f(n) \ll n^{(k-1)/2}$, on average at least (resp., $a_f(n) \ll 1$ on average).

That this is actually individually true was first conjectured by Ramanujan for the τ function, in the strikingly precise form

$$|\tau(n)| \leqslant d(n)n^{11/2} \tag{4.9}$$

($d(n) = \sigma_0(n)$ is the number of divisors of n).

Petersson generalized this conjecture to other modular forms (in a less precise form); one should have

$$a_f(n) \ll_\varepsilon n^{(k-1)/2+\varepsilon}, \text{ for } n \geqslant 1, \tag{4.10}$$

for any $f \in S_k(q, \chi)$, the implied constant depending on f and $\varepsilon > 0$.

Indeed, besides the evidence on average described above, one can justify this expectation quickly by applying the "square-rooting" philosophy to the sum of Kloosterman sums in (4.4); since the Poincaré series span $S_k(q, \chi)$ (for $k > 2$ at least), one derives very strong evidence for (4.10). However, no proof has been found along those lines and it is through a remarkable application of the Riemann Hypothesis for varieties over finite fields (involving the link between modular forms and Galois representations or "motives") that P. Deligne [De] proved the Ramanujan–Petersson conjecture for cusp forms of weight $\geqslant 2$, in the precise form (4.10) for the τ function, or, more generally,

$$|a_f(n)| \leqslant d(n)n^{(k-1)/2}$$

for the coefficients of a primitive form (see Section 5 below). For weight 1, the analogue was proved by Deligne and Serre.

One should not neglect the approach through (4.4), however; its analogues still apply for instance in the case of Maass forms or of half-integral weight forms for which the corresponding conjectures remain open (see [I1, 5.3] for the half-integral case which has remarkable applications to quadratic forms in 3 variables).

4.3 *L-series attached to modular forms*

Let f be a modular form of weight k or Maass form, so that is has the Fourier expansion (4.1) or (4.3) at ∞ (one can work at the other cusps also). To study the properties of the coefficients, consider the associated L-series (it seems better to reserve the terminology "L-function" to Dirichlet series with Euler products)

$$L(f, s) = \sum_{n\geqslant 1} a_f(n)n^{-s} \text{ (for } f \text{ holomorphic)} \tag{4.11}$$

$$L(f, s) = \sum_{n\neq 0} a_f(n)|n|^{-s} \text{ (for Maass forms).} \tag{4.12}$$

Hecke (resp., Maass[7]) showed that those L-series still carried most analytic properties of the L-functions considered in the previous chapters. First observe

[7] If the Maass form f is odd, i.e., $f(-\bar{z}) = -f(z)$, this L-series vanishes; a slight modification still works which we will not discuss.

that by the trivial bound $a_f(n) \ll n^k$ (resp., $a_f(n) \ll |n|$) for Fourier coefficients, the L-series converges absolutely for $\mathrm{Re}(s)$ large enough.

Let

$$W = \begin{pmatrix} 0 & -1/\sqrt{q} \\ \sqrt{q} & 0 \end{pmatrix} \in SL(2, \mathbf{R})$$

acting by $z \mapsto -1/(qz)$. This W normalizes $\Gamma_0(q)$ (it is in $SL(2, \mathbf{Z})$ for $q = 1$), so a simple computation shows that if $f \in M_k(q, \chi)$, then $g = f\,|_k\, W \in M_k(q, \overline{\chi})$ (resp., if $f \in M^\lambda(q, \chi)$, then $g = f \,|\, W \in M^\lambda(q, \overline{\chi})$).

Proposition 4.4. *Let*

$$\Lambda(f, s) = (2\pi)^{-s}\Gamma(s)L(f, s)$$

for f holomorphic,

$$L(f, s) = \pi^{-s}\Gamma\Big(\frac{s + it}{2}\Big)\Gamma\Big(\frac{s - it}{2}\Big)L(f, s) \tag{4.13}$$

for Maass forms with eigenvalue $\lambda = 1/4 + t^2$.

Then $\Lambda(f, s)$ extends to a meromorphic function on $\mathbf{C}$. If f is a cusp form it is entire, otherwise it has poles at $s = 1$ and $s = 0$. Moreover we have

$$\Lambda(f, s) = i^k q^{k/2-s}\Lambda(g, k - s)$$

where $g = f\,|_k\, W$ for f holomorphic or

$$\Lambda(f, s) = q^{1/2-s}\Lambda(g, 1 - s)$$

where $g = f \,|\, W$ for f a Maass form.

Proof. Consider the holomorphic case. The Maass form case is similar, except one has to appeal to the formula for the Mellin transform of the Bessel K-functions. Also we assume that f is a cusp form to avoid dealing with the constant term (which is not hard either).

We proceed much as in proving the analytic continuation of Hecke L-functions, f replacing the theta function. One has for $\mathrm{Re}(s) > 1$

$$(2\pi)^{-s}\Gamma(s)n^{-s} = \int_0^{+\infty} e^{-2\pi ny}y^s\frac{dy}{y},$$

hence we find for $\mathrm{Re}(s)$ large enough that

$$\Lambda(f, s) = \int_0^{\infty} f(iy)y^s\frac{dy}{y}. \tag{4.14}$$

Again we split the integral at $\alpha > 0$ to get

$$\begin{aligned}\Lambda(f,s) &= \int_0^{\alpha} f(iy)y^s\frac{dy}{y} + \int_{\alpha}^{\infty} f(iy)y^s\frac{dy}{y}\\ &= \int_{1/(q\alpha)}^{+\infty} f\Big(\frac{i}{qu}\Big)(qu)^{-s}\frac{du}{u} + \int_{\alpha}^{\infty} f(iy)y^s\frac{dy}{y}\\ &= \int_{1/(q\alpha)}^{+\infty} q^{k/2}i^k u^k g(iu)(qu)^{-s}\frac{du}{u} + \int_{\alpha}^{\infty} f(iy)y^s\frac{dy}{y}\end{aligned}$$

since $g(iu) = q^{-k/2}i^{-k}u^{-k}f(i/(qu))$

$$= i^k q^{k/2-s}\int_{1/(q\alpha)}^{+\infty} g(iu)u^{k-s}\frac{du}{u} + \int_{\alpha}^{\infty} f(iy)y^s\frac{dy}{y}.$$

This already yields the analytic continuation, and the functional equation follows if $\alpha = q^{-1/2}$. □

Compared to the various results seen in the first two chapters about analytic continuation and functional equation, one notices that the gamma factors are different (especially for Maass forms), and the functional equation relates s to $k-s$, putting the center of the critical strip at $s = k/2$. Moreover, the functional equation relates f to $g = f\,|_k\, W$. If $q = 1$, then $W \in SL(2,\mathbf{Z})$ so that $f = g$, but in general there is no reason for such a relation to hold.

In analytic number theory, it is often convenient to renormalize the coefficients, putting

$$a_f(n) = n^{(k-1)/2}\lambda_f(n)$$

and using

$$\sum_{n\geqslant 1}\lambda_f(n)n^{-s}$$

instead of $L(f,s)$, so that the functional equation for this other series relates s to $1-s$.

5 Hecke operators and applications

Proposition 4.4 gives us more examples of Dirichlet series with functional equations. However, there does not seem to be any reason to have an Euler product. However, Ramanujan also conjectured that the τ function is multiplicative (i.e., $\tau(mn) = \tau(m)\tau(n)$ if $(m,n) = 1$) and more precisely that

$$L(\Delta,s) = \sum_{n\geqslant 1}\tau(n)n^{-s} = \prod_p (1-\tau(p)p^{-s} + p^{11-2s})^{-1}. \tag{5.1}$$

Similarly, we know that the Fourier coefficient of $\theta(2z)^2$ is $4r(n)$ where $r(n)$ is the coefficient of the Dedekind zeta function of $\mathbf{Q}(i)$, so

$$\sum_{n\geqslant 1} r(n)n^{-s} = \zeta(s)L(\chi_4, s) = \prod_p (1-p^{-s})^{-1}(1-\chi_4(p)p^{-s})^{-1}$$
$$= \prod_p (1-(1+\chi_4(p))p^{-s}+p^{-2s})^{-1}. \tag{5.2}$$

Notice in both cases the denominator is of degree 2 in p^{-s} (except for $p=2$ in the second case, when $\chi_4(2)=0$).

Mordell first proved Ramanujan's multiplicativity conjecture, but it was Hecke who did the most to create a coherent theory, revealing a really remarkable arithmetic structure for the Fourier coefficients of holomorphic forms. However, he could not obtain a completely satisfactory answer to some problems and it was not until Atkin and Lehner developed the theory of "newforms" that the situation became really clarified. The adèlic theory of Jacquet–Langlands also throws much light on these matters.

Hecke's idea is to obtain "good" arithmetic modular forms by finding an algebra acting on spaces of modular forms in such a way that it is diagonalizable. The eigenfunctions then inherit much of the structural property of the algebra. This had no "classical" counterpart for Dirichlet characters although analogues can be constructed a posteriori.

The *Hecke operators* can be defined in a number of ways. We mention a few:

- One can define an abstract algebra generated by *double cosets* $\Gamma\gamma\Gamma$ for some γ in the *commensurator* of Γ. For congruence subgroups, the latter is larger than Γ. Then this algebra is shown to act on $M_k(q,\chi)$ in an appropriate way, essentially

$$f\,|_k\,\Gamma\gamma\Gamma = \sum_{\alpha_i} f\,|_k\,\alpha_i$$

where $\Gamma\gamma\Gamma = \cup_i \Gamma\alpha_i$ is a decomposition of the double coset. By local-global principles (the Chinese Remainder Theorem), the algebra is shown to admit generators

$$T_p = \Gamma\begin{pmatrix} p & 0\\ 0 & 1\end{pmatrix}\Gamma \text{ for prime } p$$
$$R_p = \Gamma\begin{pmatrix} p & 0\\ 0 & p\end{pmatrix}\Gamma \text{ for prime } p$$

and those satisfy simple multiplicativity relations (see below). See [Sh] for a detailed study based on this approach; to study the interactions between Hecke operators defined on various subgroups, it is often the most precise.

- If one sees modular forms as functions on the space of lattices, one can define the Hecke operator T_p as follows:

$$(f\,|_k\,T_p)(\Lambda) = \sum_{[\Lambda:\Lambda']=p} f(\Lambda'),$$

where the sum is over all sublattices of index p in Λ (see [Se1, VII-5]).

- To get a quick definition, although not a very practical one for proving the properties of the operators, one can just give the action on the Fourier coefficients, using the fact that the expansion at ∞ suffices to recover modular forms. We will take this approach for simplicity.

Fix q, the weight k and the character χ. For p prime we let

$$T(p) = \frac{1}{p} \sum_{b \,(\mathrm{mod}\, p)} \begin{pmatrix} 1 & b \\ 0 & p \end{pmatrix} + \chi(p)p^{k-1} \begin{pmatrix} p & 0 \\ 0 & 1 \end{pmatrix}$$

(in the rational group ring of $GL(2, \mathbf{R})^+$). For p dividing q, notice that the second term vanishes (even for the trivial character, χ is seen as defined modulo q). When dealing with Maass forms, a different definition must be taken, where the factor p^{k-1} is removed; the same change applies to the various formulae below.

For a given weight k, we let these $T(p)$ act on 1-periodic functions on $\mathbf{H}$ in the obvious way (with the $|_k$ action or the $|$ action), and we denote this $f\,|_k\,T(p)$ or $f\,|\,T(p)$. (Notice $T(p)$ still depends on χ and q).

Lemma 5.1. *The above operator $T(p)$ on a 1-periodic function induces an endomorphism on $M_k(q, \chi)$, resp., on $M^\lambda(q, \chi)$, and preserves cusp forms. If $a_f(n)$ are the Fourier coefficients of f, then the nth Fourier coefficient of $f\,|_k\,T(p)$ is equal to*

$$a_f(n/p) + \chi(p)p^{k-1}a_f(np)$$

with the convention $a_f(x) = 0$ for $x \in \mathbf{Q}$ not an integer.[8]

The first part of this lemma is, of course, the crucial assertion and cannot be proved without, in effect, relating the ad hoc definition of $T(p)$ we have given to one of the more intrinsic definitions. The second part, on the other hand, is quite simple (orthogonality of the characters of $\mathbf{Z}/p\mathbf{Z}$).

Now we extend the definition to define $T(n)$, $n \geqslant 1$ by multiplicativity and induction

$$\begin{aligned} T(nm) &= T(n)T(m) \text{ for } (n, m) = 1 \\ T(p^{i+1}) &= T(p)T(p^i) + \chi(p)p^{k-1}T(p^{i-1}), \text{ for } i \geqslant 0 \\ T(1) &= 1. \end{aligned}$$

A simple computation shows that the mth Fourier coefficient of $f\,|_k\,T(n)$ is

$$\sum_{d|(n,m)} \chi(d)d^{k-1}a_f\Big(\frac{nm}{d^2}\Big); \tag{5.3}$$

in particular, the first Fourier coefficient of $f\,|_k\,T(n)$ is equal to $a_f(n)$, the nth coefficient of f and the 0th Fourier coefficient is $\sigma_{k-1}(n)a_f(0)$.

---

[8] As previously this formula is for holomorphic forms and must be changed (remove p^{k-1}) for Maass forms.

Lemma 5.2. *The subalgebra of* End $M_k(q,\chi)$ *generated by the Hecke operators* $T(p)$ *or* $T(n)$ *is commutative and one has*

$$\sum_{n\geqslant 1} T(n)n^{-s} = \prod_p (1 - T(p)p^{-s} + \chi(p)p^{k-1-2s})^{-1}.$$

Compare the shape of this Euler product with that conjectured by Ramanujan for the τ function (5.1) and also with that in (5.2). In fact, we have the following.

Corollary 5.3. *The Euler product* (5.1) *holds for* Re(s) *large enough.*

Proof. Consider the Hecke operators for weight 12, level 1 and $\chi = 1$. One has $\dim S_{12}(1) = 1$, so $S_{12}(1) = \Delta\mathbf{C}$. It follows that $\Delta\,|_{12}\,T(n) = \lambda(n)\Delta$ for any n. By the lemma, we have tautologically

$$\sum_{n\geqslant 1} \lambda(n)n^{-s} = \prod_p (1 - \lambda(p)p^{-s} + p^{11-2s})^{-1}.$$

Comparing the first Fourier coefficient using (5.3), one finds

$$\tau(n) = \lambda(n)\tau(1) = \lambda(n)$$

hence the result. □

For the same reason, there will be an Euler product for any f which is an eigenfunction of all Hecke operators $T(n)$. But do these exist? When the weight or the level is large there are many linearly independent holomorphic modular forms so the simple argument of the corollary cannot extend much. However, Hecke proved the following.

Lemma 5.4. *Let* $n \geqslant 1$ *be coprime with* q. *Then the operator* $T(n)$ *acting on* $S_k(q,\chi)$, *resp.,* $S^\lambda(q,\chi)$, *is* normal *with respect to the Petersson inner product* (4.5), *resp., with respect to the inner product in* $L^2(\Gamma_0(q)\backslash\mathbf{H})$, *and, in fact, the adjoint of* $T(n)$ *is*

$$T(n)^* = \overline{\chi}(n)T(n).$$

The proof of this lemma is actually quite subtle. Most importantly, the condition $(n,q) = 1$ is necessary: the operators $T(p)$ for $p \mid n$ are usually *not* normal. In any case Hecke could deduce:

Corollary 5.5. *There is an orthonormal basis of* $S_k(q,\chi)$, *resp.,* $S^\lambda(q,\chi)$, *consisting of forms* f *which are eigenfunctions of all Hecke operators* $T(n)$ *with* $(n,q) = 1$. *Such a modular form is called a* Hecke form, *and we denote by* $\lambda_f(n)$ *its Hecke eigenvalues.*

For a Hecke form f, we derive from the equation

$$f\,|_k\,T(n) = \lambda_f(n)f \text{ for } (n,q) = 1$$

that $a_f(n) = \lambda_f(n)a_f(1)$ for $(n, q) = 1$. If $a_f(1) \neq 0$, we deduce from the Euler product for the Hecke operators that the form $g = a_f(1)^{-1} f$ is such that

$$L_q(g, s) = \sum_{(n,q)=1} a_g(n)n^{-s} = \sum_{(n,q)=1} \lambda_f(n)n^{-s}$$
$$= \prod_{p\nmid q} (1 - \lambda_f(p)p^{-s} + \chi(p)p^{k-1-2s})^{-1}.$$

Two difficulties remain: one might have $a_f(1) = 0$, in which case the reasoning breaks down, and we would prefer an Euler product involving all primes for $L(f, s)$ itself.

Hecke was able to show that in some cases a basis consisting of eigenfunctions of all the $T(n)$ existed, in which case the L-functions of forms in this basis would have an Euler product (after normalizing the first coefficient to be 1). In particular if the character χ is primitive modulo q, this is so, but this excludes the important case $\chi = 1$ for $q > 1$. Hecke's idea was to prove that there is "multiplicity one" for the Hecke algebra: the space of modular forms with given eigenvalues $\lambda(n)$, $(n, q) = 1$, is at most 1-dimensional. If this is true, then since Hecke operators $T(p)$ with $p \mid q$ act on this common eigenspace (they commute with all the other Hecke operators), it would follow that $f |_k T(p)$ is a multiple of f for any p, and the theory would proceed as before. However, multiplicity one fails in general.

Atkin and Lehner [AL] showed how to correct Hecke's theory by introducing an analogue of the notion of primitive character. This is reasonable since we know that only primitive Dirichlet characters satisfy a nice functional equation.

The starting point of the theory is the following way of inducing modular forms to a higher level: let $q \geqslant 1$, χ a character modulo q and $a > 1$. Let $f \in S_k(q, \chi)$. Then for any integer $d \mid a$, the function

$$g_d(z) = f(dz)$$

is an element of $S_k(aq, \chi')$, χ' being induced modulo aq from χ, and if f is a Hecke form, then g_d is a Hecke form *with the same eigenvalues for* $(n, aq) = 1$. This result is very simple to check. In particular, f and g_d provide an example showing the failure of multiplicity one. (Note also that

$$g_d(z) = \sum_{n\equiv 0 \pmod d} a_f(n/d)e(nz),$$

so the first Fourier coefficient of g_d is $= 0$ if $d > 1$.) This example motivates the following:

Definition. Let $q \geqslant 1$, χ a character modulo q and k an integer. Let $q' \mid q$ be the conductor of χ, χ' the primitive character inducing χ. The *old space* of $S_k(q, \chi)$, denoted $S_k(q, \chi)^\flat$, is the subspace spanned by all functions of the form $f(az)$, where

(i) $f \in S_k(q'', \chi')$ for some q'' such that $q' \mid q'' \mid q$, $q'' \neq q$;

(ii) $a \mid q/q''$.

The *new space* $S_k(q,\chi)^*$ is the orthogonal of the old-space in $S_k(q,\chi)$ for the Petersson inner product.

Example 5.6. (i) If $q = 1$ or if χ is primitive, then $S_k(q,\chi) = S_k(q,\chi)^*$.

(ii) If $k < 12$ and $q = p$ is prime, then since $S_k(1) = 0$, it follows that $S_k(p)^* = S_k(p)$. If $k \geqslant 12$, then

$$S_k(p) = S_k(1) \oplus pS_k(1) \oplus S_k(p)^*,$$

denoting $pS_k(1)$ the space of cusp forms of type $f(pz)$ with $f \in S_k(1)$. In general, Möbius inversion can be used to compute the dimension of the new-space.

One shows easily that all the Hecke operators $T(n)$ with $(n,q) = 1$ act on the old-space and the new-space.

Theorem 5.7 (Atkin–Lehner). *The multiplicity one principle holds for $S_k(q,\chi)^*$, i.e., if $(\lambda(n))$ is any sequence of complex numbers defined for $(n,q) = 1$, the space of $f \in S_k(q,\chi)^*$ such that*

$$f \mid_k T(n) = \lambda(n) f \text{ for } (n,q) = 1$$

is at most 1*-dimensional.*

There is a unique orthonormal basis of $S_k(q,\chi)^*$ made of primitive forms *(also called newforms), i.e., forms f such that $a_f(1) = 1$ and $f \mid_k T(n) = a_f(n) f$ for any $n \geqslant 1$.*

The first part is the crucial statement and the second part follows readily following the sketch above.

Corollary 5.8. *Let f be a primitive form. The L-function $L(f,s)$ has an Euler product*

$$L(f,s) = \prod_p (1 - a_f(p)p^{-s} + \chi(p)p^{k-1-2s})^{-1}. \tag{5.4}$$

Moreover there exists $\eta(f) \in \mathbf{C}$ with modulus 1 *such that $f \mid_k W = \eta(f)\bar{f} \in S_k(q,\overline{\chi})^*$, where*

$$\bar{f}(z) = \sum_{n \geqslant 1} \overline{a_f(n)} e(nz),$$

hence $L(f,s)$ satisfies the functional equation

$$\Lambda(f,s) = \varepsilon(f) q^{k/2-s} \Lambda(\bar{f}, k-s)$$

with $\varepsilon(f) = i^k \eta(f)$.

The functional equation is now perfectly analogous to that of (3.3) for primitive Dirichlet characters. The argument $\varepsilon(f)$ is a very subtle invariant for which there is no simple formula in general. If q is squarefree, then one can relate $\eta(f)$ to the

qth Fourier coefficient of f. If $q = 1$, for example, $f|_k W = f$ and the sign of the functional equation is i^k.

It also follows from this that in fact for any Hecke form f of level q, there is a unique primitive form g, of some lower level in general, with the same eigenvalues for the operators $T(n)$, $(n, q) = 1$. Hence it is indeed possible to use the primitive forms to gain information about the whole space of cusp forms, using primitive forms of lower levels if necessary.

Also, one may find that the original definition of a primitive form, involving as it does that $f \in S_k(q, \chi)^*$, which is defined "negatively," is not very convenient. However, W. Li has shown that the converse of the corollary holds: if f is such that $f|_k W = \eta f$ and $L(f, s)$ has an Euler product (it automatically also satisfies the required functional equation), then f is a primitive form. This agrees perfectly with the larger "automorphic" philosophy that "good" L-functions are associated to the correct "primitive" objects.

6 Other "openings"

We close this chapter as we opened it, with three brief remarks about further topics in this classical theory which have had a great influence on the evolution of the subject of L-functions and automorphic forms, and that will be presented, more or less transmogrified, in the other chapters.

6.1 *Converse theorem*

The simple steps leading to the proof of Proposition 4.4 can be obviously reversed and lead to a criterion for two 1-periodic holomorphic functions $f(z)$ and $g(z)$ to be related by the Fricke involution, $g = f|_k W$ (similarly for Maass forms with f and g having Fourier expansions of type (4.3)).

For level 1, where $W \in SL(2, \mathbf{Z})$, it is also known that the matrices

$$\begin{pmatrix} 1 & 1 \\ 0 & 1 \end{pmatrix} \text{ and } \begin{pmatrix} 0 & -1 \\ 1 & 0 \end{pmatrix}$$

generate $SL(2, \mathbf{Z})$; hence one deduces a *characterization* of modular forms of level 1 from the functional equation of their L-functions.

One would like to have a similar characterization for higher levels, but the group theory is not so simple. It was Weil who found the correct generalization, using the functional equations not only of the L-function of f itself, but of the *twists* of f by Dirichlet characters. Roughly speaking, one wants information not only about the L-series of f, but also about the Dirichlet series generated by the coefficients $a_f(n)$ of f in a given congruence class (although those are seen "dually" using Dirichlet characters).

We discuss the holomorphic case, the nonholomorphic one is similar, with functional equations of the type (4.13) to "specify" the eigenvalue.

Let $f \in S_k(q,\chi)$ for instance, and let ψ be a Dirichlet character modulo m. One defines

$$(f \otimes \psi)(z) = \sum_{n \geqslant 1} a_f(n)\psi(n)e(nz).$$

Decomposing ψ into additive characters (which introduces the Gauss sum of ψ), one shows that this is again a modular form, precisely $f \otimes \psi \in S_k(qm^2, \chi\psi^2)$. If $(q,m) = 1$, computing $(f \otimes \psi)\,|_k\, W$ yields the following functional equation for $L(f \otimes \psi, s)$, where $g = f\,|_k\, W$:

$$\Lambda(f \otimes \psi, s) = \varepsilon(f \otimes \psi)(qm^2)^{k/2-s}\Lambda(g \otimes \overline{\psi}, k-s),$$

with

$$\varepsilon(f \otimes \psi) = i^k \chi(m)\psi(q)\tau(\psi)^2 r^{-1}. \tag{6.1}$$

Weil's theorem says that having such functional equations for "enough" twists ensures that a Dirichlet series comes from a modular form. For simplicity we state the version for cusp forms. For a proof, see, e.g., [I1, 7.4].

Theorem 6.1. *Let*

$$L_1(s) = \sum_{n \geqslant 1} a(n)n^{-s} \quad \textit{and} \quad L_2(s) = \sum_{n \geqslant 1} b(n)n^{-s}$$

be two Dirichlet series absolutely convergent for $\mathrm{Re}(s) > C$ *for some* $C > 0$.

Assume that there exists integers $k \geqslant 1$, $q \geqslant 1$ *and* $M > 0$ *such that for any* ψ *primitive modulo* m, *with* $(m, Mq) = 1$, *the Dirichlet series*

$$L(f \otimes \psi, s) = \sum_{n \geqslant 1} a(n)\psi(n)n^{-s} \quad \textit{and} \quad L(g \otimes \psi, s) = \sum_{n \geqslant 1} b(n)\psi(n)n^{-s}$$

admit analytic continuation to entire functions bounded in vertical strips such that the functions

$$\Lambda(f \otimes \psi, s) = (2\pi)^{-s}\Gamma(s)L(f \otimes \psi, s)$$

and

$$\Lambda(g \otimes \psi, s) = (2\pi)^{-s}\Gamma(s)L(g \otimes \psi, s)$$

are entire and satisfy the functional equation

$$\Lambda(f \otimes \psi, s) = \varepsilon(f \otimes \psi)(qm^2)^{k/2-s}\Lambda(g \otimes \overline{\psi}, k-s),$$

with $\varepsilon(f \otimes \psi)$ *given by* (6.1).

Then there exists a cusp form $f \in S_k(q,\chi)$ *for some* χ *modulo* q *such that* $L(f,s) = L_1(s)$ *and* $L(f\,|_k\, W, s) = L_2(s)$.

Generalizations of this converse theorem have a long history and have been of great importance in many developpments of automorphic forms and for the functoriality conjectures of Langlands in particular; one may mention the construction of the symmetric square of $GL(2)$-forms by Gelbart and Jacquet. The more recent results of Cogdell and Piatetski-Shapiro are also used in Lafforgue's proof of the Global Langlands Correspondence for $GL(n)$ over function fields.

6.2 *Rankin–Selberg L-function*

Let f be a weight k cusp form of level q. We mentioned in (4.8) that the upper bound

$$\sum_{n\leqslant N} |a_f(n)|^2 \ll N^k \text{ for } N \geqslant 1,$$

is quite easy to obtain. Rankin and Selberg independently proved that it is indeed sharp; as for the proof of the Euler product (5.1), their method had considerable influence on the later development of the theory of L-functions and automorphic forms.

Given two modular forms f and g (of arbitrary weights and levels, one or both being possibly nonholomorphic), define first the (naive) *Rankin–Selberg L-function*

$$L(f\times g,s)=\sum_{n\geqslant 1} a_f(n)a_g(n)n^{-s},$$

so that $L(f\times \bar{f},s)$ has $|a_f(n)|^2$ as coefficients.

By the polynomial bound for Fourier coefficients, the series converges for $\mathrm{Re}(s)$ large enough. The upshot of the work of Rankin and Selberg is that this new type of L-function satisfies much of the good analytic properties of Hecke L-functions (abelian) and automorphic L-functions; namely, they possess analytic continuation, functional equations, and in privileged cases, an Euler product expansion. We state here only the simplest case:

Proposition 6.2. *Let $f\in S_k(1)$ and $g\in S_\ell(1)$ with $k\geqslant 2$ and $\ell\geqslant 2$ be two holomorphic modular forms. In addition to $L(f\times g,s)$ define*

$$L(f\otimes g,s)=\zeta(2s-(k+\ell)+2)L(f\times g,s)$$

and

$$\Lambda(f\otimes g,s)=(2\pi)^{-2s}\Gamma(s)\Gamma(s-(k+\ell)/2+1)L(f\otimes g,s).$$

Then $\Lambda(f\otimes g,s)$ admits analytic continuation to $\mathbf{C}$, entire if $f\neq\bar{g}$ and with simple poles at $s=0$ and $s=k+\ell$ otherwise, satisfying the functional equation

$$\Lambda(f\otimes g,s)=\Lambda(\bar{f}\otimes\bar{g},k+\ell-s).$$

Hence $L(f \otimes g, s)$ is entire if $f \neq \bar{g}$ and has a simple pole at $s = k = \ell$ if $f = \bar{g}$. The residue is given by

$$\mathrm{Res}_{s=k} L(f \times g, s) = \frac{3}{\pi} \frac{(4\pi)^k}{\Gamma(k)} <f, f>.$$

Sketch of proof. The proof depends crucially on the nonholomorphic Eisenstein series $E(z; s)$ defined in (3.3); indeed, the key is the *Rankin–Selberg integral representation*

$$(4\pi)^{-s}\Gamma(s)L(f \times g, s) \\ = \int_{\Gamma_0(1)\backslash \mathbf{H}} f(z)g(z)y^{(k+\ell)/2}E(z; s - (k+\ell)/2 + 1)d\mu(z)$$

involving the nonholomorphic Eisenstein series. To prove this, notice that

$$(4\pi)^{-s}\Gamma(s)n^{-s} = \int_0^{+\infty} e^{-4\pi ny} y^s \frac{dy}{y}$$

and

$$\int_0^1 f(z)g(z)dx = \sum_{n\geqslant 1} a_f(n)a_g(n)e^{-4\pi ny};$$

hence since $]0, 1[\times]0, +\infty[\subset \mathbf{R}^2$ is a fundamental domain for the stabilizer Γ_∞ of $+\infty$ in $SL(2, \mathbf{Z})$, we obtain

$$(4\pi)^{-s}\Gamma(s)L(f \times g, s) \\ = \int_{\Gamma_\infty\backslash \mathbf{H}} f(z)g(z)y^{s-1}dxdy \\ = \sum_{\gamma\in\Gamma_\infty\backslash SL(2,\mathbf{Z})} \int_{SL(2,\mathbf{Z})\backslash \mathbf{H}} f(z)g(z)y^{(k+\ell)/2}(\mathrm{Im}\,\gamma z)^{s-(k+\ell)/2+1}d\mu(z),$$

using the fact that $z \mapsto f(z)y^{k/2}$ and $z \mapsto g(z)y^{\ell/2}$ are both $SL(2, \mathbf{Z})$-invariant functions. Exchanging the order of summation, the integral formula follows from the definition of $E(z; s)$.

Next multiply both sides by $\pi^{-s}\Gamma(s - (k+\ell)/2 + 1)\zeta(2s - (k+\ell) + 2)$, finding $\Lambda(f \otimes g, s)$ on the left-hand side. On the right-hand side, the functional equation (4.7) of $E(z; s)$ means that we obtain an expression invariant under $s \mapsto (k+\ell)/2 - s$, hence the functional equation for $\Lambda(f \otimes g, s)$.

Since $E(z, s)$ at a simple pole at $s = 1$ with residue equal to $1/\mathrm{Vol}(SL(2, \mathbf{Z})\backslash\mathbf{H}) = 3/\pi$, the other statements are also consequences of this integral representation.

Note in particular that it is indeed $L(f \otimes g, s)$ which has "good" analytic properties. The simpler $L(f \times g, s)$ is indeed meromorphic on $\mathbf{C}$ but it has infinitely many poles at the points $s = \rho/2 + (k+\ell)/2 + 1$ where ρ is a zero of the Riemann zeta function. □

Corollary 6.3. *Let $f \in S_k(1)$ be a modular form. We have*

$$\sum_{n \leqslant X} |a_f(n)|^2 \sim c(f) X^k$$

as $X \to +\infty$, where

$$c(f) = \mathrm{Res}_{s=k} L(f \times \bar{f}, s).$$

Proof. Apply the standard methods of contour integration to $L(f \times \bar{f}, s)$: the simple pole at $s = k$ gives the leading term. □

Remark 6.4. In addition to the above, Selberg observed that if f and g have multiplicative Fourier coefficients, then the coefficients of $L(f \times g, s)$ are also multiplicative, hence this L-function also has an Euler product expansion. If $f \in S_k(1)$ and $g \in S_\ell(1)$ are primitive, write (using (5.4))

$$a_f(p^k) = \alpha_1^k + \alpha_2^k, \text{ and } a_g(p^k) = \beta_1^k + \beta_2^k,$$

where (α_1, α_2) (resp., (β_1, β_2)) are the roots of the quadratic equation

$$1 - a_f(p)X + p^{k-1}X^2 = 0 \text{ (resp., } 1 - a_g(p)X + p^{\ell-1}X^2 = 0).$$

Then

$$\begin{aligned}\sum_{k \geqslant 0} a_f(p^k) a_g(p^k) X^k &= \frac{1}{1 - \alpha_1\beta_1 X} + \frac{1}{1 - \alpha_1\beta_2 X} + \frac{1}{1 - \alpha_2\beta_1 X} + \frac{1}{1 - \alpha_2\beta_2 X} \\ &= \frac{1 - p^{\ell+k-2}X^2}{(1 - \alpha_1\beta_1 X)(1 - \alpha_1\beta_2 X)(1 - \alpha_2\beta_1 X)(1 - \alpha_2\beta_2 X)};\end{aligned}$$

hence $L(f \otimes g, s)$ has the following Euler product expansion valid for $\mathrm{Re}(s)$ large enough

$$\begin{aligned}L(f \otimes g, s) = \prod_p \big((1 - \alpha_{1,p}\beta_{1,p}p^{-s})(1 - \alpha_{1,p}\beta_{2,p}p^{-s}) \\ \cdot (1 - \alpha_{2,p}\beta_{1,p}p^{-s})(1 - \alpha_{2,p}\beta_{2,p}p^{-s})\big)^{-1}.\end{aligned}$$

In the general case, if f and g are primitive forms, the same formal computation shows that $L(f \otimes g, s)$ has an Euler product expansion, with all but finitely many Euler factors of the same form. However, the functional equation becomes quite a delicate matter if one proceeds classically; since it should be inherited from the Eisenstein series, and for these, whenever the group involved has more than one cusp, the functional equation takes a matrix form, involving the Eisenstein series at all cusps. In the adèlic language, the results are again much cleaner and the proofs conceptually probably simpler.

Remark 6.5. In keeping with the analogy with abelian L-functions, one should think of $L(f \otimes g, s)$ as the analogue of the L-function $L(\chi_1\chi_2, s)$ for χ_i some Dirichlet characters. Of course, in the latter case, this remains an abelian L-function since $1 \times 1 = 1$. In particular one should think of $L(f \otimes g, s)$ as giving a way of measuring the "distance" between two modular forms. That the order of the pole at $s = 1$ gives a way of deciding whether or not $f = \bar{g}$ is very useful, as is the fact that the residue at $s = k$ is related to the Petersson norm of f (this is the way the Petersson norm of primitive forms can often be best studied, for instance).

A difference is that the notation $f \otimes g$ is rather formal. It is only recently that Ramakrishnan proved that the Rankin–Selberg L-functions are also attached to some automorphic object, namely an automorphic form on $GL(2) \times GL(2)$, as predicted by the Langlands functoriality conjectures.

6.3 *Theta functions and quadratic fields, revisited*

It has already been mentioned that the kind of theta functions used to prove the analytic properties of abelian L-functions are themselves modular forms. Hecke and later Maass considered a particularly interesting special case which can be seen as a first instance of relating automorphic forms on $GL(1)$ over a quadratic field K to automorphic forms on $GL(2)$ over $\mathbf{Q}$. The result can be stated as follows, in the special case of class group characters (i.e., trivial weight and modulus $\mathfrak{m} = \mathcal{O}$):

Proposition 6.6. *Let $K/\mathbf{Q}$ be a quadratic extension with discriminant D. Let χ be a character of the ideal class group of K. There exists a primitive modular form f_χ of level $|D|$ with nebentypus ε_D equal to the quadratic character associated to K, which is holomorphic of weight* 1 *if $D < 0$ and a Maass form with eigenvalue $\lambda = 1/4$ if $D > 0$, such that*

$$L(f_\chi, s) = L(\chi, s),$$

and this holds even locally for every p-factor. Moreover, f_χ is a cusp form if and only if the character χ is not real.

One can construct f as a theta function with character, given by

$$f(z) = \sum_{\mathfrak{a} \neq 0} \chi(\mathfrak{a}) e(zN\mathfrak{a})$$

(if χ is not real and K is imaginary).

Or, one can easily see that the functional equation of the L-function of χ is indeed compatible with the existence of f. Expanding

$$L(\chi, s) = \sum_{n \geqslant 1} a_\chi(n) n^{-s}$$

one can then apply Weil's converse theorem: the functional equations of abelian L-functions over K will show that the hypothesis is satisfied. The primitivity of f_χ is easy to derive since by construction $L(f_\chi, s)$ has an Euler product.

In the case of real fields, this gives essentially the only known explicit examples of Maass forms. Notice they have algebraic Fourier coefficients. It is conjectured that all primitive Maass forms with eigenvalue $\lambda = 1/4$ must arise from real quadratic fields.

A similar argument can be applied with characters of other weights, yielding in particular the analytic continuation and functional equation of the Hasse-Weil zeta functions of elliptic curves *with complex multiplication* over number fields (Deuring, Weil; see, e.g., [I2, 8] for a specific case in complete details).

References

[AL] A. Atkin and J. Lehner, Hecke operators on $\Gamma_0(m)$, *Math. Ann.*, **185**(1970), 134–160.

[Bu] D. Bump, *Automorphic Forms and Representations*, Cambridge University Press, Cambridge, UK, 1996.

[CF] J. W. S. Cassels and A. Fröhlich, eds., *Algebraic Number Theory*, Academic Press, New York, 1967.

[CP] J. Cogdell and I. Piatetski-Shapiro, *The Arithmetic and Spectral Analysis of Poincaré Series*, Perspectives in Mathematics 13, Academic Press, New York, 1990.

[Co] H. Cohen, *Advanced Topics in Computational Number Theory*, Graduate Texts in Mathematics 193, Springer-Verlag, New York, 2000.

[CS] B. Conrey and K. Soundararajan, Real zeros of quadratic Dirichlet L-functions, *Invent. Math.*, **150**-1(2002), 1–44.

[De] P. Deligne, Les conjectures de Weil I, *Inst. Hautes Études Sci. Publ. Math.*, **43**(1974), 273–300.

[Di] P. G. Dirichlet, Beweis des Satzes, dass jede unbegrenzte arithmetische Progression, deren erstes Glied und Differenz ganze Zahlen ohne gemeinschaftlichen Factor sind, unendlich viele Primzahlen enthält, in *Collected Works*, Vol. I, Chelsea, New York, 1969, 313–342.

[Du] W. Duke, The dimension of the space of cusp forms of weight one, *Internat. Math. Res. Notices*, **2**(1995), 99–109.

[Ge] S. Gelbart, *Automorphic Forms on Adèle Groups*, Annals of Mathematical Studies 83, Princeton University Press, Princeton, NJ, 1975.

[Go] D. Goldfeld, The class number of quadratic fields and the conjectures of Birch and Swinnerton–Dyer, *Ann. Scuola Norm. Sup. Pisa* 3, **4**(1976), 623–663.

[I1] H. Iwaniec, *Topics in Classical Automorphic Forms*, Graduate Studies in Mathematics 17, American Mathematical Society, Providence, 1997.

[I2] H. Iwaniec, *Introduction to the Spectral Theory of Automorphic Forms*, Biblioteca de la Revista Matemática Iberoamericana, Madrid, 1995.

[IR] K. Ireland and M. Rosen, *A Classical Introduction to Modern Number Theory*, 2nd ed., Graduate Texts in Mathematics 84, Springer-Verlag, New York, 1990.

[KS] N. Katz and P. Sarnak, *Random Matrices, Frobenius Eigenvalues, and Monodromy*, Colloquium Publications, American Mathematical Society, Providence, 1998.

[L] E. Landau, Bemerkungen zum Heilbronnschen Satz, *Acta Arithmetica*, **1**(1936), 1–18.

[La] S. Lang, *Algebraic Number Theory*, Graduate Texts in Mathematics 110, Springer-Verlag, New York, 1986.

[LRS] W. Luo, Z. Rudnick, and P. Sarnak, On Selberg's eigenvalue conjecture, *Geometric Functional Anal.*, **5**(1994), 387–401.

[Mi] T. Miyake, *Modular Forms*, Springer-Verlag, New York, 1989.

[PS] R. Phillips and P. Sarnak, On cusp forms for cofinite subgroups of $PSL(2, \mathbf{R})$, *Invent. Math.*, **80**(1985), 339–364.

[Rie] B. Riemann, Über die Anzahl der Primzahlen unter einer gegebenen Grösse, in *Monatsberichte der Berliner Akademie*, Berliner Akademie, Berlin, 1859, 671–680; reprinted in R. Narasimhan, ed., *Gesammelte mathematische Werke, wissenschaftlicher Nachlass und Nachtrage*, Springer-Verlag, Berlin, 1990, 177–185.

[RS] Z. Rudnick and P. Sarnak, Zeros of principal L-functions and random matrix theory, *Duke Math. J.*, **81**(1996), 269–322.

[Sa] P. Sarnak, Quantum chaos, symmetry and zeta functions, *Curr. Dev. Math.*, **3**(1997), 84–115.

[Se1] J.-P. Serre, *Cours d'arithmétique*, 3rd ed., Presses Université de France, Paris, 1988.

[Se2] J.-P. Serre, Modular forms of weight one and Galois representations, in A. Frölich, ed., *Algebraic Number Fields*, Academic Press, New York, 1977, 193–268.

[Sh] G. Shimura, *Introduction to the Arithmetic Theory of Automorphic Functions*, Iwanami Shoten and Princeton University Press, Princeton, NJ, 1971.

[Si] C. L. Siegel, Über die Classenzahl quadratischer Zahlkörper, *Acta Arithmetica*, **1**(1936), 83–86.

[Ta] J. Tate, Fourier analysis in number fields and Hecke's zeta function, in J. W. S. Cassels and A. Fröhlich, eds., *Algebraic Number Theory*, Academic Press, New York, 1967, 305–347 [CF].

[Tu] J. Tunnell, Graduate course, Rutgers University, New Brunswick, NJ, 1995–1996.

[Wa] L. Washington, *Cyclotomic Fields*, 2nd ed., Graduate Texts in Mathematics 83, Springer-Verlag, New York, 1997.

[We] A. Weil, Prehistory of the zeta function, in K. E. Aubert, E. Bombieri, and D. Goldfeld, eds., *Number Theory, Trace Formulas and Discrete Groups*, Academic Press, New York, 1989.

[Ti] A. C. Titchmarsh, *The Theory of the Riemann Zeta-Function*, 2nd ed. (revised by D. R. Heath-Brown), Oxford University Press, Oxford, UK, 1986.

4
Artin L Functions

Ehud de Shalit*

1 Introduction

Let $\chi : (\mathbb{Z}/m\mathbb{Z})^\times \to \mathbb{C}^\times$ be a primitive Dirichlet character modulo m. Let $K = \mathbb{Q}(\zeta)$, where $\zeta = e^{2\pi i/m}$. The identification $G = \mathrm{Gal}(K/\mathbb{Q}) \simeq (\mathbb{Z}/m\mathbb{Z})^\times$ allows us to attach to χ a character $\chi_{\mathrm{Gal}} : G \to \mathbb{C}^\times$ satisfying

$$\chi_{\mathrm{Gal}}(\sigma_p) = \chi(p) \tag{1.1}$$

if $(p, m) = 1$ and σ_p is the *Frobenius automorphism* at p (the canonical generator of the decomposition group of p in G, which induces on the residue field of any prime of K above p the automorphism $x \mapsto x^p$.) The *Kronecker–Weber theorem* (Kronecker 1853, Weber 1886) asserts that every 1-dimensional character of $G_{\mathbb{Q}} = \mathrm{Gal}(\bar{\mathbb{Q}}/\mathbb{Q})$ is of the form χ_{Gal} for an appropriate χ.

The Dirichlet L-function $L(\chi, s)$ may be accordingly written

$$L(\chi, s) = \prod_p \left(1 - \chi_{\mathrm{Gal}}(\sigma_p)p^{-s}\right)^{-1}. \tag{1.2}$$

In a famous paper from 1923 ("Über eine neue Art von L-Reihen"), E. Artin extended this definition to *arbitrary* complex-valued characters of $G_{\mathbb{Q}}$.

2 Representations of finite groups

Let G be a finite group. The category Rep_G of complex-valued finite-dimensional representations of G is *semisimple*: every representation is a direct sum of irreducible ones. If $\rho : G \to GL(V)$ is a representation, its *character* $\chi = \chi_\rho$ is the function

$$\chi(g) = \mathrm{tr}\, \rho(g), \qquad g \in G. \tag{2.1}$$

The character determines the representation up to isomorphism. This is a consequence of the *orthogonality relations*, which state that the characters of the irreducible representations, $\chi_1, \ldots, \chi_s$, make up an orthonormal basis for the space

*A disclaimer: This is an expository work based on a talk given at the instructional conference in Jerusalem in March 2001. None of the results—but all the mistakes—are due to the author.

of *class functions* on G (functions constant on conjugacy classes) under the inner product

$$\langle u, v\rangle = |G|^{-1} \sum_{g\in G} u(g)\bar{v}(g). \tag{2.2}$$

The number of the (isomorphism types of) irreducible representations, s, is therefore the number of conjugacy classes in G. Their *dimensions* $n_i = \chi_i(1)$ satisfy

$$\sum_{i=1}^{s} n_i^2 = |G|. \tag{2.3}$$

2.1 *Operations on representations*

- Direct sum: $\chi_{\rho_1\oplus\rho_2} = \chi_{\rho_1} + \chi_{\rho_2}$
- Tensor product: $\chi_{\rho_1\otimes\rho_2} = \chi_{\rho_1}\chi_{\rho_2}$
- Contragredient: $\chi_{\rho^\vee} = \bar{\chi}_\rho$, where $\rho^\vee(g) = {}^t\rho(g^{-1})$
- Restriction: $\chi_{\mathrm{Res}_H^G \rho} = \chi_\rho|_H$, where $\mathrm{Res}_H^G \rho \in \mathrm{Rep}_H$ is the restriction of ρ to a subgroup H.
- Induction: If $\rho : H \to GL(W)$ is in Rep_H, then $R = \mathrm{Ind}_H^G \rho : G \to GL(V)$ is the representation whose space is

$$V = \{f : G \to W;\ f(hg) = \rho(h)f(g)\ \forall h \in H,\ g \in G\} \tag{2.4}$$

  and where R is the representation by right translation:

$$(R(g)f)(g') = f(g'g). \tag{2.5}$$

  The formula for $\chi_R = \mathrm{Ind}_H^G \chi_\rho$ is a bit complicated:

$$(\mathrm{Ind}_H^G \chi)(g) = \sum_{\gamma\in G/H \text{ s.t. } \gamma^{-1}g\gamma\in H} \chi(\gamma^{-1}g\gamma). \tag{2.6}$$

Induction and Restriction are dual with respect to the inner products on G and H (*Frobenius Reciprocity*):

$$\langle\psi, \chi|_H\rangle_H = \langle\mathrm{Ind}_H^G \psi, \chi\rangle_G. \tag{2.7}$$

Another consequence of Frobenius reciprocity (which is easy to prove directly too) is the *"projection formula"*

$$\mathrm{Ind}_H^G(\psi \cdot \mathrm{Res}_H^G \chi) = \mathrm{Ind}_H^G(\psi)\cdot\chi. \tag{2.8}$$

The (right) *regular representation* is $\mathrm{Ind}_1^G 1$, the representation of G by right translation in the space of complex valued functions on G. By Frobenius reciprocity, the regular representation contains every irreducible representation with a multiplicity equal to its dimension. A representation of the form $\mathrm{Ind}_H^G \rho$, where ρ is one-dimensional, is called *monomial*.

A *virtual representation* ρ is a formal sum of irreducible representations with coefficients from $\mathbb{Z}$. Its character can be defined in the obvious way, and still determines ρ. The virtual representations (equivalently, their characters) form a *commutative ring* R_G with respect to direct sum and tensor product (equivalently, ordinary sum and multiplication of class functions). The restriction (resp., induction) functor extends to a ring (resp., group) homomorphism $R_G \to R_H$ (resp., $R_H \to R_G$), and Frobenius reciprocity and the projection formula still hold.

REFERENCES

[1] J.-P. Serre, *Linear Representations of Finite Groups*, Graduate Texts in Mathematics 42, Springer-Verlag, New York, 1977.

3 Artin L-functions

3.1 *Galois groups of global fields*

Let k be a *global field.* Thus k is either a finite extension of $\mathbb{Q}$ (a number field) or the field of functions of a smooth, projective, geometrically irreducible curve X over the finite field $\mathbb{F}_q$ (a function field). Let K be a finite Galois extension of k and $G = \mathrm{Gal}(K/k)$.

If v is a *prime* of k (in the number field case, a nonzero prime of $\mathcal{O}_k$; in the function field case, a prime divisor on X) and if $\bar{v}$ is a prime of K above v, we let $G_{\bar{v}}$ denote the *decomposition group* and $I_{\bar{v}}$ the *inertia group* of $\bar{v}$ in G. We denote by $\sigma_{\bar{v}}$ the Frobenius automorphism of $G_{\bar{v}}/I_{\bar{v}}$, and by σ_v any of the $\sigma_{\bar{v}}$ for $\bar{v}|v$ (they are all conjugate).

In the number field case, let v be an archimedean place of k, and $\bar{v}$ a place of K above it. If both are real or both are complex, let $G_{\bar{v}} = \{1\}$. If v is real but $\bar{v}$ is complex, let $G_{\bar{v}} = I_v$ be the group of order 2, generated by the unique $\sigma_{\bar{v}} \in G$ which exchanges the two (conjugate) complex embeddings corresponding to the place $\bar{v}$.

The primes (or archimedean places) v where $I_v \neq 1$ are called *ramified*. There are only finitely many ramified primes.

3.2 *The Artin L-function*

Let $\rho : G \to GL(V)$ be a representation, and χ its character. For every prime v the quotient $G_{\bar{v}}/I_{\bar{v}}$ acts on $V^{I_{\bar{v}}}$, so the following is well defined:

$$L(\chi, s) = \prod_v L_v(\chi, s), \tag{3.1}$$

where the product extends over all the primes, and the *Euler factor at* v is

$$L_v(\chi, s) = \det\left(1 - \sigma_{\bar{v}} q_v^{-s} | V^{I_{\bar{v}}}\right)^{-1} \tag{3.2}$$

(q_v is the cardinality of the residue field at v). The *degree* of the Euler factor as a polynomial in q_v^{-s} is $\dim V^{I_{\bar{v}}}$. For almost all v it is equal to $\dim \rho$. The prime v is said to be *ramified in* ρ (or "*bad*") if $V^{I_{\bar{v}}} \neq V$. If ρ is faithful, this is the same as being ramified in K/k.

The Euler factors at the bad primes, apart from being natural, are dictated by the desire to have a nice looking functional equation, and other nice properties such as additivity in χ (see below). If we omit them, they will reappear as fudge factors in the functional equation. If ρ is one-dimensional, $V^{I_{\bar{v}}} = 0$ for bad primes, so the Euler product covers only the good ones, but in general bad Euler factors might be different from 1.

The equation

$$\log \det(1 - A)^{-1} = \sum_{n=1}^{\infty} \frac{\operatorname{tr}(A^n)}{n} \tag{3.3}$$

(A: a matrix) gives the following useful formula. Let

$$\chi(\sigma_{\bar{v}}^n) = |I_{\bar{v}}|^{-1} \sum_{\sigma \bmod I_{\bar{v}} = \sigma_{\bar{v}}^n} \chi(\sigma) \tag{3.4}$$

be the average of χ over the coset $\sigma_{\bar{v}}^n$. Then

$$\log L_v(\chi, s) = \sum_{n=1}^{\infty} \frac{\chi(\sigma_{\bar{v}}^n)}{n q_v^{ns}}. \tag{3.5}$$

The Euler product converges for $Re(s) > 1$, uniformly on compact sets, hence represents a nowhere vanishing analytic function in that region, which is furthermore bounded in any closed half-plane of the form $Re(s) \geq 1 + \epsilon$.

3.3 *The function field case*

(See the survey paper by Serre [3] cited on p. 79.) The analytic properties of the Artin L-function in this case are well understood, thanks to Weil's work on curves over finite fields, and shed some light on what should be expected in the number field case. Assume that $\mathbb{F}_q$ is algebraically closed in K (the extension K/k does not involve a constant field extension). In this case K is the function field of a unique smooth projective geometrically irreducible curve Y over $\mathbb{F}_q$, and the extension K/k corresponds to a finite *Galois covering* $Y \to X$ with covering group G. Let $\bar{Y}$ and $\bar{X}$ be the corresponding curves over $\bar{\mathbb{F}}_q$. Assume that the representation ρ of G is irreducible and nontrivial (otherwise we are studying the Zeta function of k, $\zeta_k(s)$).

Recall that the *Jacobian variety* $\mathrm{Jac}(\bar{Y})$ is a group variety, whose $\bar{\mathbb{F}}_q$-rational points are divisor classes of degree 0 on $\bar{Y}$ modulo principal divisors. If l is a prime different from $p = \mathrm{char}(\mathbb{F}_q)$, its l^r-torsion points are

$$\mathrm{Jac}(\bar{Y})[l^r] \cong (\mathbb{Z}/l^r\mathbb{Z})^{2g}, \tag{3.6}$$

where $g = genus(\bar{Y})$. Under multiplication by l these groups form an inverse system whose projective limit is called the l-adic *Tate module* of $\mathrm{Jac}(\bar{Y})$. Let $H^1(\bar{Y}, \mathbb{Q}_l)$ denote its $\mathbb{Q}_l$-dual,

$$H^1(\bar{Y}, \mathbb{Q}_l) = \mathrm{Hom}\left(\lim_{\leftarrow} \mathrm{Jac}(\bar{Y})[l^r], \mathbb{Q}_l\right). \tag{3.7}$$

It is a vector space of dimension $2g$ over $\mathbb{Q}_l$, which carries commuting actions of G and $\mathrm{Gal}(\bar{\mathbb{F}}_q/\mathbb{F}_q)$. Let $H^1(\bar{Y}, \mathbb{Q}_l)(\rho)$ be the *ρ-isotypical component*. To define it we first regard ρ as a representation in a vector space V over $\bar{\mathbb{Q}}_l$ (rather than over the complex numbers), and then define

$$H^1(\bar{Y}, \mathbb{Q}_l)(\rho) = \mathrm{Hom}_G\left(V, H^1(\bar{Y}, \mathbb{Q}_l) \otimes \bar{\mathbb{Q}}_l\right). \tag{3.8}$$

Since the actions of G and of $\mathrm{Gal}(\bar{\mathbb{F}}_q/\mathbb{F}_q)$ commute, this space still carries the action of $\mathrm{Gal}(\bar{\mathbb{F}}_q/\mathbb{F}_q)$.

Let F be the Frobenius automorphism of order q, which is a topological generator of $\mathrm{Gal}(\bar{\mathbb{F}}_q/\mathbb{F}_q)$. The expression for $\log L_v(\chi, s)$ recorded in the previous section allows one to express $L(\chi, s)$ in terms of the number $\Lambda(gF^n)$ of *fixed points* of gF^n on $Y(\bar{\mathbb{F}}_q)$ for $n = 1, 2, \ldots$ and $g \in G$:

$$\log L(\chi, s) = \sum_{n=1}^{\infty} \left\{ |G|^{-1} \sum_{g \in G} \chi(g^{-1}) \Lambda(gF^n) \right\} \frac{q_v^{-ns}}{n}$$

This is elementary and only involves some algebra. Notice that every fixed point of gF^n necessarily lies over a point of $X(\mathbb{F}_{q^n})$, since G acts in the fibers of $\bar{Y} \to \bar{X}$.

A. Weil's theory (*Courbes algébriques et variétés abéliennes*, 1948) and, in particular, his analogue of the *Lefschetz fixed point formula* now imply that

$$L(\chi, s) = Z(X, \rho, q^{-s}), \tag{3.9}$$

where

$$Z(X, \rho, t) = \det\left(1 - Ft | H^1(\bar{Y}, \mathbb{Q}_l)(\rho)\right) \tag{3.10}$$

is a polynomial in t over $\bar{\mathbb{Q}}$.

One therefore has the first of the following properties (*analytic continuation*). The second one (*functional equation*) follows from Poincaré duality (the existence of the Weil pairing and its functoriality in G). The third—the *Riemann Hypothesis* (RH) for function fields—is one of Weil's major achievements, and lies much deeper. Note however that because of (i), and because of the formula relating the product of the Artin L-series to Zeta functions (see below), every zero of $L(\chi, s)$ appears as a zero of $\zeta_K(s)$, so the RH for Artin L-functions follows from the RH for Zeta functions.

Theorem 3.1 (A. Weil). *Assume that ρ is irreducible and nontrivial. Then we have the following:*

(i) *The Artin L-series $L(\chi, s)$ is a polynomial in q^{-s}, and is therefore analytic in the whole complex plane.*

(ii) *One has a functional equation*

$$Z(X, \rho, t) = \epsilon Z(X, \rho^{\vee}, q^{-1}t^{-1}) \tag{3.11}$$

(equivalently, relating $L(\chi, 1-s)$ to $L(\bar{\chi}, s)$), where $\epsilon = at^{d(\rho)}$, $d(\rho) = \dim H^1(\bar{Y}, \mathbb{Q}_l)(\rho)$, and $a = (-1)^{d(\rho)} \det(F|H^1(\bar{Y}, \mathbb{Q}_l)(\rho))$.

(iii) *All the zeroes of $L(\chi, s)$ lie on the line $Re(s) = 1/2$.*

These results were extended by Dwork, Grothendieck and Deligne to varieties over finite fields of arbitrary dimension, using (Grothendieck and Deligne) l-adic étale cohomology. The corresponding questions in the number field case are among the most intriguing conjectures in number theory.

3.4 *First properties of $L(\chi, s)$*

- **Additivity:** $L(\chi_1 + \chi_2, s) = L(\chi_1, s)L(\chi_2, s)$.
- **Induction:** If $K/k'/k$ is an intermediate extension, and χ is the character of a representation of $H = \mathrm{Gal}(K/k')$, then

$$L(\mathrm{Ind}_H^G \chi, s) = L(\chi, s). \tag{3.12}$$

  Note that the base field is k on the left, but k' on the right.
- **Inflation:** If $K'/K/k$ is an over-extension, which is also Galois over k, and χ' is the character of $G' = \mathrm{Gal}(K'/k)$ obtained by first projecting to G, and then evaluating χ,

$$L(\chi', s) = L(\chi, s). \tag{3.13}$$

- **Class field theory:** Assume that G is *abelian* and χ irreducible (one-dimensional). Class field theory associates to χ a character χ_{CFT} on a *generalized ideal class group* $\mathrm{Cl}(\mathfrak{f})$. (In the number field case, ideals prime to $\mathfrak{f}$, modulo principal ideals generated by totally positive elements which are congruent to 1 mod $\mathfrak{f}$; in the function field case, a similar group constructed from a generalized Jacobian of X.) For example, if K is the Hilbert class field of k, so K/k is unramified, then we may take $\mathfrak{f} = 1$, and $\mathrm{Cl}(\mathfrak{f})$ is the (narrow) class group of k. The correspondence is determined by

$$\chi_{\mathrm{CFT}}([v]) = \chi(\sigma_v). \tag{3.14}$$

The two facts: (i) that χ_{CFT} so defined factors through $\mathrm{Cl}(\mathfrak{f})$, and (ii) that $\chi \longleftrightarrow \chi_{\mathrm{CFT}}$ is a bijection, are a summary of Class Field Theory in a nutshell. Class Field Theory was developed in the first half of the twentieth century, and there are many good textbooks exposing it, using different approaches (analytic versus cohomological, local versus global). We shall not dwell on this subject here.

From the definition,

$$L(\chi, s) = L(\chi_{\mathrm{CFT}}, s) \tag{3.15}$$

and the latter admits analytic continuation (if χ is nontrivial) and a functional equation (Hecke).

The correspondence $\chi_{\mathrm{Gal}} \longleftrightarrow \chi$ of the introduction is a particular case of the correspondence $\chi \longleftrightarrow \chi_{\mathrm{CFT}}$ discussed here. In general, Class Field Theory is the "automorphic representation theory of $GL(1)$," and the correspondence between Galois characters and ideal class group characters is the first (and motivating) example for the Langlands' conjectures relating motivic L functions and automorphic L functions.

Example. A consequence of the first two properties, and of the well-known decomposition of the regular representation, is that

$$\zeta_K(s) = \zeta_k(s) \prod_{\chi \neq 1} L(\chi, s)^{\chi(1)}. \tag{3.16}$$

From now on we concentrate on the number field case.

References

[1] J. Martinet, Character theory and Artin L-functions, in *Algebraic Number Fields: Proceedings of the Durham Symposium*, Academic Press, New York, 1977.

[2] J. Tate, *Les conjectures de Stark sur les fonctions L d'Artin en s = 0: Notes d'un cours a Orsay*, Progress in Mathematics 47, Birkhäuser, Basel, 1984, Chapter 0,

[3] J.-P. Serre, Zeta and L functions, in O. F. G. Schilling, ed., *Arithmetic Algebraic Geometry*, Harper and Row, New York, 1965.

4 Brauer's theorem, meromorphic continuation, and the functional equation

4.1 *Archimedean Euler factors*

It is well known, and first seen in the product formula, that in the analogy between number fields and function fields, the finite primes of a number field correspond to the *open* curve, and to get the analogue of the *complete* curve one must consider also the archimedean places. The consequence for L-functions, as exemplified already by the Riemann Zeta function, is that L-functions of number fields should be augmented by *Euler factors at the archimedean places* to obtain nice looking functional equation. The exact shape of these factors was derived from studying many particular cases. Only Tate's thesis (to be discussed by S. Kudla) gave a uniform derivation of the finite and infinite Euler factors from first principles, in the special case of abelian L-functions. Nowadays, general principles from Hodge theory allow one to predict the correct shape of these Euler factors, even when the analytic continuation and functional equation are unknown. We shall now give them for Artin L-functions.

Define

$$\begin{aligned}\Gamma_{\mathbb{R}}(s) &= \pi^{-s/2}\Gamma(s/2),\\ \Gamma_{\mathbb{C}}(s) &= \Gamma_{\mathbb{R}}(s)\Gamma_{\mathbb{R}}(s+1) = 2(2\pi)^{-s}\Gamma(s),\end{aligned} \tag{4.1}$$

and write $\Gamma_v(s)$ for $\Gamma_{\mathbb{R}}(s)$ or $\Gamma_{\mathbb{C}}(s)$, depending on whether v is real or complex. Let

$$n_+(\chi, v) = \dim V^{G_{\bar{v}}}, \qquad n_-(\chi, v) = \dim V/V^{G_{\bar{v}}}, \tag{4.2}$$

and

$$L_v(\chi, s) = \Gamma_v(s)^{n_+(\chi,v)}\Gamma_v(s+1)^{n_-(\chi,v)}. \tag{4.3}$$

4.2 *The Artin conductor*

The *Artin conductor* $\mathfrak{f}(\chi)$ is an ideal of k which measures the ramification of ρ. For example, if χ is a primitive Dirichlet character modulo m, its conductor is m. The Artin conductor is divisible precisely by the primes which are ramified in ρ. For any prime v of k let $I_{\bar{v}} = G_0 \supset G_1 \supset G_2 \supset \cdots$ be the sequence of higher ramification groups of $\bar{v}/v$. Let

$$f(\chi, v) = |G_0|^{-1}\sum_{i=0}^{\infty}|G_i| \operatorname{codim} V^{G_i}. \tag{4.4}$$

(The sum is finite, and vanishes for a good v.) It is a deep result of Artin, which in the abelian case boils down to the Hasse–Arf theorem in local class field theory, that this is an integer. Let

$$\mathfrak{f}(\chi) = \prod_v v^{f(\chi,v)}. \tag{4.5}$$

4.3 *Meromorphic continuation and functional equation*

Let d_k be the discriminant of k, and define

$$\Lambda(\chi, s) = \left\{|d_k|^{\chi(1)}\mathbb{N}\mathfrak{f}(\chi)\right\}^{s/2} \prod_{v|\infty} L_v(\chi, s) \cdot L(\chi, s). \tag{4.6}$$

This is the *augmented* L-series. It can be checked that it satisfies the same functoriality with respect to χ (additivity, induction, inflation) as was observed for $L(\chi, s)$. This requires simple formulae for the Γ-functions, and knowing how the Artin conductor and the discriminant behave in towers of fields.

Theorem 4.1. $\Lambda(\chi, s)$ *admits a meromorphic continuation to the whole complex plane. It satisfies the functional equation*

$$\Lambda(\chi, 1-s) = W(\chi)\Lambda(\bar{\chi}, s) \tag{4.7}$$

with a constant $W(\chi) \in \mathbb{C}^\times$ *of absolute value* 1 (*the* Artin root number).

The Artin conductor is defined also in the function field case, and one can show that the formula above is analogous to the precise form of the functional equation in the function field case. However, contrary to that case, there is no direct approach, and no (infinite dimensional ?) "arithmetic cohomology" which explains the theorem, as Weil's theory does there, although the search for such a theory is a very hot subject. Instead, the theorem is proved in two steps.

Step 1. Using *Class Field Theory* ("automorphic representation theory for $GL(1)$") the theorem is proved when ρ is one-dimensional. This is classical, and due to Dirichlet (over $\mathbb{Q}$), Hecke (in general), and Iwasawa and Tate (in the modern approach using harmonic analysis on idele groups). We shall not pursue this any further, because it is the subject of Kudla's chapter on Tate's thesis. We only remark that working with the character χ_{CFT}, which is defined on objects built from k alone, *and not on an a priori unknown Galois group*, allows one to reduce the proof to geometric considerations for the embedding of k in Euclidean space (in Hecke's approach) or to harmonic analysis on some locally compact groups constructed from k and its completions (in Tate's approach), which are concrete and accessible objects.

Step 2. Using a theorem of Brauer in the theory of representations of finite groups, the general case is reduced to that of one-dimensional χ.

4.4 *Brauer's induction theorem*

Theorem 4.2 (R. Brauer, 1947). *Let G be a finite group. Then every character of G is a $\mathbb{Z}$-linear combination of monomial characters.*

Recall that a monomial character is of the form $\mathrm{Ind}_H^G \chi$, where χ is one-dimensional. It is in general impossible to gurantee that all the coefficients are *positive*, even if we allow them to be rational, and as we shall see, this is the source of the difficulty in Artin's conjecture. Indeed, if χ is a character which is a linear combination of monomial characters with positive rational coefficients, then some multiple $m\chi$ is monomial itself, and there are examples (e.g., when $G = \mathfrak{A}_5$), where this is impossible.

We sketch a proof of Brauer's theorem given in 1951 by Brauer and Tate (following an idea of Roquette). First one observes that *any character of a nilpotent group is monomial*. For if G is nilpotent and nonabelian (the abelian case being trivial, of course) it contains a normal abelian subgroup A properly containing the center. Let (V, ρ) be an irreducible representation of G and consider its restriction to A. If ρ is not faithful, induction on $|G|$ may be applied to deduce the result. If ρ is faithful, since A is noncentral, $\rho(A)$ does not act through homotheties, so A acts on V through several distinct characters. Each eigenspace of A is preserved by G, since A is normal in G, and G permutes the A-eigenspaces transitively, since ρ is irreducible. But this means that ρ is induced from a representation σ of the stabilizer of one of these eigenspaces, and again induction may be applied to finish the proof.

A subgroup H of G is called *p-elementary* (p a prime) if it is a product of a p-group by a cyclic group. Such a group is nilpotent, so all its characters are monomial. It is enough to prove that every virtual character of G is a $\mathbb{Z}$-linear combination of characters induced from elementary subgroups. More precisely, one considers the map

$$\mathrm{Ind}_p : \oplus_{H:\ p\text{-elementary}} R_H \to R_G \tag{4.8}$$

which is the sum of the group homomorphisms Ind_H^G for H a p-elementary subgroup. Write $|G| = p^a l$ where $(p, l) = 1$. The strategy is to prove that (the constant function) l is in the image of Ind_p. Applying this for all the p's dividing $|G|$, we conclude that 1 is in the image of $\mathrm{Ind} = \sum_p \mathrm{Ind}_p$, which is what we wanted, because by the projection formula, the image of Ind is an *ideal* in the ring R_G.

Instead of $\mathbb{Z}$-linear combinations of characters, one considers A-linear combinations where $A = \mathbb{Z}[\zeta]$, ζ a $|G|$-th primitive root of unity. The advantage of this ring is that all the characters of G are A-valued, and it is enough to show that l is in the image of $A \otimes \mathrm{Ind}_p$. By explicit construction, one proves two things.

(1) Each $\mathbb{Z}$-valued class function on G with values divisible by $|G|$ is in the image of $A \otimes \mathrm{Ind}_p$, and we may even restrict the H's here to *cyclic* subgroups of G. To show this, one first constructs the constant function $|G|$, and then uses the projection formula, and the fact that we allowed coefficients from A.

(2) Let x be an element of G of order prime to p. Let $C = \langle x\rangle$, let P be a p-Sylow subgroup of the centralizer $Z(x)$, and $H = C \times P$. Then there exists a function $\psi \in A \otimes R_H$, which is $\mathbb{Z}$-valued, such that $\psi' = \mathrm{Ind}_H^G(\psi)$ satisfies

(a) $\psi'(x) \not\equiv 0 \bmod p$

(b) $\psi'(s) = 0$ for elements s of order prime to p not conjugate to x.

Using (2) it is easy to construct a χ in the image of $A \otimes \mathrm{Ind}_p$ with values in $\mathbb{Z}$, such that $\chi(g)$ is not congruent to 0 mod p for any element g of order prime to

p, and hence (this is easy) for any $g \in G$. Using a suitable linear combination of elements obtained in (1) and (2), we get the function l.

Brauer's theorem implies the meromorphic continuation and functional equation as follows. Pick an irreducible χ, and write it as $\sum n_i \operatorname{Ind}(\chi_i)$ where χ_i are characters of dimension 1 of various subgroups, and the n_i are integers. Then

$$\Lambda(\chi, s) = \prod \Lambda(\chi_i, s)^{n_i}. \tag{4.9}$$

Since by Step 1, the theorem is known for χ_i, it follows for χ. Note that the χ_i's are attached to various base field k_i between k and K. The fact that we cannot guarantee the positivity of the n_i is responsible for the fact that we cannot deduce by this method an analytic continuation, only a meromorphic one.

Another conequence of Brauer's theorem is that the GRH (the Riemann Hypothesis for all Dedekind Zeta functions) implies that the zeroes (or poles) of every $\Lambda(\chi, s)$ lie on $Re(s) = 1/2$, and this without assuming the Artin conjecture.

5 Artin's conjecture

Conjecture 5.1. *If χ is a character of $G = \mathrm{Gal}(K/k)$ which does not contain the trivial character, Artin's L-function $L(\chi, s)$ is analytic in the whole complex plane.*

As we have seen, this conjecture is true for one-dimensional χ's, or more generally for monomial characters. It is thus true for nilpotent groups G. It is also true in the function field case. Furthermore, since we know that $L(\chi, s)$ is meromorphic, it is enough to prove the conjecture for $L(m\chi, s) = L(\chi, s)^m$ for some m.

Corollary 5.2. *If $\chi = \sum r_i \operatorname{Ind}_{H_i}^G \chi_i$ where $0 < r_i \in \mathbb{Q}$ and χ_i are one-dimensional and nontrivial, then Artin's conjecture holds for χ.*

Example 5.1 (Aramata–Brauer). The augmentation representation of G (the regular representation minus the trivial one) is of this sort. Thus for Galois K/k, $\zeta_K(s)/\zeta_k(s)$ is holomorphic. The corresponding theorem for non-Galois K/k is open, although it would follow from the conjecture.

Artin's conjecture today is incorporated into the general Langlands program, and follows from the general Langlands conjectures. We shall say a few words on this in the last section, in the case of representations into $GL(2)$.

6 Local constants

The *Artin root number* is a very interesting invariant. For example, for a Dirichlet character χ of conductor f it is given by

$$W(\chi) = \frac{\tau(\chi)}{\sqrt{f} i^{\delta}}, \tag{6.1}$$

where $\delta = 0$ or 1 depending on whether $\chi(-1) = 1$ or -1, and $\tau(\chi)$ is a Gauss sum

$$\tau(\chi) = \sum_{a=1}^{f} \chi(a) e^{2\pi i a/f}. \tag{6.2}$$

Some elementary properties of the Artin root number are:

- $W(\chi_1 + \chi_2) = W(\chi_1)W(\chi_2)$,
- $W(\mathrm{Ind}_H^G \chi) = W(\chi)$,
- $W(\bar{\chi}) = \overline{W(\chi)}$.

Note the consequence of the last property: If χ is *real*, then $\Lambda(\chi, 1-s) = W(\chi)\Lambda(\chi, s)$ and $W(\chi) = \pm 1$.

Real characters are associated with representations which are isomorphic to their contragredient. This is equivalent to saying that V has a nondegenerate G-invariant bilinear form. In particular χ is real if there is an underlying real structure $V_{\mathbb{R}}$ invariant under ρ. It is well known that ρ is then *orthogonal* (may be realized in orthogonal matrices). In this case, a theorem of Fröhlich–Queyrut says that $W(\chi) = 1$.

In his derivation of the functional equation for Hecke L-series, Tate found a canonical decomposition of $W(\chi)$ into a product over all the places v of k:

$$W(\chi) = \prod_v W_v(\chi). \tag{6.3}$$

The $W_v(\chi)$ depend only on the restriction of χ to G_v (recall that G is abelian now, and χ one-dimensional). If v is archimedean, $W_v(\chi) = i^{-\delta}$, where $\delta = 0$ or 1, depending on whether $\chi|_{G_v}$ is trivial or not. For nonarchimedean v, let

$$\mathfrak{f}_v(\chi) = v^{f(\chi, v)} \tag{6.4}$$

be the v-part of the conductor of χ. Then

$$W_v(\chi) = \frac{\tau(\chi_v)}{\sqrt{\mathbb{N}\mathfrak{f}_v(\chi)}}, \tag{6.5}$$

and $\tau(\chi_v)$ is a certain *generalized Gauss sum* associated with the character $\chi_v : k_v^\times \to \mathbb{C}^\times$ corresponding to $\chi|_{G_v}$ via local class field theory (for details, see Tate's article cited above). Compare with the case of Dirichlet L-functions.

Langlands (1970) noticed that as a consequence of his conjectures attaching automorphic L-functions on $GL_n(k)$ to degree-n Artin L-functions, the Artin root number $W(\chi)$ should be expressible as a product of canonically defined *local root numbers* $W_v(\chi)$ in the nonabelian case as well. This was announced by him, without a proof. Deligne (1974) found a short global proof, building upon some earlier results of Dwork (1956), and Tate's exposition is based on Deligne's.

Any way of expressing χ as a $\mathbb{Z}$-linear combination of monomial characters

$$\chi = \sum n_i \operatorname{Ind}_{H_i}^G \chi_i \tag{6.6}$$

yields the identity

$$W(\chi) = \prod_i W(\chi_i)^{n_i}, \tag{6.7}$$

and suggests a similar definition of $W_v(\chi)$ in terms of the $W_v(\chi_i)$. Here, if k_i is the fixed field of H_i, by $W_v(\chi_i)$ we mean the product of Tate's local constants $W_w(\chi_i)$, over all the places w of k_i lying above v. The difficulty is in showing that this is independent of the chosen expression (for the global root number we know it only because $W(\chi)$ can be alternatively defined by the functional equation). In fact, this is not quite true (here lies the difference between *extendibility* and *strong extendibility*). However, it *is* true if the χ's are virtual characters of dimension 0 (formal differences of characters of the same dimension), and this is enough because one then defines

$$W_v(\chi) = W_v(\chi - \dim\chi \cdot 1) W_v(1)^{\dim\chi}. \tag{6.8}$$

Thus the main part of the proof is to show that if

$$\sum n_i \operatorname{Ind}_{H_i}^G \chi_i = \sum n'_j \operatorname{Ind}_{H'_j}^G \chi'_j, \tag{6.9}$$

then

$$\prod_i W_v(\chi_i)^{n_i} W_v(1_i)^{-n_i} = \prod_j W_v(\chi'_j)^{n'_j} W_v(1_j)^{-n'_j}. \tag{6.10}$$

The local root numbers appearing in this equation are the abelian ones, and 1_i is the trivial character of k_i, the fixed field of H_i.

REFERENCES

[1] J. Tate, Local constants, in *Algebraic Number Fields: Proceedings of the Durham Symposium*, Academic Press, New York, 1977.

7 Relation with modular forms (for $GL(2)$)

The Artin conjecture was one of the first instances where the relation between motivic and automorphic L-functions was tested. Although the results, to this day, are partial, we could not leave this subject out of our survey. We shall assume familiarity with classical modular forms for $GL(2)$ over $\mathbb{Q}$.

In this section, $k = \mathbb{Q}$, we fix the representation $\rho : \mathrm{Gal}(K/\mathbb{Q}) \to GL(2, \mathbb{C})$, and assume throughout that it is *odd*, meaning that $\epsilon = \det(\rho)$ sends complex conjugation to -1. Note that ϵ can be viewed as an odd Dirichlet character.

Modular forms are a very important source of Galois representations. Historically, the l-adic representations attached to eigenforms of weight 2 were discovered first (by Eichler and Shimura), then those attached to eigenforms of weight ≥ 2 (by Deligne). Only in 1974 did Deligne and Serre find the simpler complex representations attached to weight 1 eigenforms.

Theorem 7.1 (Deligne–Serre). *Let*

$$f = \sum_{n=1}^{\infty} a_n q^n \qquad (q = e^{2\pi i \tau}) \tag{7.1}$$

be a normalized ($a_1 = 1$) weight 1 newform on $\Gamma_0(N)$, with nebentypus ϵ (an odd Dirichlet character modulo N). Then there exists an irreducible representation $\rho : \mathrm{Gal}(K/\mathbb{Q}) \to GL(2, \mathbb{C})$ of conductor N and determinant ϵ, whose Artin L-function is

$$L(\rho, s) = L(f, s) = \sum a_n n^{-s}. \tag{7.2}$$

These representations satisfy the Artin conjecture, because it is known that $L(f, s)$ is analytic in the whole complex plane.

The proof of Deligne–Serre goes roughly as follows. Assume for simplicity that the Fourier coefficients of f lie in $\mathbb{Z}$. Our purpose is to construct a representation with $\mathrm{tr}(\rho(\sigma_p)) = a_p$ and $\det(\rho(\sigma_p)) = \epsilon(p)$. Now, representations attached to eigenforms of weight 2 are pretty well understood, and arise from the l-adic Tate modules of the Jacobians of the modular curves. They may be reduced modulo l to obtain representations in $GL(2, \mathbb{F}_l)$. Starting with f, and multiplying it with a certain Eisenstein series of weight 1, whose q-expansion is congruent to 1 modulo l, Deligne and Serre were able to obtain weight 2 eigenforms whose q-expansions were congruent to that of f modulo l. They were therefore able to construct, not yet the desired representation ρ, but a "mod l" approximation: a representation $\bar{\rho}_l : \mathrm{Gal}(\bar{\mathbb{Q}}/\mathbb{Q}) \to GL(2, \mathbb{F}_l)$ satisfying $\mathrm{tr}(\bar{\rho}_l(\sigma_p)) = a_p \bmod l$, and $\det(\bar{\rho}_l(\sigma_p)) = \epsilon(p) \bmod l$. This they were able to do for infinitely many l's.

The next step was to use an analytic result of Rankin to show that the a_p's are finite in number, at least if we exclude a set of primes p of (arbitrarily) small density. It is here that weight 1 is crucial. Using this result one obtains a uniform bound on the image of $\bar{\rho}_l$.

The last step is to "glue" the $\bar{\rho}_l$: to obtain a representation ρ into $GL(2, \mathcal{O}_E)$ for some ring of integers $\mathcal{O}_E$ which would reduce to $\bar{\rho}_l$ for infinitely many l. This is possible thanks to the bound on the image of $\bar{\rho}_l$ obtained in the second step. It is then relatively easy to show that ρ has the desired properties.

In the converse direction, one has the following important theorem, due (in this case) to A. Weil, and dating back to 1967. It is known as "Weil's converse theorem," and is the first example of more theorems of the same type.

Theorem 7.2 (Weil). *Let ρ be an odd, irreducible, two-dimensional complex representation of* $\mathrm{Gal}(\bar{\mathbb{Q}}/\mathbb{Q})$. *Let N be its conductor and ϵ its determinant. Assume that there exists an M such that for* every *Dirichlet character χ of conductor prime to M, the Artin L-function $L(\rho \otimes \chi, s)$ is analytic in the whole complex plane. Then there exists a normalized newform of level N, weight* 1 *and nebentypus ϵ such that*

$$L(\rho, s) = L(f, s). \tag{7.3}$$

An excellent source for the proof of this theorem is Ogg's book.

The two theorems point to the tight link between the Artin conjecture and the question of *modularity of Galois representations.* In fact, they show that for ρ as in this section, the two questions are equivalent. In view of this, some authors call the *strong Artin conjecture* the assertion that ρ is obtained from an f as above.

The next big step was taken by Langlands, and complemented by Tunnel. The finite subgroups G of $GL(2, \mathbb{C})$ are classified by their image $\tilde{G}$ in $PGL(2, \mathbb{C})$. It is a classical theorem of Jordan that $\tilde{G}$ can be cyclic, dihedral, A_4 (the Tetrahedral case), S_4 (the Octahedral case) or A_5 (the Icosahedral case).

The next theorem is very deep, and even a sketch (to be found in the paper of Gelbart cited above) is outside the scope of this survey.

Theorem 7.3 (Langlands–Tunnel). *Assume that ρ is odd, irreducible and two-dimensional, and that $G = Im(\rho)$ is of one of the first four types listed above ($\tilde{G}$ is cyclic, dihedral, tetrahedral or octahedral—in fact, the cyclic case does not occur, since then ρ cannot be irreducible). Then $L(\rho, s)$ satisfies the strong Artin conjecture, i.e., ρ is modular.*

We close this very sketchy and partial survey on Artin's L functions with the remark that the icosahedral case of Artin's conjecture for two-dimensional representations has been studied in recent years by R. Taylor and his collaborators. For a short account on Taylor's program (written however before most of the results were obtained) see *Icosahedral Galois Representations*, by R. Taylor, in *Pacific J. Math.*, 1997, 337–347.

References

[1] J.-P. Serre, Modular forms and Galois representations, *Algebraic Number Fields: Proceedings of the Durham Symposium*, Academic Press, New York, 1977.

[2] A. Ogg, *Modular Forms and Dirichlet Series*, Benjamin, New York, 1969.

[3] S. Gelbart, Three lectures on the modularity of $\bar{\rho}_{E,3}$ and the Langlands reciprocity conjecture, in G. Cornell, J. H. Silverman, and G. Stevens, eds., *Modular Forms and Fermat's Last Theorem*, Springer-Verlag, New York, 1997.

5
L-Functions of Elliptic Curves and Modular Forms

Ehud de Shalit

1 Elliptic curves

In the previous chapter we examined representations of Galois groups of global fields into $GL(n, \mathbb{C})$, and their L-functions. Such representations are necessarily of finite image. We saw at the end that an important example is supplied by modular forms of weight 1.

However, most of the representations of $G_{\mathbb{Q}} = \mathrm{Gal}(\bar{\mathbb{Q}}/\mathbb{Q})$ that arise in arithmetic are not of finite image, i.e., they do not factor through a finite extension of $\mathbb{Q}$. Roughly speaking, to every finite collection of polynomial equations with coefficients from $\mathbb{Q}$, defining an algebraic variety of dimension d, and for every index $0 \leq i \leq 2d$, one can attach by the theory of *étale cohomology*, a *system* of l-adic representations (representations of $G_{\mathbb{Q}}$ into $GL_n(\mathbb{Q}_l)$, where $\mathbb{Q}_l$ is the field of l-adic numbers), one for each l. The Artin representations correspond to $d = 0$. They are of finite image, essentially independent of l, and thus may be realized in any algebraically closed field of characteristic 0, in particular $\mathbb{C}$. The same holds true for the Galois group of any global field instead of $\mathbb{Q}$.

As in the previous chapter, it is possible to attach to every such l-adic representation an L-function. It is a Dirichlet series, given by an Euler product, convergent in some right half-plane, which should have algebraic coefficients, depending on the index i, but not on l. In the number field case, these L-functions are augmented by Euler factors at the archimedean places. They should satisfy analytic continuation and functional equation similar to those of the Artin L-functions. In general these are grand open conjectures.

In these chapters we treat the first interesting case, beyond the Artin L-series. We restrict the base field to $\mathbb{Q}$ (abandoning the very important example of a function field), and take for our algebraic variety a *curve* ($d = 1$), in fact the simplest curve which still produces nontrivial results: an elliptic curve (a curve of genus 1). For the index i we take 1 ($i = 0$ or 2 give $\zeta(s)$ and $\zeta(s+1)$; in general only the range $0 < i < 2d$ is interesting, and in many cases—e.g., hypersurfaces in projective space—the only interesting index is $i = d$).

The l-adic representations attached to elliptic curves defined over $\mathbb{Q}$ are *odd, two-dimensional, and irreducible*. As in the case of Artin L-functions, an important source of such representations is supplied by modular forms, this time of weight 2, and Weil's converse theorem predicts that they all come from modular forms. This is the famous Shimura–Taniyama–Weil conjecture, which can be phrased also in terms of modular uniformization of elliptic curves. It is now a proven theorem, thanks to the work of A. Wiles and Taylor and Wiles and the later complementary work of F. Diamond, R. Taylor, B. Conrad, and C. Breuil.

1.1 *Elliptic curves over* $\mathbb{C}$

A *complex elliptic curve* is a compact Riemann surface E of genus 1. By the uniformization theorem it is isomorphic to

$$E_\tau = \mathbb{C}/(\mathbb{Z} + \mathbb{Z}\tau) \tag{1.1}$$

for a $\tau \in \mathfrak{H}$, the complex upper half plane, and τ is unique up to the action of $SL(2, \mathbb{Z})$ given by

$$\begin{pmatrix} a & b \\ c & d \end{pmatrix} : \tau \mapsto \frac{a\tau + b}{c\tau + d}. \tag{1.2}$$

The field of meromorphic functions on E (= elliptic functions), $\mathbb{C}(E)$, is of transcendental degree 1. Weierstrass wrote down generators. Let $\mathcal{L} = \mathbb{Z} + \mathbb{Z}\tau$, or, more generally, any lattice in $\mathbb{C}$ (which is of course homothetic to some lattice of this kind) and define

$$\wp(z, \mathcal{L}) = \frac{1}{z^2} + \sum_{\omega \in \mathcal{L} - \{0\}} \left(\frac{1}{(z-\omega)^2} - \frac{1}{\omega^2} \right). \tag{1.3}$$

Then $\wp$ converges, is meromorphic, $\mathcal{L}$-periodic, its only poles on E are double poles at $0 (\mathrm{mod}\ \mathcal{L})$, and together with its derivative $\wp'$ it generates $\mathbb{C}(E) = \mathbb{C}(\wp, \wp')$. Since this is a field of transcendence degree 1, $\wp$ and $\wp'$ are algebraically dependent. Indeed, they satisfy the *Weierstrass equation*, which can be viewed as a differential equation for $\wp$:

$$\wp'^2 = 4\wp^3 - g_2(\mathcal{L})\wp - g_3(\mathcal{L}), \tag{1.4}$$

where the constants are given by the *Eisenstein series* of weights 4 and 6:

$$g_2(\mathcal{L}) = 60 \sum_{\omega \in \mathcal{L} - \{0\}} \omega^{-4}, \quad g_3(\mathcal{L}) = 140 \sum_{\omega \in \mathcal{L} - \{0\}} \omega^{-6}. \tag{1.5}$$

As an abstract field,

$$\mathbb{C}(E) \simeq \mathbb{C}(X)[Y]/(Y^2 - 4X^3 + g_2 X + g_3). \tag{1.6}$$

The *discriminant* of the cubic in X is

$$\Delta(\mathcal{L}) = g_2^3 - 27 g_3^2 \neq 0. \tag{1.7}$$

Conversely, every Weierstrass equation with nonzero discriminant arises from some lattice $\mathcal{L}$. (Warning: Different equations may represent isomorphic elliptic curves if the lattices are homothetic).

The embedding $E - \{0\} \hookrightarrow \mathbb{C}^2 \subset \mathbb{P}^2(\mathbb{C})$, given by

$$z \mapsto (\wp(z), \wp'(z)) = (\wp(z) : \wp'(z) : 1), \tag{1.8}$$

extends to an embedding of E as a *smooth projective curve* in $\mathbb{P}^2(\mathbb{C})$. The unique point "at infinity" $(0 : 1 : 0)$ is the image of $0_E = 0 \bmod \mathcal{L}$.

The description of E as a complex torus makes it evident that it carries a structure of an *abelian group*, with 0_E as its origin. It is instructive to compute the *group law* in terms of the projective embedding. If P and P' are two points, elementary considerations from algebraic geometry give the following interpretation of the group law: let Q be the third point of intersection of the line through P and P' (or of the tangent at P if $P = P'$), with the cubic E (recall the Bézout theorem!). Then $P + P'$ is the point obtained by reflecting Q in the x-axis.

1.2 *Elliptic curves over arbitrary fields*

Over an arbitrary field F, an *elliptic curve* is a smooth, projective, geometrically irreducible curve E of genus 1, *equipped with an F-rational point* 0_E. It follows from the Riemann–Roch theorem that E carries a structure of an abelian group variety, where the operations are defined over F, and where 0_E is the neutral element.

If F is a subfield of $\mathbb{C}$, then a complex elliptic curve is definable over F, if its Weierstrass equation can be chosen to have coefficients in F. But *note* that nonisomorphic elliptic curves over F may become isomorphic over $\mathbb{C}$! The fact that the group operations are F-rational is clear in this case from its geometric description given above, in terms of the cubic. More generally, elliptic curves over fields of characteristic different from 2 or 3 have plane projective models given by Weierstrass equations with $g_2, g_3 \in F$. Tate has worked out "Weierstrass models" over fields of characteristic 2 and 3 as well.

For any extension K of F, the K-rational points of E, denoted $E(K)$, make up an abelian group. If K is finite, it is a finite group. If K is a global field (more generally, a finitely generated field over its prime field), it is a finitely generated abelian group. (This is fairly deep: it is the *Mordell–Weil theorem.*) Its (free) rank is called the Mordell–Weil rank of $E(K)$.

An elliptic curve is canonically isomorphic to its Jacobian, and also to its *Picard variety* (the variety classifying line bundles of degree 0, or what is the same, divisors of degree 0 modulo principal divisors). The isomorphism with the Picard variety is effected by sending $x \in E(\bar{F})$ to the class (modulo principal divisors) of $(x) - (0)$. The *Abel–Jacobi theorem* asserts that a divisor $D = \sum n_i(x_i)$ is principal if and only if $\deg(D) = \sum n_i = 0$ and $s(D) = \sum n_i x_i = 0_E$. Over $\mathbb{C}$ it may be proved by analytical means, i.e., Cauchy's integral formula, and theta functions. The algebraic proof follows from the Riemann–Roch theorem.

An *isogeny* of elliptic curves is a nonconstant homomorphism of group varieties $\alpha : E \to E'$. Its *degree*, $\deg(\alpha)$ is defined as the number of points in the fiber above a generic point of E', counted with multiplicites (if α is inseparable), or equivalently

$$\deg(\alpha) = [F(E) : \alpha^* F(E')]. \tag{1.9}$$

The degree is a positive integer. We denote $\mathrm{Ker}(\alpha)$ by $E[\alpha]$. It is a finite group scheme, which is reduced if and only if α is separable. In the purely inseparable case, it consists just of the origin "thickened by nilpotents." The degree of α is equal to its order as a group scheme. In the separable case, it is the actual number of points of $E[\alpha]$ in $\bar{F}$.

If $\alpha \in E \to E'$ is an isogeny, the *dual isogeny* $\check{\alpha} : E' \to E$ is the isogeny obtained from α by Picard functoriality, using the fact that an elliptic curve is its own Picard variety. It is also characterized uniquely by

$$\check{\alpha} \circ \alpha = \deg(\alpha) \cdot 1_E. \tag{1.10}$$

1.3 *The endomorphism ring*

The *endomorphism ring* of E over F, $\mathrm{End}(E)$, is by definition the ring of all algebraic group endomorphisms of E. (They form a ring because E is commutative.) In general, it will not be larger than $\mathbb{Z}$. If it is, E is said to have *complex multiplication* (CM). If $\mathrm{char}(F) = 0$, so we may assume $F \subset \mathbb{C}$ and attach to E a point $\tau \in \mathfrak{H}$ as before (modulo the action of $SL(2, \mathbb{Z})$) then $\mathrm{End}(E)$ is either $\mathbb{Z}$ or an order in the quadratic imaginary field $\mathbb{Q}(\tau)$, in case the latter is indeed a quadratic imaginary field. This can be seen immediately from the complex uniformization, if we observe that $\mathrm{End}(E/\mathbb{C})$ is the set of complex numbers α such that $\alpha\mathcal{L} \subset \mathcal{L}$. It also explaine the terminology "complex multiplication"!

If $\mathrm{char}(F) = p > 0$, the endomorphism ring may be even larger, and non-commutative. It is then a maximal order in the unique quaternion algebra over $\mathbb{Q}$, which is ramified at p and ∞ and split elsewhere. Such elliptic curves are called *supersingular*, and they are finite in number. Otherwise E is called *ordinary*. In any case, as an abelian group, the endomorphism ring is free of rank 1, 2, or 4. These results are due to Deuring, and date to the 1930s. They are proved by referring to the l-adic Tate module of E (see the next section).

It can be shown that deg is a *quadratic function* on the (free) abelian group $\mathrm{End}(E)$. This requires some effort, especially if $\mathrm{char}(F) > 0$, and the proper way to do it is to use *the theorem of the cube* (see Mumford's book *Abelian varieties,* which discusses the more general case of endomorphism rings of abelian varieties). Thus $\deg(n \cdot 1 - \alpha)$ is a monic, quadratic polynomial in n with integral coefficients, which we write

$$n^2 - \mathrm{tr}(\alpha)n + \deg(\alpha). \tag{1.11}$$

It is called the *characteristic polynomial* of α. For example, if F is of characteristic 0, and E has complex multiplication by an order $\mathrm{End}(E) \subset \mathbb{Q}(\tau)$, the degree coincides with the *norm* from $\mathbb{Q}(\tau)$ to $\mathbb{Q}$.

A very important case is when $F = \mathbb{F}_q$ is a finite field. Thinking in terms of Weierstrass equations, for example, we see that $\varphi : E \to E$,

$$\varphi(x, y) = (x^q, y^q) \tag{1.12}$$

is an endomorphism of the curve (caution: although it is bijective on $E(\bar{\mathbb{F}}_q)$, it is not an automorphism; it is a purely inseparable endomorphism of degree q). This special endomorphism is called the *Frobenius endomorphism of E* (of degree q).

$\mathrm{End}(E)$ and $\mathrm{Gal}(\bar{F}/F)$ act on $E(\bar{F})$, and the two actions commute. In general they are very different; for example, if F is a number field, the endomorphisms act continuously (in the complex topology), but the Galois elements, except for complex conjugations, act noncontinuously.

Nevertheless, if $F = \mathbb{F}_q$ is *finite,* the geometric action of φ is the same as that of σ, the generator of $\mathrm{Gal}(\bar{F}/F)$ which raises every field element to the power q. Since by Galois theory, the fixed points of σ^n are the points of $E(\mathbb{F}_{q^n})$, we obtain

$$E(\mathbb{F}_{q^n}) = \text{ fixed points of } \varphi^n = E(\bar{F})[\varphi^n - 1]. \tag{1.13}$$

The endomorphism $\varphi^n - 1$ is separable, so its degree is the number of elements in its kernel. From the quadratic formula for the degree we obtain:

Corollary 1.1. *Suppose that $F = \mathbb{F}_q$. Then*

$$\#E(\mathbb{F}_{q^n}) = 1 - (\lambda^n + \lambda'^n) + q^n, \tag{1.14}$$

where λ and λ' are the two roots of the characteristic polynomial of φ.

For the roots of the characteristic polynomial of φ^n are λ^n and λ'^n (this follows once the connection between the characteristic polynomial and the l-adic representation is established; see below) and its degree is q^n.

A very important observation is that the quadratic form $\deg(n - m\varphi)$ is positive definite, unless φ is an integer, where it is positive semidefinite. In the latter case $\lambda = \lambda' = \sqrt{q}$ is an integer. If not, λ and λ' should be complex conjugate to each other, generating an imaginary quadratic field. In particular $|\lambda| = |\lambda'| = \sqrt{q}$. We then have the following.

Theorem 1.2.

$$|1 + q - \#E(F_q)| \leq 2\sqrt{q}. \tag{1.15}$$

This is *Hasse's thesis* (1934), conjectured by E. Artin. As we shall see, it is equivalent to the Riemann Hypothesis for $E/\mathbb{F}_q$.

2 The l-adic representation

2.1 *The Tate module*

Over any field F, if l is prime to $\mathrm{char}(F)$, multiplication by l is a separable endomorphism, so $E[l]$ is an abelian group of order l^2, and of exponent l. It is therefore isomorphic to $(\mathbb{Z}/l\mathbb{Z})^2$. In the same way one shows that

$$T_l E = \varprojlim E[l^n] \simeq \mathbb{Z}_l^2. \tag{2.1}$$

This module is called the *Tate module* of E. If $p = \text{char}(F) > 0$, we use the same definition for $T_p E$, and it is not hard to see that in this case $T_p E \simeq \mathbb{Z}_p$ if E is ordinary, and 0 if E is supersingular.

Both $\text{End}(E)$ and $G_F = \text{Gal}(\bar{F}/F)$ act continuously on $T_l E$, and the actions commute. Incidently, this is how Deuring showed his structure theorem for $\text{End}(E)$. Note that the endomorphism ring acts *faithfully*, even on $T_p E$ if it is not 0, because a nonzero endomorphism cannot have an infinite kernel.

About the action of $\text{End}(E)$ we have the following.

Theorem 2.1. *Let $\alpha \in \text{End}(E)$, and let $T_l(\alpha)$ be its matrix acting on $T_l E$. Then*

$$\det T_l(\alpha) = \deg(\alpha). \tag{2.2}$$

Corollary 2.2. *The characteristic polynomial of $T_l(\alpha)$ is equal to the characteristic polynomial of α. In particular, it has integral coefficients, and is independent of l.*

It is customary to denote by $\rho_l : G_F \to GL(T_l E)$ the representation of the Galois group. It is called the *l-adic representation* of E. (Sometimes this name is reserved for the representation of the endomorphism algebra as well.)

The image of the l-adic representation is the subject of a lot of important research. For example, Serre proved that if F is a number field, and E does not have CM over $\bar{F}$, the image is open in $GL(T_l E) \simeq GL(2, \mathbb{Z}_l)$, and for all but finitely many l's it is the full group. The CM case is easier to understand, and dates back to the first half of the 20th century. If E has CM over F, the image of ρ_l is abelian.

More generally, if we replace E by an abelian variety (so the endomorphism ring can get more complicated), Tate conjectured, and Faltings proved, that the *Lie algebra* of the image of G_F is the full commutant of $\text{End}(E/\bar{F})$. This deep theorem is the key step in Faltings' proof of Mordell's conjecture. On the other hand, it is an example of a vast array of conjectures due to Tate on algebraic cycles. This fascinating subject is somewhat orthogonal to our work, so we leave it here.

2.2 *The Weil pairing*

Weil found an important G_F- and $\text{End}(E)$-compatible, skew-symmetric, nondegenerate pairing

$$\langle \cdot, \cdot \rangle : T_l E \times T_l E \to T_l \mu, \tag{2.3}$$

where by the latter we mean

$$T_l \mu = \varprojlim \mu_{l^n} \tag{2.4}$$

and μ_r is the group of rth roots of unity in $\bar{F}$. Over the complex numbers, if $E = \mathbb{C}/\mathcal{L}$, where

$$\mathcal{L} = \mathbb{Z} + \mathbb{Z}\tau, \tag{2.5}$$

it may be normalized by the requirement that

$$\langle l^{-n}, l^{-n}\tau \rangle = e^{2\pi i/l^n}. \tag{2.6}$$

In general, given $a \in E[N]$, there are functions f and $g \in \bar{F}(E)$ whose divisors are

$$\begin{aligned} div(f) &= N(a) - N(0), \\ div(g) &= N^*((a) - (0)) = \sum_{Na'=a} (a') - \sum_{Nu=0} (u). \end{aligned} \tag{2.7}$$

(This follows from the Abel–Jacobi theorem.) One easily sees that $f \circ N = g^N$ up to a constant, because they have the same divisor. If $b \in E[N]$, define

$$e_N(a, b) = g(x)/g(x + b). \tag{2.8}$$

That this function is an Nth root of unity follows from the fact that $g(x)^N = f(Nx)$. The Weil pairing on $T_l E$ is defined from the e_N-pairing taking the limit over $N = l^n$. It is proved then that it has all the desired properties. It is Galois equivariant, and the *dual* with respect to the pairing *of an endomorphism* α is the dual endomorphism $\check{\alpha}$.

An immediate consequence of the existence of the Weil pairing for the l-adic representation is that

$$\det(\rho_l) = \chi_l \tag{2.9}$$

is the cyclotomic character $G_F \to \mathbb{Z}_l^\times$. If $F = \mathbb{Q}$, since χ_l sends complex conjugation to -1, we see that ρ_l is an *odd, two-dimensional representation.*

Little is lost by tensoring with $\mathbb{Q}$: one defines $V_l = T_l E \otimes \mathbb{Q} \simeq \mathbb{Q}_l^2$, and views

$$\rho_l : G_F \to GL(V_l) \simeq GL(2, \mathbb{Q}_l), \tag{2.10}$$

suppressing the information about the integral lattice $T_l E$. Integral lattices play a role once again when one wants to reduce the representation ρ_l modulo l (as is crucial in Wiles' work), but for the purpose of constructing the L-function, they are irrelevant.

The Weil pairing implies that the contragredient $\check{V}_l$ of V_l is isomorphic to $V_l(-1) = V_l \otimes_{\mathbb{Z}_l} (T_l\mu)^\vee$.

2.3 *Good and bad reduction*

Let us start with an elliptic curve $E/\mathbb{Q}$. It is said to have *good reduction* at a prime p if there exists an *elliptic curve* $\mathcal{E}$ over the local ring $\mathbb{Z}_{(p)}$, whose generic fiber is E. The definition of an elliptic curve over a general ring R is similar to that over a field: it is a proper, smooth scheme $\mathcal{E}$ over Spec R, whose geometric fibers are all of dimension 1, equipped with a section s : Spec $R \to \mathcal{E}$. In such a case $\mathcal{E}$ is unique. The *special fiber* $\mathcal{E} \times_{\mathbb{Z}_{(p)}} \mathbb{F}_p$ is sometimes called the *reduction* of E modulo p and denoted by $\tilde{E}$.

If E is given by a Weierstrass planar model as above, where g_2 and g_3 are p-adic integers, and Δ is not divisible by p, then clearly E has good reduction at p, and $\tilde{E}$ is obtained by "reading the Weierstrass equation modulo p." Incidently, this shows that E has good reduction at almost all the primes p. Now it is true that the converse holds, too: if E has good reduction, there exists a *minimal* (generalized) Weierstrass equation, whose Δ is a p-adic unit. However, this is a matter of good luck concerning the numerics of Weierstrass models, and does not generalize to other arithemetic varieties, even curves of higher genus, so the scheme-theoretic definition of good reduction is a better one.

E is said to have *bad reduction* if it does not have good reduction. One can still ask for a best possible model, which is again describable by the minimal Weierstrass equation. The computations here, especially in characteristics 2 and 3, become lengthy. One says that E has *multiplicative reduction* (or *semistable reduction*) if there is a proper and flat model $\mathcal{E}/\mathbb{Z}_{(p)}$ which is not smooth, but whose special fiber contains only one singular point, which is a node. Otherwise one says that E has *additive reduction*, and a model may be found where there is a unique singular point in the special fiber, which is a cusp. The names derive from the fact the the set of nonsingular points on the special fiber in these cases forms an algebraic group which is the multiplicative or the additive group in one variable.

In terms of the minimal Weierstrass equations, and in characteristics different from 2 or 3, multiplicative reduction means that two out of the three roots of the cubic coincide modulo p, and additive reduction means that they all coincide modulo p.

Good or bad reduction is reflected in the ramification in the (infinite) extension of $\mathbb{Q}$ "cut by" ρ_l. In other words $\mathrm{Ker}(\rho_l) \subset G_{\mathbb{Q}}$ is a closed subgroup, and its fixed field is $K_l = \mathbb{Q}(E[l^\infty])$, the field obtained by adjoining to $\mathbb{Q}$ the coordinates of all the l^n-torsion points of E.

Theorem 2.3 (criterion of Ogg–Néron–Shafarevich). *A prime $p \neq l$ is ramified in ρ_l (i.e., ramified in K_l) if and only if it is a prime of bad reduction.*

Only recently (in the last 15 years) it became clear how to decide whether p is good or bad, looking at the representation ρ_p, which will be always ramified at p (K_p contains the p^n roots of unity by the Weil pairing), but this is a different, much more complicated story!

One direction of the criterion of Ogg–Néron–Shafarevich can be explained relatively easily. If there is good reduction, points in $E[l^n]$ reduce injectively modulo primes above p (this involves studying the kernel of the reduction map, which is a pro-p group, so contains no l-torsion). But this means that reduction modulo a prime above p induces an isomorphism

$$T_l E \simeq T_l \tilde{E}. \tag{2.11}$$

Since the inertia group acts trivially on the residue field, it acts trivially on $T_l E$. The other direction is more difficult, and relies on the theory of *Néron models*. In fact, one has the following precise answer. Recall that $p \neq l$, and I_p is the inertia group of some prime above p in K_l.

- If p is a prime of good reduction, $I_p = 0$.
- If p is a prime of multiplicative reduction, $\dim (V_l)_{I_p} = 1$.
- If p is a prime of additive reduction, $\dim (V_l)_{I_p} = 0$.

As a final step in the classification, one distinguishes two sub-cases of the multiplicative case. One says that E has *split* multiplicative reduction if the two tangents at the node of $\tilde{E}$ are defined over $\mathbb{F}_p$. Otherwise, they are defined over a quadratic extension, and the reduction is nonsplit. The theory of Néron models allows us to compute the effect of the *Frobenius* class $\sigma_p \in G_p/I_p$ on $(V_l)_{I_p}$. It acts trivially in the split case, and by -1 in the nonsplit case.

3 The Hasse–Weil L-function

3.1 *Definition*

Let E be an elliptic curve over $\mathbb{Q}$. The L-function of $E/\mathbb{Q}$ is defined from the l-adic representations $\rho_l : G_{\mathbb{Q}} \to GL(V_l E)$, by the same recipe that was used in the definition of the Artin L-function. For each prime p choose an $l \neq p$, and let

$$L_p(E, s) = \det(1 - p^{-s}\sigma_p | (V_l)_{I_p})^{-1}. \tag{3.1}$$

Claim: this is the inverse of a polynomial in p^{-s} with integral coefficients, which is independent of l.

If p is a prime of good reduction, we have seen that V_l is unramified, so we can substitute $V_l\tilde{E}$ for $V_l E$. But we have noticed earlier that over $\mathbb{F}_p$ the action of σ_p is the same as that of $\varphi_p \in \mathrm{End}(\tilde{E})$. If λ and λ' are the roots of the characteristic polynomial of φ_p, it is customary to write

$$a_p = \mathrm{tr}(\varphi_p) = \lambda + \lambda', \quad \lambda\lambda' = p. \tag{3.2}$$

Then $a_p \in \mathbb{Z}$ and

$$L_p(E, s) = (1 - a_p p^{-s} + p^{1-2s})^{-1}. \tag{3.3}$$

Recall that by *Hasse's thesis*

$$|a_p| \leq 2\sqrt{p}. \tag{3.4}$$

If p is a prime of multiplicative reduction, we have seen that

$$L_p(E,s) = (1 - a_p p^{-s})^{-1} \tag{3.5}$$

and $a_p = 1$ in the split case, -1 in the nonsplit case.

Finally if p is a prime of additive reduction, $a_p = 0$ and

$$L_p(E,s) = 1. \tag{3.6}$$

The L-function is the Euler product

$$L(E,s) = \prod L_p(E,s). \tag{3.7}$$

It converges uniformly on compact sets in $\mathrm{Re}(s) > 3/2$. This is a consequence of the estimates on a_p established by Hasse.

Remarks.

(a) This is the first example of L-functions attached to *étale cohomology* groups. In general, independence of l for the good Euler factors is a consequence of Deligne's proof of the Weil conjectures, but for the bad Euler factors, this is still an open question.

(b) The Euler factor at a good prime p is related to the Zeta function of the reduction. The latter is defined as

$$Z(\tilde{E},t) = \exp\left(-\sum_{n=1}^{\infty} \frac{\nu_n}{n} t^n\right), \tag{3.8}$$

where $\nu_n = \#\tilde{E}(\mathbb{F}_{p^n})$. From the expression for ν_n obtained before we get

$$Z(\tilde{E},t) = \frac{(1-t)(1-pt)}{(1-\lambda t)(1-\lambda' t)} = \frac{(1-t)(1-pt)}{(1-a_p t + pt^2)}, \tag{3.9}$$

so

$$Z(\tilde{E},p^{-s}) = \frac{(1-p^{-s})(1-p^{1-s})}{L_p(E,s)}. \tag{3.10}$$

(c) When $F = \mathbb{F}_q(X)$, the function field of a smooth projective curve over a finite field, and E is an elliptic curve over F, the same procedure can be used to define $L(E,s)$, taking an Euler product over all the prime divisors of the global field F. In this case we can associate to E a *surface* $\mathcal{E}$ fibered over X, whose generic fiber is E. This $\mathcal{E}$ can be taken to be regular (although not smooth over X). A similar relation exists then between the Zeta function of the *surface* $\mathcal{E}$ and $L(E,s)$. This relation, together with the work of Grothendieck and Deligne on Zeta functions of varieties over finite fields, can be exploited to prove analytic continuation and functional equation of $L(E,s)$, which for a number field F as the base are still conjectural (with the exception of $F = \mathbb{Q}$!). A detailed discussion of elliptic curves over function fields may be found in Tate's Bourbaki talk "On the conjecture of Birch and Swinnerton–Dyer and a geometric analogue."

(d) Since the dual space of $(V_l)_{I_p}$ is $\check{V}_l^{I_p}$, and the dual of $\rho_l(\sigma_p)$ is $\check{\rho}_l(\sigma_p^{-1})$, and since dual linear transformations have the same characteristic polynomials, we could also define

$$L_p(E,s) = \det(1 - p^{-s}\sigma_p^{-1}|\check{V}_l^{I_p})^{-1}. \tag{3.11}$$

Some authors use this. They call σ_p^{-1} the geometric Frobenius. Here $\check{V}_l^{I_p} = H^1(E_{/\bar{\mathbb{Q}}}, \mathbb{Q}_l)^{I_p}$ (invariants of étale cohomology).

3.2 *The conductor*

In a way similar to the definition of the Artin conductor, one can attach a conductor N to an elliptic curve $E/\mathbb{Q}$. It is determined by the higher ramification groups in the l-adic representation. We do not give the details, but record the following:

- If E has good reduction at p, then $p \nmid N$.
- If E has multiplicative reduction at p, $p||N$.
- If E has additive reduction at p and $p \geq 5$, $p^2||N$.
- For $p = 2, 3$, if E has additive reduction, $p^2|N$, but it is possible that the power of p in N is higher.

3.3 *The augmented L-function*

The field $\mathbb{Q}$ has one archimedean place. One defines

$$\Lambda(E,s) = N^{s/2}(2\pi)^{-s}\Gamma(s)L(E,s). \tag{3.12}$$

Conjecture 3.1 (Hasse–Weil). *$\Lambda(E,s)$ admits analytic continuation to the whole complex plane, and satisfies a functional equation*

$$\Lambda(E,s) = \pm\Lambda(E, 2-s). \tag{3.13}$$

Note that the *sign of the functional equation* should determine the parity of the order of $\Lambda(E,s)$ at $s = 1$. The "special value" $L(E,1)$ (more precisely, the leading term in the Taylor expansion around $s = 1$) is one of the most interesting special values in arithmetic. It is supposed to record subtle arithmetic invariants of E. It suffices if we say that it is conjectured that the order of vanishing of $L(E,s)$ at $s = 1$ is equal to the Mordell–Weil rank of $E(\mathbb{Q})$ (the conjecture of *Birch and Swinnerton–Dyer*)!

The Hasse–Weil conjecture for elliptic curves over $\mathbb{Q}$ is now a theorem, thanks to the work of Wiles et al. mentioned above. But the proof is indirect. It is based on the fact that L-series of modular forms satisfy these properties, and on the proof of the Shimura–Taniyama–Weil conjecture, which asserts that $L(E,s) = L(f,s)$ for a suitable modular form f.

Over general number fields, the Hasse–Weil conjecture is still wide open. Of course, the definition of $L(E,s)$ and the form of the functional equation have to be modified, but the gap between $\mathbb{Q}$ and a general number field is not technical, and has to do with the limitations of Wiles' program at the moment.

Working with the Jacobian variety of an arbitrary smooth projective curve C defined over$\mathbb{Q}$, and its l-adic representation (which is now of dimension $2g$, $g = genus(C)$), one can attach an L-function $L(C,s)$ to every curve in a way similar to the above. The good Euler factors of $L(C,s)$ will be of degree $2g$ (as polynomials in p^{-s}). In general, analytic continuation and functional equation for $L(C,s)$ are unknown, if $g > 1$, even over $\mathbb{Q}$.

3.4 *Elliptic curves with CM*

There is one other case where the Hasse–Weil conjecture has been known for quite some time, over an arbitrary number field, and this is the case of E's *admitting CM*. For simplicity, assume again that E is defined over $\mathbb{Q}$, and has (over $\bar{\mathbb{Q}}$) CM by the full ring of integers $\mathcal{O}_K$ in the quadratic imaginary field K. The endomorphisms are then defined already over K. We normalize the isomorphism

$$\mathcal{O}_K \simeq \mathrm{End}(E/K) \tag{3.14}$$

so that the effect of the endomorphism α on the tangent space at 0 is multiplication by α (rather than by $\bar{\alpha}$).

The *theory of complex multiplication* shows that in this case the extension generated by adjoining the torsion points of E to K is *abelian* over K. One associates to E a *grössencharacter* ψ of K. This is a character from $I(\mathfrak{f})$, the group of all fractional ideals relatively prime to $\mathfrak{f}$, to $K^\times$, which (in this case) satisfies

$$\psi((\alpha)) = \alpha \tag{3.15}$$

if $\alpha \equiv 1 \bmod \mathfrak{f}$. Its defining property is this: If $\mathfrak{p}$ is a prime ideal of K not dividing $\mathfrak{f}$, and l is relatively prime to $\mathfrak{p}$, $K_l = K(E[l^\infty])$ is unramified over K at $\mathfrak{p}$, and the Frobenius automorphism $\sigma_\mathfrak{p}$ of $\mathfrak{p}$ acts via $\psi(\mathfrak{p})$:

$$\rho_l(\sigma_\mathfrak{p})(x) = \psi(\mathfrak{p})(x) \text{ for } x \in T_l E. \tag{3.16}$$

The conductor $\mathfrak{f}$ of ψ is related to the conductor N of $E/\mathbb{Q}$ by

$$N = d_K \mathbf{N}\mathfrak{f}. \tag{3.17}$$

The *Hecke* L-function of ψ is defined as the sum over all the ideals of K:

$$L(\psi, s) = \sum \psi(\mathfrak{a}) \mathbf{N}\mathfrak{a}^{-s}. \tag{3.18}$$

Comparing Euler factors one obtains the equality

$$L(E/\mathbb{Q}, s) = L(\psi, s). \tag{3.19}$$

Hecke has established (for any grössencharacter, and arbitrary base fields) the analytic continuation and functional equation of $L(\psi, s)$. This will be explained in Kudla's chapter on Tate's thesis. The key point is that grössencharacters, like characters of generalized ideal class groups of finite order, have an interpretation as *idele class group characters*. In the case of E, it boils down to the Hasse–Weil conjecture for $L(E, s)$.

4 Modular curves and their Jacobians

In this section we assume familiarity with the analytic theory of modular forms, their L-series, and Hecke operators, as discussed by Kowalski.

4.1 *Modular curves*

The complex *modular curve* $X_0(N)$ is defined as the compactification of the open Riemann surface

$$Y_0(N) = \Gamma_0(N)\backslash\mathfrak{H} \tag{4.1}$$

where $\mathfrak{H}$ is the upper half plane, and

$$\Gamma_0(N) = \left\{ \begin{pmatrix} a & b \\ c & d \end{pmatrix} \in SL(2, \mathbb{Z});\ c \equiv 0 \bmod N \right\}. \tag{4.2}$$

The points of $X_0(N) - Y_0(N)$ are called the *cusps*, and play a special role in the theory of modular forms.

Consider the map that associates to $\tau \in \mathfrak{H}$ the couple

$$(E_\tau, H_\tau) \tag{4.3}$$

where $E_\tau = \mathbb{C}/\mathcal{L}_\tau$, $\mathcal{L}_\tau = \mathbb{Z}\tau + \mathbb{Z}$, and $H_\tau = \langle \frac{1}{N} \bmod \mathcal{L}_\tau \rangle$. This couple consists of an elliptic curve and a cyclic subgroup of order N. An isomorphism of two such couples is an isomorphism of the elliptic curves, mapping the cyclic subgroups to each other. It is then an easy exercise to check that the pairs corresponding to τ and τ' are isomorphic if and only if τ and τ' lie in the same orbit of $\Gamma_0(N)$. Thus $Y_0(N)$ is the *moduli space* of couples (E, H) as above. An appropriate definition of what one calls a "generalized (or degenerate) elliptic curve" allows a moduli-theoretic interpretation of the cusps as well.

Algebraic geometry supplies the tools to study moduli problems of this sort over arbitrary rings, and to represent them by algebraic varieties or schemes.

This is how one obtained a canonical model of $X_0(N)$, first over $\mathbb{Q}$ (Kronecker, Shimura), then over $\mathbb{Z}[\frac{1}{N}]$ (Igusa), and finally over $\mathbb{Z}$ (Deligne–Rapoport). The modular curve $X_0(N)$ and similar modular curves built from other congruence subgroups of $SL(2,\mathbb{Z})$, are fascinating objects. Because of their moduli interpretation they come equipped with lots of symmetries—more precisely, with a large ring of correspondences (the *Hecke algebra*)—and this makes their geometry richer. Because they are defined, at least over $\mathbb{C}$, in a group theoretic manner—recall that $\mathfrak{H} = SL(2,\mathbb{R})/SO(2)$—their L-functions coincide with *automorphic* L-functions, for which one can prove analytic continuation and functional equation.

We are interested in a relation between $X_0(N)$ and elliptic curves of conductor N, which at first sight is rather surprising. One confusing point should be cleared up at the beginning: this relation, which is the subject of our chapter, has nothing to do with the fact that $X_0(N)$ itself is a moduli space of elliptic curves with level structure. Thus elliptic curves intervene in two completely unrelated ways!

For the study of *modular forms*, the modular curves are important because they are the spaces over which modular forms live. More precisely, the correct geometric interpretation of modular forms, which has a lot of arithmetic consequences (such as the Ramanujan estimates for the growth of their Fourier coefficients) is as *sections of certain line bundles over the modular curve*. In this chapter we shall be interested exclusively in modular forms of weight 2. If

$$f \in S_2(\Gamma_0(N)) \tag{4.4}$$

is a *cusp-form* on $\Gamma_0(N)$,

$$\omega_f = 2\pi i f(z)dz \tag{4.5}$$

becomes a $\Gamma_0(N)$-invariant differential form on $\mathfrak{H}$, i.e., a *holomorphic differential form* on $X_0(N)$. (The cuspidality condition precisely guarantees that ω_f is holomorphic at the cusps as well.) It is well known that the space $H^0(X,\Omega)$ of holomorphic differential forms on a Riemann surface X is of dimension $g = genus(X)$, and may be identified with the cotangent space at 0 to the Jacobian variety of X, $\mathrm{Jac}(X)$, which is a complex torus of dimension g. We therefore proceed to study $J_0(N)$, the Jacobian of $X_0(N)$, and its endomorphism ring.

4.2 *The modular Jacobian*

We have just said that $J_0(N)$ is a complex torus of dimension g. Moreover, it has the structure of a smooth projective group variety—an abelian variety—which is defined over $\mathbb{Q}$. As such it has a Tate module, similar to the Tate module of an elliptic curve,

$$T_l J_0(N) = \varprojlim J_0(N)[l^n] \simeq \mathbb{Z}_l^{2g}, \tag{4.6}$$

on which $G_{\mathbb{Q}}$ and $\mathrm{End}(J_0(N))$ act commutingly. For some small values of N, e.g., $N = 11$, $g = 1$ and $X_0(N) = J_0(N)$ is itself an elliptic curve. In general, this is a higher-dimensional analogue of what we have discussed at length for elliptic curves.

Since $X_0(N)$ has good reduction over $\mathbb{Z}[\frac{1}{N}]$, so does its Jacobian. By the criterion of Ogg–Néron–Shafarevitch, the l-adic representation

$$\rho_l : G_{\mathbb{Q}} \to GL(V_l J_0(N)) \tag{4.7}$$

is unramified at every prime not dividing Nl.

4.3 *The Hecke algebra*

A *correspondence* between two smooth projective curves X and X' is a (perhaps reducible, but complete) curve Γ on the surface $X \times X'$, which does not contain any vertical or horizontal components. One should think of Γ as the graph of a multivalued function from X to X' or vice versa. Correspondences form naturally an abelian group w.r.t. formal addition, and if $X = X'$, taking composition as a product, they form a ring (composition is best understood when we think of correspondences as multivalued functions, and use the ordinary composition of functions). Any correspondence Γ between X and X' induces a homomorphism of abelian varieties

$$\gamma : \mathrm{Jac}(X) \to \mathrm{Jac}(X'). \tag{4.8}$$

For example, if we think of the Jacobian as being identified with the Picard variety, and if p and p' are the two projections from Γ to X and X', take a divisor of degree 0 on X, call it D, denote by $[D]$ its class modulo principal divisors, and define

$$\gamma([D]) = [p'_* p^* D]. \tag{4.9}$$

This is well defined. Moreover, $\Gamma \mapsto \gamma$ is additive, and in case $X = X'$, it is a *ring homomorphism* from the ring of correspondences of X into $\mathrm{End}(\mathrm{Jac}(X))$.

The modular curve $X_0(N)$ has, in view of its moduli interpretation, many interesting correspondences. For each prime p not dividing N take a point $x \in X_0(N)$, represent it by (E, H) as above, and let $C_0, \ldots, C_p$ be the $p+1$ cyclic subgroups of E of order p. Let T_p be the correspondence sending x to the formal sum $x_0 + \cdots + x_p$, where x_i is represented by $(E/C_i, H + C_i \bmod C_i)$. With some care, similar correspondences are defined for $p|N$ too. The T_p are called the *Hecke correspondences.* The endomorphisms that they induce on $J_0(N)$, or by functoriality on its cotangent space, which is the space of cusp forms of weight 2, are called the *Hecke operators.* Their effect on q-expansions of cusp forms is given by the formulae recorded in Kowalski's chapter.

The ring they generate, $\mathbb{Z}[\ldots, T_p, \ldots]$ is called the *Hecke algebra*. It is commutative. For general N it may not be semisimple, a fact that leads to the *Atkin–Lehner theory of old and new forms.* To avoid this technical complication, assume from now on, at least when we deal with the Hecke algebra, that

$$N \textit{ is prime.} \tag{4.10}$$

This implies that the (rational) Hecke algebra is semisimple, a product of fields, and therefore $S_2(\Gamma_0(N))$ *has a basis of eigenforms for all the Hecke operators.*

5 The Eichler–Shimura congruence relation

5.1 *The proof of the congruence relation*

If p is a good prime of $X = X_0(N)$ (a prime not dividing N), X extends to a smooth projective curve $\mathfrak{X}$ over $\mathbb{Z}_{(p)}$, which classifies the same couples (E, H) as before over $\mathbb{Z}_{(p)}$-algebras. In particular we may consider its reduction modulo p, $\tilde{X}$. The correspondence T_l, being a curve on $X \times X$, undergoes a similar reduction $\tilde{T}_l$. We would like to understand it in terms of a correspondence on the moduli problem in characteristic p. For $l \neq p$ there is nothing difficult. For T_p, however, there is a problem: an elliptic curve in characteristic p does not have $p+1$ cyclic subgroups of order p. If E is ordinary, there is one, and if E is supersingular there are none!

To overcome the difficulty one should talk about *finite group schemes* of order p in E. In characteristic p there is always a special one, the kernel of the Frobenius isogeny of degree p,

$$\varphi_E : E \to E^{(p)}, \tag{5.1}$$

where $E^{(p)}$ is the curve obtained by raising the coefficients of the equations defining E to power p. (If E is defined over $\mathbb{F}_q$, and $q = p^f$, then the fth iterate of φ_E is the endomorphism of E which we considered before.) If E is *ordinary*, which is the generic case, the dual isogeny

$$E^{(p^{-1})} \leftarrow E : \check{\varphi}_{E^{(p^{-1})}} \tag{5.2}$$

is separable, and its kernel is an "actual" (reduced, to be precise) cyclic subgroup of E of order p. If E is supersingular, $E^{(p^{-1})} = E^{(p)}$, and $\check{\varphi}_{E^{(p^{-1})}}$ is equal, up to an automorphism of the curve, to φ_E, so their kernels are equal. In any case, these are the only subgroup schemes in characteristic p.

If E is an elliptic curve in characteristic 0 (representing the point x on X), with good reduction at p (meaning that x does not reduce to a cusp on $\tilde{X}$), one of the $p+1$ subgroups of order p, say C_0, lies close to the origin (in the p-adic topology) and reduces to $\mathrm{Ker}(\varphi_{\tilde{E}})$. The other p subgroups all reduce to Ker $\check{\varphi}_{\tilde{E}^{(p^{-1})}}$. When we divide by the C_i's we therefore get once $\tilde{E}^{(p)}$ and p times $\tilde{E}^{(p^{-1})}$. The same applies to the auxiliary level-N structure. It is customary to call the correspondence on $\tilde{X}$ which sends $x = [(E, H)]$ (in characteristic p) to $x^{(p)}$ by *Frobenius*, and denote it by F. The dual correspondence (obtained by reversing the roles of the source and the target) is called *Verschiebung*, and is denoted by V. In the notation used before, these are the two correspondences obtained from

$$\Gamma = \left\{(x, x^{(p)}) \in \tilde{X} \times \tilde{X}\right\}. \tag{5.3}$$

Note that they are special to characteristic p and do not lift to characteristic 0. Note also that the first projection $\Gamma \to \tilde{X}$ is of degree 1 (an isomorphism), but the second projection is inseparable of degree p. Thus F is of degree 1, and V purely inseparable of degree p. The effect on points is

$$\begin{aligned} F &: x \mapsto x^{(p)} \\ V &: x \mapsto px^{(p^{-1})}. \end{aligned} \tag{5.4}$$

We have sketched the proof of the following.

Theorem 5.1 (Eichler–Shimura congruence relation). *The reduction modulo p of the correspondence T_p is*

$$\tilde{T}_p = F + V. \tag{5.5}$$

Although F and V separately do not lift to characteristic 0, their sum does! This important theorem has its seeds already in the work of Kronecker on the modular equation.

5.2 *Consequences for L-functions of modular curves*

As we observed in the case of elliptic curves, to study the action of σ_p, a Frobenius automorphism in $G_{\mathbb{Q}}$, in the l-adic representation of $J_0(N)$ (p good), we may reduce modulo p, so that the *arithmetically defined* σ_p becomes the *geometrically defined* F. Weil's work from 1948 on the RH for curves over finite fields (generalizing Hasse's thesis for elliptic curves!) tells us that the eigenvalues of F in its action on $T_l J_0(N)$ are algebraic integers $\lambda_1, \ldots, \lambda_g, \bar{\lambda}_1, \ldots, \bar{\lambda}_g$ of absolute value $\sqrt{p}$ (in every complex embedding). The eigenvalues of V (in the same basis) are $\bar{\lambda}_1, \ldots, \bar{\lambda}_g, \lambda_1, \ldots, \lambda_g$. Thus the eigenvalues of T_p are $a_{p,1}, \ldots, a_{p,g}$, each with multiplicity 2, where $a_{p,i} = \lambda_i + \bar{\lambda}_i$.

The action of T_p on $T_l J_0(N)$ is isomorphic to its action on the singular homology of the Jacobian, which is the sum of its representation on the cotangent space and on the complex conjugate of the cotangent space. But the cotangent space is identified (as a Hecke module) with $S_2(\Gamma_0(N))$, and the eigenvalues of T_p on cusp forms are real. We therefore conclude that $a_{p,1}, \ldots, a_{p,g}$ are just the eigenvalues of T_p in its action on $S_2(\Gamma_0(N))$.

The Euler factor at p of the L-function of X over $\mathbb{Q}$ is

$$\prod(1 - \lambda_i p^{-s})(1 - \bar{\lambda}_i p^{-s}) = \prod_{i=1}^{g}(1 - a_{p,i} p^{-s} + p^{1-2s}). \tag{5.6}$$

Corollary 5.2 (Shimura, Carayol). *Let $f_1, \ldots, f_g$ be a basis of $S_2(\Gamma_0(N))$ consisting of normalized eigenfunctions of the Hecke operators (recall that we assume N to be prime now, so all the f_i are newforms). Let*

$$f_i = \sum_{n=1}^{\infty} a_{n,i} q^n \tag{5.7}$$

be their q-expansions, so that $T_p f_i = a_{p,i} f_i$. Let

$$L(f_i, s) = \sum_n a_{n,i} n^{-s} = \prod_p (1 - a_{p,i} p^{-s} + p^{1-2s})^{-1} \tag{5.8}$$

be the corresponding L-functions. Then

$$L(X_0(N), s) = \prod_i L(f_i, s). \tag{5.9}$$

The good Euler factors (for p not dividing N) match, as we have just checked. The bad Euler factors match too, but this was proved only later by Carayol, andit is more difficult. Note that to establish analytic continuation, it is enough to have the L-function up to a finite number of Euler factors. It is only the precise form of the functional equation that requires knowing them all.

An important consequence of the corollary is that *the Hasse–Weil conjecture for the modular curve $X_0(N)$* is true. This relies of course on the analytic continuation and the functional equation of the $L(f_i, s)$ discussed in Kowalski's chapter. We are ready to consider elliptic curves over $\mathbb{Q}$.

6 Uniformization of elliptic curves and the Shimura–Taniyama–Weil conjecture

6.1 *Elliptic curves that are quotients of $J_0(N)$*

To simplify the exposition, keep the assumption that N is prime. The space $S_2(\Gamma_0(N))$ has a nice interpretation as the space of *linear maps* from the Hecke algebra to $\mathbb{C}$. The *eigenforms* are just those linear maps which are also *algebra* homomorphisms. Suppose that $f = \sum a_n q^n$ is a normalized cusp form, which is an eigenform of all the Hecke operators, and assume that its Fourier coefficients are rational: $a_n \in \mathbb{Z}$. Interpreting f as an algebra homomorphism, we see that the *rational* Hecke algebra

$$\mathcal{H} = \mathbb{Q}[\ldots, T_p, \ldots] \subset \mathrm{End}(J_0(N)) \otimes \mathbb{Q} \tag{6.1}$$

admits a quotient isomorphic to $\mathbb{Q}$ in which T_p is mapped to a_p. The corresponding idempotent e_f is a projector in the rational endomorphism algebra, some integral multiple of which is an honest endomorphism projecting $J_0(N)$ onto an *elliptic curve* quotient (the dimension can be determined from the cotangent space, as usual). Call it E_f. It is defined over $\mathbb{Q}$, since both $J_0(N)$ and the projector e_f are.

A variant of the arguments given above for the whole modular Jacobian, implies now that a_p (for good p) is the trace of Frobenius on E_f. As usual, the bad Euler factors match too, although this is more difficult. We obtain the following.

Theorem 6.1. *$L(E_f, s) = L(f, s)$ satisfies the Hasse–Weil conjecture.*

Ogg has proved that the conductor of E_f is N. (*The arithmetic conductor is equal to the analytic conductor.*)

6.2 *The Shimura–Taniyama–Weil conjecture*

Sometimes in the late 50's the following conjecture travelled from Japan to Princeton, and caught the attention of A. Weil:

Conjecture 6.2. *Every elliptic curve E over $\mathbb{Q}$ is isogenous to an E_f. More precisely, if N is the conductor of E, there exists a newform of level N and weight 2 such that E is isogenous to E_f.*

There was little evidence for the conjecture at the beginning, and nobody dared to put it on paper. But in 1967 Weil proved his *converse theorem*, discussed in our earlier chapter. It had the following implication.

Theorem 6.3 (Weil). *Suppose that $L(E, s) = \sum a_n n^{-s}$, together with all the twists*

$$L(E, \chi, s) = \sum a_n \chi(n) n^{-s}, \tag{6.2}$$

for every Dirichlet character χ of conductor prime to a certain integer M, admit analytic continuation and a functional equation (similar to the one predicted for $L(E, s)$). Then $E = E_f$ is modular.

As in the case of Artin L-functions, the modularity of E (or of the associated l-adic representation, or of the L-series, they are all equivalent) became a necessary and sufficient condition for the Hasse–Weil conjecture for all the $L(E, \chi, s)$.

In 1995 Taylor and Wiles proved the conjecture for square-free N, among other, but not all, cases. Further work by Diamond, Taylor, Conrad, and Breuil eliminated the restrictions on N, so today the conjecture is proved for all N.

It is interesting to note that the conjecture can be formulated purely as a conjecture in complex function theory, for E is isogenous to a factor of $J_0(N)$ if and only if there exists a nonconstant map of Riemann surfaces from $X_0(N)$ to E.

We shall not attempt to expose the work of Taylor and Wiles. Excellent expository papers exist, but they all require more background than what we wanted to assume in this book.

References

There are many excellent textbooks on the subject of these two chapters. We list a few. The references on Fermat's Last Theorem also contain very good survey chapters on the previous topics.

On elliptic curves:

[1] A. Robert, *Elliptic Curves*, Springer-Verlag, New York, 1973.

[2] S. Lang, *Elliptic Functions*, Addison–Wesley and Springer-Verlag, New York and Reading, MA, 1987.

[3] J. Silverman, *The Arithmetic of Elliptic Curves*, Springer-Verlag, New York, 1986.

[4] J. Silverman, *Advanced Topics in the Arithmetic of Elliptic Curves*, Springer-Verlag, New York, 1994.

On l-adic representations:

[5] J.-P. Serre, *Abelian l-Adic Representations and Elliptic Curves*, Benjamin, New York, 1968.

On modular forms, modular curves, and the relation to elliptic curves:

[6] G. Shimura, *Introduction to the Arithmetic Theory of Automorphic Forms*, Princeton University Press, Princeton, NJ, 1971.

[7] S. Lang, *Introduction to Modular Forms*, Springer-Verlag, New York, 1976.

[8] A. Knapp, *Elliptic Curves*, Princeton University Press, Princeton, NJ, 1992.

On Wiles's work on the Taniyama–Weil conjecture:

[9] H. Darmon, F. Diamond, and R. Taylor, Fermat's last theorem (a survey), *Curr. Dev. Math.*, **1**(1995), 1–157.

[10] G. Cornell, J. Silverman, and G. Stevens, eds., *Modular Forms and Fermat's Last Theorem*, Springer-Verlag, New York, 1997.

[11] K. Murty, ed., *Seminar on Fermat's Last Theorem*, CMS Conference Proceedings 17, American Mathematical Society, Providence, 1995.

6
Tate's Thesis

Stephen S. Kudla*

Tate's thesis, *Fourier Analysis in Number Fields and Hecke's Zeta-Functions* (Princeton, 1950) first appeared in print as Chapter XV of the conference proceedings *Algebraic Number Theory*, edited by Cassels and Frölich (Thompson Book Company, Washington, DC, 1967). In it, Tate provides an elegant and unified treatment of the analytic continuation and functional equation of the L-functions attached by Hecke to his Größencharaktere in his pair of papers [7]. The power of the methods of abelian harmonic analysis in the setting of Chevalley's adèles/idèles provided a remarkable advance over the classical techniques used by Hecke.[1] A sketch of the analogous interpretation of classical automorphic forms in terms of *nonabelian* harmonic analysis—automorphic representations—is given in the next chapter. The development of the theory of automorphic representations for arbitrary reductive groups G is one of the major achievements of mathematics of the later part of the twentieth century. And, of course, this development is still under way, as is evident from the later chapters [6]. In hindsight, Tate's work may be viewed as giving the theory of automorphic representations and L-functions of the simplest connected reductive group $G = \mathrm{GL}(1)$, and so it remains a fundamental reference and starting point for anyone interested in the modern theory of automorphic representations.

These articles have two main goals. The first is to give a unified treatment, following Tate, of the analytic properties, analytic continuation, functional equation, etc., of the L-function $L(s, \chi)$ attached to any Hecke character χ for any number field. The second is to introduce the point of view and techniques of representation theory of adèle groups in the simplest case. For example, the progression from local results to global results and the role played by uniqueness theorems for eigendistributions is rather typical in this business. The following chapters provide only a sketch and the reader should consult more extended sources for details.

References

Of course, the best reference is Tate's thesis itself. In addition, there are many other expositions, including Weil [13, Chapter VII], Bump [2, Chapter 3.1], [5], [9], and

---

*This chapter is a slightly expanded version of two lectures on Tate's thesis given at the school on Automorphic Forms, L-Functions, and Number Theory at Hebrew University in Jerusalem in March of 2001.

[1]Tate gives some interesting historical comments in the thesis and at the end of [3, Chapter XV].

the recent book of Valenza and Ramakrishnan [11]. In the present condensed survey, I more or less follow the approach sketched in Weil's 1966 Bourbaki talk *Fonction zeta et distributions* [12, 14], which places a greater emphasis on the equivariant distributions which are only implicit in Tate's original work. Weil's commentary on [14, Volume III, p. 448] is also of interest.

1 Adèles, idèles, and Dirichlet characters

As a motivation for the role played by the idèles and quasicharacters, consider the following description of classical Dirichlet characters.

For a positive integer N, a classical Dirichlet character $\underline{\chi}_N : \mathbb{Z} \to \mathbb{C}$ modulo N is a function obtained from a character $\chi_N : (\mathbb{Z}/N\mathbb{Z})^\times \to \mathbb{C}^\times$ by first extending by 0 on $\mathbb{Z}/N\mathbb{Z}$ and then pulling back to $\mathbb{Z}$. We will call χ_N a Dirichlet character. If $N \mid M$, then χ_N defines a character $\chi_M = \chi_N \circ \mathrm{pr}_{M,N}$ of $(\mathbb{Z}/M\mathbb{Z})^\times$, by pulling back under the projection

$$\mathrm{pr}_{M,N} : (\mathbb{Z}/M\mathbb{Z})^\times \longrightarrow (\mathbb{Z}/N\mathbb{Z})^\times, \qquad N \mid M,$$

and there is an associated classical Dirichlet character $\underline{\chi}_M$ modulo M as well. The inverse limit of the system $\{(\mathbb{Z}/N\mathbb{Z})^\times, \mathrm{pr}_{M,N}\}$ is the compact, totally disconnected, topological group[2]

$$\widehat{\mathbb{Z}}^\times := \varprojlim_N (\mathbb{Z}/N\mathbb{Z})^\times,$$

and every Dirichlet character χ_N can be viewed as a *continuous* character χ of $\widehat{\mathbb{Z}}^\times$. The *conductor* of χ is the smallest N_0 for which χ is trivial on the kernel of the projection $\mathrm{pr}_{N_0} : \widehat{\mathbb{Z}}^\times \to (\mathbb{Z}/N_0\mathbb{Z})^\times$. Then $N_0 \mid N$ and χ is the pullback of a unique Dirichlet character χ_{N_0} of $(\mathbb{Z}/N_0\mathbb{Z})^\times$. The collection of classical Dirichlet characters $\underline{\chi}_M$ for $M \mid N_0$ all correspond to χ, where $\underline{\chi}_{N_0}$ is the unique *primitive* classical Dirichlet character, and the others are *imprimitive*. An analogous phenomena takes place in the dictionary between classical holomorphic modular forms and automorphic representations of GL(2), where the normalized newforms play the role of the primitive classical Dirichlet characters.

[2]Recall that this is, by definition,

$$\widehat{\mathbb{Z}}^\times = \{ (x_N) \in \prod_N (\mathbb{Z}/N\mathbb{Z})^\times \mid \mathrm{pr}_{M,N}(x_M) = x_N \}.$$

The product of the finite groups $(\mathbb{Z}/N\mathbb{Z})^\times$ is compact and hence so is the closed subgroup $\widehat{\mathbb{Z}}^\times$. The projection maps

$$\mathrm{pr}_N : \widehat{\mathbb{Z}}^\times \to (\mathbb{Z}/N\mathbb{Z})^\times$$

are all surjective and continuous.

Recall that, for a fixed prime p,

$$\begin{aligned}\mathbb{Z}_p &:= \varprojlim_n \mathbb{Z}/p^n\mathbb{Z}\\ &= \{\,\alpha = a_0 + a_1 p + a_2 p^2 + \cdots \mid 0 \le a_i \le p-1\,\},\\ \mathbb{Z}_p^\times &= \{\,\alpha \in \mathbb{Z}_p \mid a_0 \neq 0\,\},\\ \mathbb{Q}_p &:= \{\,\alpha = \sum_i a_i p^i\,\}\end{aligned}$$

and so on.

Recall that the adèle ring $\mathbb{A}$ of $\mathbb{Q}$ is the restricted product[3]

$$\mathbb{A} = \mathbb{R} \times \prod_p{}' \mathbb{Q}_p$$

with respect to the subrings $\mathbb{Z}_p$. Also $\mathbb{Q} \hookrightarrow \mathbb{A}$, $\alpha \mapsto (\ldots, \alpha, \alpha, \alpha, \ldots)$, embedded diagonally, is a discrete subring. Similarly, the idèle group is the restricted product

$$\mathbb{A}^\times = \prod_v{}' \mathbb{Q}_v^\times$$

with respect to the subgroups $\mathbb{Z}_p^\times$, there is a diagonal embedding $\mathbb{Q}^\times \hookrightarrow \mathbb{A}^\times$, and there is a decomposition

$$\mathbb{A}^\times \simeq \mathbb{Q}^\times \times \mathbb{R}_+^\times \times \widehat{\mathbb{Z}}^\times, \qquad x = \alpha t u. \tag{1.1}$$

From this last decomposition, we see that any Dirichlet character χ defines a continuous character

$$\omega : \mathbb{Q}^\times\backslash\mathbb{A}^\times \longrightarrow \mathbb{C}^\times, \qquad \omega(x) = \omega(\alpha t u) = \chi(u).$$

The most general such *quasicharacter*[4] has the form $\omega \cdot |\ |^s$, for $s \in \mathbb{C}$, where $|x| = |\alpha t u| = t$. Thus, the Dirichlet characters are precisely the quasicharacters of finite order.

Note that, since

$$(\mathbb{Z}/N\mathbb{Z})^\times = \prod_p (\mathbb{Z}/p^n\mathbb{Z})^\times, \qquad n = n_p = \mathrm{ord}_p(N),$$

any Dirichlet character χ_N has a factorization

$$\chi_N = \otimes_p (\chi_N)_p,$$

where $(\chi_N)_p$ is a character of $(\mathbb{Z}/p^n\mathbb{Z})^\times$. This gives rise to a *factorization*

$$\chi = \otimes_p \chi_p$$

[3]Almost all components of an element (x_p) must lie in the subring $\mathbb{Z}_p$.

[4]A continuous homomorphism to $\mathbb{C}^\times$ is called a character if it takes values in $\mathbb{C}^1$, i.e., if it is unitary.

of the corresponding character χ of $\widehat{\mathbb{Z}}^\times$, where each χ_p is the character of $\widehat{\mathbb{Z}}_p^\times$ associated to $(\chi_N)_p$. Note that, for each place v, i.e., for each $\mathbb{Q}_p$ and for $\mathbb{Q}_\infty = \mathbb{R}$, we have an inclusion

$$\mathbb{Q}_v^\times \hookrightarrow \mathbb{Q}^\times\backslash\mathbb{A}^\times \qquad x_v \mapsto (\dots, 1, 1, x_v, 1, 1, \dots).$$

Thus, any quasicharacter ω determines quasicharacters ω_v of each $\mathbb{Q}_v^\times$, and there is a factorization

$$\omega = \otimes_v \omega_v \tag{1.2}$$

compatible with the previous one.

Exercise: (i) Check that an *odd* classical Dirichlet character $\underline{\chi}$ mod N (i.e., $\underline{\chi}(-a) = -\underline{\chi}(a)$) yields a character ω such that $\omega_\infty(x) = \operatorname{sgn}(x) = x/|x|$.

(ii) For ω associated to a classical Dirichlet character $\underline{\chi}_N$, modulo N, determine $\omega_p(p)$ for primes $p \nmid N$ and for primes $p \mid N$.

2 L-functions for quasicharacters

For k a general number field (or function field over a finite field),[5] there is a similar picture. We will use the following notation:

$$\begin{aligned}
\mathcal{O} &= \text{ring of integers in } k,\\
\mathcal{O}_v &= \text{its completion at a nonarchimedean place } v \ (= \text{a dvr}),\\
\mathcal{P}_v &= \text{the maximal ideal of } \mathcal{O}_v,\\
\varpi_v &= \text{a generator of } \mathcal{P}_v,\\
\mathbb{F}_v &= \mathcal{O}_v/\mathcal{P}_v = \text{the residue field at } v,\\
q_v &= |\mathcal{O}_v/\mathcal{P}_v| = \text{its order},\\
\operatorname{ord}_v &: k_v \longrightarrow \mathbb{Z}\cup\{\infty\}, \quad \text{the valuation, with } \operatorname{ord}_v(\varpi_v) = 1,\\
|x|_v &:= q_v^{-\operatorname{ord}_v(x)}, \quad \text{the normalized absolute value},\\
\mathcal{O}_v^\times &= \text{the group of units},
\end{aligned}$$

and so, $\mathcal{O}_v^\times \times \mathbb{Z} \xrightarrow{\sim} k_v^\times$, $(u, n) \mapsto u\varpi_v^n$. For an archimedean place v with $k_v \simeq \mathbb{R}$ (resp., $k_v \simeq \mathbb{C}$), let $|x|_v$ denote the usual absolute value (resp., the square of the usual absolute value).

Then we have restricted products, the adèles,

$$\mathbb{A} = \prod_v{}' k_v \qquad \text{and the idèles,} \qquad \mathbb{A}^\times = \prod_v{}' k_v^\times,$$

with diagonal embeddings $k \hookrightarrow \mathbb{A}$ and $k^\times \hookrightarrow \mathbb{A}^\times$. Here restricted means that for $x = (\dots, x_v, \dots)$ in $\mathbb{A}$ (resp., $\mathbb{A}^\times$), almost all components x_v lie in $\mathcal{O}_v$ (resp.,

[5] We will exclude this case, simply to avoid the extra side comments it would require.

$\mathcal{O}_v^\times$). There is an absolute value

$$| \ |_{\mathbb{A}} : \mathbb{A}^\times \longrightarrow \mathbb{R}_+^\times, \qquad |x| = \prod_v |x_v|_v,$$

which is trivial on $k^\times$ (the product formula). The decomposition (1.1) becomes more complicated due to (i) global units and (ii) the nontriviality of the ideal class group. These are precisely the difficulties which Hecke had to work hard to overcome and which, by contrast, the adèlic formalism handles so beautifully. In any case, there are inclusions

$$i_v : k_v^\times \hookrightarrow k^\times \backslash \mathbb{A}^\times, \qquad x_v \mapsto (\dots, 1, 1, x_v, 1, 1, \dots),$$

as before.

Definition 2.1. A *quasicharacter* or Hecke character is a continuous complex character $\omega : \mathbb{A}^\times \to \mathbb{C}^\times$, trivial on $k^\times$.

An explanation of the relation between this definition and the classical notion of Hecke character described in Kowalski's article [8] is given in Section 5 below.

There is a factorization

$$\omega = \otimes_v \omega_v,$$

where $\omega_v = \omega \circ i_v$.

Definition 2.2.

(i) For a nonarchimedean place v, a quasicharacter $\omega_v : k_v^\times \to \mathbb{C}^\times$ is *unramified* if it is trivial on $\mathcal{O}_v^\times$, and hence can be written in the form

$$\omega_v(x) = t_v^{\mathrm{ord}_v(x)}$$

for $t_v = t_v(\omega_v) \in \mathbb{C}^\times$.

(ii) A quasicharacter ω of $k^\times \backslash \mathbb{A}^\times$ is unramified at v if its local component $\omega_v = \omega \circ i_v$ at v is unramified.

For any ω, there is a finite set of places $S = S(\omega)$, including all archimedean places[6] such that ω_v is unramified for all $v \notin S$. For example, in the case $k = \mathbb{Q}$ discussed in the previous section, if ω is associated to a primitive classical Dirichlet character $\underline{\chi}_N$ modulo N, then $S(\omega) = \{\infty\} \cup \{p \mid p \mid N\}$. Thus, associated to ω is a collection of complex numbers $\{t_v(\omega)\}_{v \notin S}$. For example, for $s \in \mathbb{C}$, the quasicharacter $\omega_s(x) = |x|_{\mathbb{A}}^s$ is everywhere unramified and determines the set $\{q_v^{-s}\}$.

Definition 2.3. The (partial) *L-function* associated to a quasicharacter ω is the Euler product

$$L^S(s, \omega) = \prod_{v \notin S} (1 - t_v(\omega) q_v^{-s})^{-1}, \tag{EP}$$

[6]Where the notion of "unramified" is not defined.

where $S \supset S(\omega)$. The factors

$$L_v(s, \omega_v) := (1 - t_v(\omega)\, q_v^{-s})^{-1}$$

are the *local L-factors* associated to the ω_v's.

Of course,

$$L^S(s, \omega) = L^S(0, \omega\omega_s),$$

so we could dispense with s in the notation. Usually, we will prefer to keep s and to assume that ω is a character instead. In this case, the Euler product is absolutely convergent in the half plane $\mathrm{Re}(s) > 1$.

The main goals are to prove the analytic properties of these functions, i.e.,

1. to "complete" the partial L-function $L^S(s, \omega)$ by including additional local L-factors for the primes $v \in S$, for example, for the archimedean places, and

2. to prove the meromorphic analytic continuation and functional equation of the completed L-function.

These goals will be achieved by interpreting the local L-factors (resp., the global Euler product) as constants of proportionality between two naturally constructed basis elements of a one-dimensional complex vector space of *distributions*!

3 Local theory

Eigendistributions

We fix a place v of k and, to streamline notation, we write $F = k_v$. Let dx (resp., $d^\times x$) be a Haar measure on F (resp., $F^\times$). Observe that, for $a \in F^\times$,

$$d(ax) = |a|\, dx$$

for the normalized absolute value, as defined above. Note that one then has $d^\times x = \mu\, |x|^{-1}\, dx$ for some positive constant μ. Also, for $s \in \mathbb{C}$, let

$$\omega_s(x) = |x|^s$$

be the associated character. Of course, ω_0 is the trivial character. If v is nonarchimedean, we write $\mathcal{O} = \mathcal{O}_v$, $\mathcal{P} = \mathcal{P}_v$, $q = q_v$, etc. In this case, the character ω_s is unramified and only depends on the coset

$$s + \frac{2\pi i}{\log(q)}\mathbb{Z}.$$

Let

$$S(F) = \text{the space of Schwartz-Bruhat functions on } F.$$
$$S(F)' = \text{the space of tempered distributions on } F\text{, i.e.,}$$

the space of continuous linear functionals $\lambda : S(F) \to \mathbb{C}$.

Examples.

(i) For F nonarchimedean, a complex valued function f on F is in $S(F)$ if and only if there is an integer $r \geq 0$, depending on f, such that

$$\operatorname{supp}(f) \subset \mathcal{P}^{-r} \qquad \text{(compact support)}$$

and

$$f = \text{ constant on cosets of } \mathcal{P}^r. \qquad \text{(locally constant)}$$

The space $S(F)$ is simply the complex vector space of all such functions and the tempered distributions are arbitrary $\mathbb{C}$-linear functionals on $S(F)$.

(ii) For $F = \mathbb{R}$ (resp., $\mathbb{C}$), $S(\mathbb{R})$ is the usual Schwartz space[7] consisting of complex valued functions which, together with all of their derivatives, are rapidly decreasing, e.g.,

$$f(x) = h(x)\, e^{-\pi x^2} \quad \left(\text{resp., } f(x) = h(x, \bar{x})\, e^{-2\pi x\bar{x}}\right),$$

for any polynomial $h \in \mathbb{C}[X]$ (resp., $h \in \mathbb{C}[X, Y]$). This is a Frechét space, and the tempered distributions are the continuous linear functionals on it.

The multiplicative group $F^\times$ acts on $S(F)$ by

$$r(a)f(x) = f(xa),$$

for $x \in F$ and $a \in F^\times$, and on $S(F)'$ by

$$\langle r'(a)\lambda, f\rangle = \langle \lambda, r(a^{-1})f\rangle,$$

where $\langle\, ,\rangle$ denotes the pairing of $S(F)'$ and $S(F)$.

Definition 3.1. For a quasicharacter ω of $F^\times$, let

$$S'(\omega) = \{\lambda \in S(F)' \mid r'(a)\lambda = \omega(a)\lambda\}$$

be the space of *ω-eigendistributions*.

The space $S'(\omega)$ can be analyzed "geometrically". Note that there are two orbits for the action of $F^\times$ on the additive group

$$F = \{0\} \cup F^\times.$$

---

[7] A nice reference for the archimedean theory is the book of Friedlander [4]. For the nonarchimedean case, see [1].

Associated to this we have an inclusion

$$C_c^\infty(F^\times) \hookrightarrow S(F),$$

where $C_c^\infty(F^\times)$ is the space of C^∞-functions with compact support in $F^\times$. By duality, there is an exact sequence of distributions

$$0 \longrightarrow S(F)'_0 \longrightarrow S(F)' \longrightarrow C_c^\infty(F^\times)' \longrightarrow 0, \tag{3.1}$$

where $S(F)'_0$ is the subspace of distributions supported at 0. This sequence is compatible with the action of $F^\times$. Taking the ω-eigenspaces, we have the sequence

$$0 \longrightarrow S'(\omega)_0 \longrightarrow S'(\omega) \longrightarrow C_c^\infty(F^\times)'(\omega), \tag{3.2}$$

where $S'(\omega)_0$ is the space of ω-eigendistributions supported at 0. First, one has the following simple uniqueness result:

Lemma 3.2. *The space $C_c^\infty(F^\times)'(\omega)$ is one dimensional and is spanned by the distribution $\omega(x)\,d^\times x$. In particular, for any $\lambda \in S'(\omega)$, there is a complex number c such that*

$$\lambda\big|_{C_c^\infty(F^\times)} = c \cdot \omega(x)\,d^\times x,$$

i.e., if $\mathrm{supp}(f) \subset F^\times$ *is compact, then*

$$\langle \lambda, f\rangle = c \cdot \int_{F^\times} f(x)\,\omega(x)\,d^\times x.$$

It remains to determine the space $S(F)'_0$ of tempered distributions supported at 0 and the subspace $S'(\omega)_0$. The delta distribution δ_0 defined by

$$\langle \delta_0, f\rangle = f(0)$$

is obviously $F^\times$-invariant and supported at 0.

Lemma 3.3.
(i) *If F is nonarchimedean, then*

$$S(F)'_0 = \mathbb{C} \cdot \delta_0 \subset S'(\omega_0),$$

where ω_0 is the trivial character. For $\omega \neq \omega_0$, $S'(\omega)_0 = 0$.
(ii) *If $F = \mathbb{R}$, then,*[8] *setting $D = \frac{d}{dx}$,*

$$S(F)'_0 = \bigoplus_{k=0}^{\infty} \mathbb{C} \cdot D^k \delta_0,$$

and

$$D^k \delta_0 \in S'(\omega),$$

[8] In the archimedean case, we give only the $F^\times$-finite distributions here.

where

$$\omega(x) = x^{-k}.$$

(iii) *If* $F = \mathbb{C}$, *then, setting* $D = \frac{\partial}{\partial x}$ *and* $\bar{D} = \frac{\partial}{\partial \bar{x}}$,

$$S(F)'_0 = \bigoplus_{k,l=0}^{\infty} \mathbb{C} \cdot D^k \bar{D}^l \delta_0.$$

Moreover,

$$D^k \bar{D}^l \delta_0 \in S'(\omega),$$

where

$$\omega(x) = x^{-k}\,\bar{x}^{-l}.$$

The fundamental local uniqueness result, whose proof we will sketch, is the following.

Theorem 3.4. *For any quasicharacter* ω *of* $F^\times$,

$$\dim S'(\omega) = 1.$$

Remarks.

(i) When $S'(\omega)_0 = 0$, i.e., when there are no ω-eigendistributions supported at 0, the reasoning above shows that the dimension of $S'(\omega)$ is at most one, so that only existence is at stake. This amounts to showing that the distribution $\omega(x)\,d^\times x$ in Lemma 3.2 can be extended to an ω-eigendistribution on $S(F)$. This will be established via local zeta integrals below.

(ii) When $S'(\omega)_0 \neq 0$, the problem will be to show that the distribution $\omega(x)\,d^\times x$ does *not* extend to an ω-eigendistribution. This more delicate fact also comes out of the analysis of local zeta integrals.

Example. Suppose that F is nonarchimedean. If ω is a *ramified* character, then $S'(\omega\omega_s)_0 = 0$, so that

$$\dim S'(\omega\omega_s) \leq 1$$

for all s. If ω is unramified with $\omega(x) = t^{\mathrm{ord}(x)}$, as above, then, by Lemma 3.3,

$$\dim S'(\omega\omega_s) \leq \begin{cases} 2 & \text{if } t \cdot q^{-s} = 1, \text{ and} \\ 1 & \text{otherwise.} \end{cases}$$

Zeta integrals

For a character ω of $F^\times$, i.e., a unitary quasicharacter, the *local zeta integral*

$$z(s, \omega; f) := \int_{F^\times} f(x)\,\omega\omega_s(x)\,d^\times x = \int_F f(x)\,\omega(x)\,\mu\,|x|^{s-1}\,dx, \qquad (3.3)$$

is absolutely convergent for all $f \in S(F)$ provided $\mathrm{Re}(s) > 0$. Here $d^\times x = \mu\,|x|^{-1}\,dx$, as before. In this range, the distribution $f \mapsto z(s, \omega; f)$ defines a nonzero element

$$z(s, \omega) \in S'(\omega\omega_s).$$

The unramified local theory

Suppose that F is nonarchimedean and suppose that ω is unramified, with $\omega(\varpi) = t$. If f has compact support in $F^\times$, then the integral

$$\int_{F^\times} f(x)\,\omega\omega_s(x)\,d^\times x$$

is entire. The idea, then, is to kill the support of an arbitrary $f \in S(F)$ by applying a suitable element of the group algebra $\mathbb{Z}[F^\times]$, more precisely, the element

$$\tau = [1] - [\varpi^{-1}], \tag{3.4}$$

where ϖ is a generator of $\mathcal{P}$. Since any $f \in S(F)$ is constant in a sufficiently small neighborhood, say $\mathcal{P}^r$, of 0, we have, for any $x \in \mathcal{P}^{r+1}$,

$$\big(r(\tau)f\big)(x) = f(x) - f(x\varpi^{-1}) = 0.$$

Thus, there is a distribution

$$z_o(s,\omega) \in S'(\omega\omega_s)$$

defined by

$$\langle z_o(s,\omega), f\rangle := \int_{F^\times} \big(r(\tau)f\big)(x)\,\omega\omega_s(x)\,d^\times x. \tag{3.5}$$

On the other hand, in the halfplane $\mathrm{Re}(s) > 0$, where it is not necessary to kill support to have convergence,

$$\begin{aligned}
\langle z_o(s,\omega), f\rangle &= \int_{F^\times} \big(f(x) - f(\varpi^{-1}x)\big)\,\omega\omega_s(x)\,d^\times x\\
&= \int_{F^\times} f(x)\,\omega\omega_s(x)\,d^\times x - \int_{F^\times} f(\varpi^{-1}x)\,\omega\omega_s(x)\,d^\times x\\
&= (1 - \omega\omega_s(\varpi))\cdot z(s,\omega; f)\\
&= (1 - t\,q^{-s})\cdot z(s,\omega; f)\\
&= L(s,\omega)^{-1}\cdot \langle z(s,\omega), f\rangle.
\end{aligned}$$

Strictly in terms of distributions, this says that, for $\mathrm{Re}(s) > 0$,

$$z(s,\omega) = L(s,\omega)\cdot z_o(s,\omega), \tag{3.6}$$

and, hence, the expression on the right-hand side gives the meromorphic analytic continuation of $z(s,\omega)$ to the whole s plane. Moreover, (3.6) provides an interpretation of the local L-factor as a constant of proportionality between natural bases for the one-dimensional space $S'(\omega\omega_s)$, away from the poles of $L(s,\omega)$.

Finally, to see why the distribution $z_o(s,\omega)$ is "natural," we normalize the multiplicative Haar measure $d^\times x$ so that the units $\mathcal{O}^\times$ have volume 1. Then, we let

$$f^o = \text{ the characteristic function of } \mathcal{O}, \tag{3.7}$$

the ring of integers in F, and compute:

$$\begin{aligned}\langle z_o(s,\omega), f^o\rangle &= \int_{F^\times} \left(f^o(x) - f^o(\varpi^{-1}x)\right)\omega\omega_s(x)\,d^\times x \\ &= \int_{\mathcal{O}^\times} d^\times x \\ &= 1,\end{aligned} \tag{3.8}$$

for all s. The point is that $r(\tau)f^o$ is the characteristic function of $\mathcal{O}^\times$. Thus, $z_o(s,\omega)$ is never zero and gives a basis vector for $S'(\omega\omega_s)$ for all s.

Remark (the L-factor as greatest common denominator). Relations (3.6) and (3.8) together show that $L(s,\omega)$ gives precisely the poles of the family of zeta integrals (3.3). More precisely, for any $f \in S(F)$, the function

$$\frac{z(s,\omega;f)}{L(s,\omega)} = \langle z_o(s,\omega), f\rangle \tag{3.9}$$

is entire and that, for any given s, there is an f (specifically f^o) for which the value of (3.9) at s is nonzero.

Example. To complete the proof of Theorem 3.4 in the unramified case, there is still one point to check, namely that the space $S'(\omega_0)$ of invariant distributions, which contains $z_o(0,\omega_0)$ and has dimension at most 2, is actually only one dimensional. Here is the argument. There is an exact sequence of distributions

$$0 \longrightarrow \mathbb{C}\cdot\delta_0 \longrightarrow S(F)' \longrightarrow C_c^\infty(F^\times)' \longrightarrow 0 \tag{3.10}$$

and, passing to ω_0-eigendistributions, i.e., $F^\times$-invariant distributions, we have

$$0 \longrightarrow \mathbb{C}\cdot\delta_0 \longrightarrow S'(\omega_0) \longrightarrow \mathbb{C}\cdot d^\times x. \tag{3.11}$$

The point now is to show that, although $d^\times x$ is in the image of the quotient map in (3.10), it is not in the image of the quotient map in (3.11), i.e., the preimage of $d^\times x$ in $S(F)'$ is *not* invariant.

Explicitly, the distribution $\lambda_0 \in S(F)'$ defined by

$$\langle\lambda_0, f\rangle := \langle d^\times x, f - f(0)f^o\rangle \tag{3.12}$$

gives a preimage of $d^\times x$ in $S(F)'$, which is clearly invariant under $\mathcal{O}^\times$ (since f^o is) and satisfies $\langle\lambda_0, f^o\rangle = 0$. The distributions $r'(\varpi)\lambda_0$ and λ_0 agree on $C_c^\infty(F^\times)$ and hence differ by a distribution supported at 0, i.e.,

$$r'(\varpi)\lambda_0 = \lambda_0 + c\,\delta_0,$$

for some constant c. To determine c (which would be 0 if λ_0 were actually $F^\times$-invariant), we evaluate on f^o, using the group algebra element τ, as above:

$$\begin{aligned} c &= \langle r'(\varpi)\lambda_0 - \lambda_0, f^o\rangle \\ &= -\langle \lambda_0, r(\tau)f^o\rangle \\ &= -\langle d^\times x, r(\tau)f^o\rangle \\ &= -\langle z_o(0,\omega_0), f^o\rangle \\ &= -1. \end{aligned} \tag{3.13}$$

Thus, on the two-dimensional subspace of $S(F)'$ spanned by λ_0 and δ_0, $F^\times$ acts by the representation

$$\rho(x) = \begin{pmatrix} 1 & -\operatorname{ord}(x) \\ & 1 \end{pmatrix}. \tag{3.14}$$

This finishes the proof of Theorem 3.4 in the unramified case.

The ramified local theory

Suppose that F is nonarchimedean and that ω is ramified, hence nontrivial on $\mathcal{O}^\times$. The conductor $c(\omega)$ of ω is the smallest integer c such that ω is trivial on $1+\mathcal{P}^c$. Since any $f \in S(F)$ is constant in a neighborhood of 0, the integral

$$\int_{F^\times - \mathcal{P}^n} f(x)\,\omega\omega_s(x)\,d^\times x$$

is independent of n for n sufficiently large. This gives the entire analytic continuation of $z(s,\omega; f)$ to the whole s plane and hence gives a basis vector

$$z_o(s,\omega) := z(s,\omega), \tag{3.15}$$

for the one-dimensional space $S'(\omega\omega_s)$ for all s. This finishes the proof of Theorem 3.4 in the ramified case. We set

$$L_v(s,\omega) = 1 \tag{3.16}$$

in this case, and we note that, for

$$f^o(x) = \begin{cases} \omega(x)^{-1} & \text{if } x \in \mathcal{O}^\times, \text{ and} \\ 0 & \text{otherwise,} \end{cases} \tag{3.17}$$

we have

$$\langle z_0(s,\omega), f^o\rangle = 1.$$

Thus the gcd property (3.9) again holds.

The archimedean case

In this case, it is the whole Taylor series at 0 of the function f which accounts for the poles of the zeta integral $z(s, \omega; f)$, rather than just the value $f(0)$ which arose in the nonarchimedean case; cf. [2, Proposition 3.1.7, p. 271]. It is easy to obtain the analytic continuation of the zeta integral $z(s, \omega; f)$ by integration by parts.

Here we just summarize the results, using the normalizations of [13]. The convention that ω is unitary is dropped in order to obtain particularly concise expressions.

If $F = \mathbb{R}$, any quasicharacter has the form $\omega\omega_s$ where $s \in \mathbb{C}$ and

$$\omega(x) = x^{-a}, \qquad a = 0, 1. \tag{3.18}$$

Let

$$L(s, \omega) = \pi^{-\frac{s}{2}} \Gamma(\frac{s}{2}). \tag{3.19}$$

If $F = \mathbb{C}$, any quasicharacter has the form $\omega\omega_s$ where $s \in \mathbb{C}$ and

$$\omega(x) = x^{-a}\bar{x}^{-b}, \qquad a,\ b \in \mathbb{Z}, \qquad \min(a, b) = 0. \tag{3.20}$$

Let

$$L(s, \omega) = (2\pi)^{1-s}\Gamma(s). \tag{3.21}$$

Proposition 3.5.
(i) *The distribution*

$$z_o(s, \omega) := L(s, \omega)^{-1}\, z(s, \omega)$$

has an entire analytic continuation to the whole s plane and, for all s, defines a basis vector for the space $S'(\omega\omega_s)$.
(ii) *If*

$$f^o(x) = \begin{cases} f_a(x) := x^a e^{-\pi x^2} & \text{if } F = \mathbb{R}, \text{ and} \\ f_{a,b}(x) := x^a\, \bar{x}^b\, e^{-2\pi x\bar{x}} & \text{if } F = \mathbb{C}, \end{cases}$$

then

$$\langle z_o(s, \omega), f^o \rangle = 1.$$

Note that we are using the measure $d^\times x = |x|^{-1}\, dx$, where $|x|$ is the usual absolute value (resp., $|x| = x\bar{x}$) and dx is Lebesque measure (resp., twice Lebesque measure) when $F = \mathbb{R}$ (resp., $F = \mathbb{C}$). For (ii), see [13, Lemma 8, p. 127].

Remark. In effect, the zeta distribution $z(s, \omega)$ has a meromorphic analytic continuation in s with simple poles at certain nonpositive integers. For example, when $F = \mathbb{R}$, the poles occur at the points $-r \in \mathbb{Z}_{\leq 0}$ with $r \equiv a \mod (2)$, and the residues are

$$\operatorname*{Res}_{s=-r} z(s,\omega) = c_r\, D^r \delta_0, \tag{3.22}$$

with a nonzero constant c_r. Here D is as in Lemma 3.3. The constant term in the Laurent expansion of $z(s,\omega)$ at $s = -r$ gives an extension to $S(F)$ of the distribution $x^{-r}\, d^{\times}x$ on $C_c^{\infty}(F^{\times})$. A simple calculation, like (3.9), shows that this extension does *not* lie in the space $S'(\omega\omega_{-r})$. In the case $F = \mathbb{C}$, the residue at $s = -r$ lies in $S'(\omega\omega_{-r})$, where $\omega\omega_{-r}(x) = x^{-a-r}\bar{x}^{-b-r}$ and is a nonzero multiple of $D^{a+r}\bar{D}^{b+r}\delta_0$.

Fourier transforms

Fix a nontrivial character ψ of the additive group $F = F^{+}$, and identify F with its topological dual character group

$$\hat{F} := \mathrm{Hom}_{\mathrm{cont}}(F, \mathbb{C}^1),$$

by

$$F \xrightarrow{\sim} \hat{F}, \qquad y \mapsto \big(x \mapsto \psi(xy)\big).$$

For F nonarchimedean, the conductor $\nu(\psi)$ of ψ is the largest integer ν such that ψ is trivial on $\mathcal{P}^{-\nu}$.

The Fourier transform

$$\hat{f}(x) = \int_F f(y)\,\psi(xy)\,dy$$

of a function $f \in S(F)$ is well defined and again lies in $S(F)$. (This is a fundamental property of the space of Schwartz–Bruhat functions.) The map $f \mapsto \hat{f}$ is an isomorphism $\hat{\ }: S(F) \xrightarrow{\sim} S(F)$. There is a unique choice of the Haar measure, the self-dual measure with respect to ψ, such that Fourier inversion gives

$$\hat{\hat{f}}(x) = f(-x).$$

Supposing ψ to be given, we fix this choice of dx from now on. The Fourier transform of a distribution is defined by

$$\langle \hat{\lambda}, f\rangle = \langle \lambda, \hat{f}\rangle.$$

Lemma 3.6. *If $\lambda \in S'(\omega)$ is an ω-eigendistribution, then*

$$\hat{\lambda} \in S'(\omega^{-1}\omega_1),$$

i.e., $\hat{\lambda}$ is an $\omega^{-1}\omega_1$-eigendistribution, where $\omega_1(x) = |x|$.

Proof. An exercise! □

By the uniqueness result, Theorem 3.4, it follows that $\widehat{z_o(s,\omega)}$ is a constant multiple (depending on s and on ψ) of $z_o(1-s,\omega^{-1})$.

Corollary 3.7 (the local functional equation).

$$\widehat{z_o(1-s,\omega^{-1})} = \epsilon(s,\omega,\psi)\, z_o(s,\omega) \tag{3.23}$$

for a nonzero constant $\epsilon(s,\omega,\psi)$, *the* local epsilon factor, *depending on* s, ω, *and* ψ.

Remark. For a given $f \in S(F)$, the local functional equation can be written as the relation

$$\frac{z(1-s,\omega^{-1};\hat{f})}{L(1-s,\omega^{-1})} = \epsilon(s,\omega,\psi)\cdot\frac{z(s,\omega;f)}{L(s,\omega)}, \tag{3.24}$$

or, alternatively, as

$$z(1-s,\omega^{-1};\hat{f}) = \epsilon(s,\omega,\psi)\frac{L(1-s,\omega^{-1})}{L(s,\omega)}\cdot z(s,\omega;f), \tag{3.25}$$

where the constant of proportionality

$$\gamma(s,\omega,\psi) = \epsilon(s,\omega,\psi)\frac{L(1-s,\omega^{-1})}{L(s,\omega)}, \tag{3.26}$$

which can now have zeroes and poles, is called the local gamma factor. This is the more traditional formulation, while that in Corollary 3.7 is emphasized in [12].

Various important properties of the ϵ-factor $\epsilon(s,\omega,\psi)$ can be obtained from relation (3.23). For example, if f^o is the standard function with respect to ω, then, evaluating both sides of (3.23) on f^o yields the useful identity

$$\epsilon(s,\omega,\psi) = \langle z_o(1-s,\omega^{-1}), \widehat{f^o}\rangle. \tag{3.27}$$

By construction, the zeta distribution satisfies

$$z(s,\omega\omega_t) = z(s+t,\omega).$$

Since the local L-function is also required to have this property, we have

$$z_o(s,\omega\omega_t) = z_o(s+t,\omega)$$

as well. The standard function f^o does not change under multiplication by ω_t. This implies that

$$\epsilon(s,\omega\omega_t,\psi) = \epsilon(s+t,\omega,\psi). \tag{3.28}$$

For $\beta \in F^\times$, let $\psi_\beta(x) = \psi(\beta x)$. If dx is the self-dual measure with respect to ψ, then $|\beta|^{\frac{1}{2}}\,dx$ is the self-dual measure with respect to ψ_β. If $\hat{f}$ is the Fourier transform of $f \in S(F)$ with respect to ψ, then $|\beta|^{\frac{1}{2}}\,r(\beta)\hat{f}$ is the Fourier transform with respect to ψ_β as always for the self-dual measure. Thus, inserting this on the right side of (3.27), we obtain

$$\epsilon(s,\omega,\psi_\beta) = |\beta|^{s-\frac{1}{2}}\omega(\beta)\,\epsilon(s,\omega,\psi). \tag{3.29}$$

The following result gives explicit values for the ϵ-factors. Using the two relations (3.28) and (3.29), it suffices to compute these factors with both ψ and ω in "standard" form.

Recall that, in the nonarchimedean case, $\nu = \nu(\psi)$ is the conductor of ψ. In the archimedean case, the "standard" choice

$$\psi(x) = \begin{cases} e(x) & \text{if } F = \mathbb{R}, \\ e(x+\bar{x}) & \text{if } F = \mathbb{C}, \end{cases} \tag{3.30}$$

where, for a real number t, $e(t) = e^{2\pi i t}$, has self-dual measures the measures described after Proposition 3.5.

Proposition 3.8.
(i) *For F nonarchimedean, ω unramified, and $\nu(\psi) = \nu$:*

$$\epsilon(s, \omega, \psi) = \omega(\varpi^{\nu}) q^{(\frac{1}{2}-s)\nu}.$$

(ii) *For F nonarchimedean, ω ramified with conductor c, and $\nu(\psi) = \nu$,*

$$\epsilon(s, \omega, \psi) = \omega(\varpi^{\nu+c}) q^{(\frac{1}{2}-s)(\nu+c)}\, \mathfrak{g}(\omega, \psi),$$

where $\mathfrak{g}(\omega, \psi)$ is the Gauss sum:

$$\mathfrak{g}(\omega, \psi) = q^{\frac{1}{2}(\nu+c)} \int_{\mathcal{O}^\times} \omega^{-1}(y)\, \psi(\varpi^{-\nu-c} y)\, dy.$$

with dy the self-dual measure on F with respect to ψ.

(iii) *For $F = \mathbb{R}$, write $\omega(x) = x^{-a}$ with $a = 0$, or 1 and let ψ be as in* (3.30). *Then*

$$\epsilon(s, \omega, \psi) = i^{a}.$$

(iv) *For $F = \mathbb{C}$, write $\omega(x) = x^{-a}\bar{x}^{-b}$, for $a, b \in \mathbb{Z}$ with $\min(a, b) = 0$ and let ψ be as in* (3.30). *Then*

$$\epsilon(s, \omega, \psi) = i^{\max(a,b)}.$$

Proof. By (3.27), one has to compute the Fourier transform of the standard function f^o for ω, where the character ψ can be taken to be standard. For example, in case (i), f^o is the characteristic function of $\mathcal{O}$, and, assuming that $\nu(\psi) = 0$, $\widehat{f^o} = f^o$. This gives $\epsilon(s, \omega, \psi) = 1$ in this case. The general unramified case is then immediate via (3.29).

In case (ii), f^o is given by (3.17). Then, following [13, Proposition 13, p. 131],

$$\widehat{f^o}(x) = \int_{\mathcal{O}^\times} \omega^{-1}(y)\, \psi(xy)\, dy$$

vanishes unless $\operatorname{ord}(x) = -\nu - c$ where $c = c(\omega)$ is the conductor of ω. In fact,

$$\widehat{f^o}(x) = \mathfrak{g}(\omega, \psi)\, q^{-\frac{1}{2}(\nu+c)}\, \overline{f^o(\varpi^{\nu+c} x)}, \tag{3.31}$$

where

$$\mathfrak{g}(\omega, \psi) = q^{\frac{1}{2}(\nu+c)} \int_{\mathcal{O}^\times} \omega^{-1}(y)\, \psi(\varpi^{-\nu-c} y)\, dy. \tag{3.32}$$

Applying (3.32) twice yields $\mathfrak{g}(\omega, \psi)\, \overline{\mathfrak{g}(\omega, \psi)} = 1$. Since $\overline{f^o}$ is the standard function for ω^{-1}, we obtain

$$\epsilon(s, \omega, \psi) = \omega(\varpi^{\nu+c})\, \mathfrak{g}(\omega, \psi)\, q^{(\frac{1}{2}-s)(\nu+c)}.$$

In case (iii), we compute, for $f^o = f_a$, as in Proposition 3.5,

$$\begin{aligned}
\widehat{f_a}(x) &= \int_{\mathbb{R}} y^a\, e^{-\pi y^2}\, e(xy)\, dy \\
&= (2\pi i)^{-a} D_x^a \left\{ \int_{\mathbb{R}} e^{-\pi y^2}\, e(xy)\, dy \right\} \\
&= (2\pi i)^{-a} D_x^a \{ e^{-\pi x^2} \} \\
&= i^a\, f_a(x).
\end{aligned} \tag{3.33}$$

Since f_a is also standard for $\omega^{-1} = \omega\, \omega_2$, we obtain (iii).

In case (iv), we have $f^o = f_{a,b}$, as in Proposition 3.5. Then, recalling that $\min(a, b) = 0$,

$$\begin{aligned}
\widehat{f_{a,b}}(x) &= \int_{\mathbb{C}} y^a \bar{y}^b\, e^{-2\pi y\bar{y}}\, e(xy + \bar{x}\bar{y})\, dy \\
&= (2\pi i)^{-a} D_x^a\, (2\pi i)^{-b} \bar{D}_x^b \left\{ \int_{\mathbb{C}} e^{-2\pi y\bar{y}}\, e(xy + \bar{x}\bar{y})\, dy \right\} \\
&= (2\pi i)^{-a} D_x^a\, (2\pi i)^{-b} \bar{D}_x^b \left\{ e^{-2\pi x\bar{x}} \right\} \\
&= i^{a+b}\, f_{b,a}(x).
\end{aligned} \tag{3.34}$$

Noting that $f_{b,a}$ is the standard function for $\omega^{-1}(x) = \bar{x}^{-a}\, x^{-b}\, |x|^{a+b}$, we obtain the claimed value. More details can be found in Tate's thesis or in [13]. □

4 Global theory

Distributions

Returning to the number field k with its adèle ring $\mathbb{A}$, we define the space $S(\mathbb{A})$ of Schwartz–Bruhat functions on $\mathbb{A}$. This space contains all functions of the form

$$f = \otimes_v f_v,$$

where $f_v \in S(k_v)$ for all v and $f_v = f_v^o$, the characteristic function of $\mathcal{O}_v$, for almost all v. Note that

$$f(x) = \prod_v f_v(x_v)$$

is then well defined, as almost all factors are 1. The finite linear combinations of such factorizable functions are dense in $S(\mathbb{A})$. Again, $S(\mathbb{A})'$ is the space of tempered distributions, i.e., continuous linear functionals on $S(\mathbb{A})$.

Lemma 4.1. *Suppose that $\{\lambda_v\}$ is a collection of local tempered distributions $\lambda_v \in S(k_v)'$ such that, for almost all v, $\langle \lambda_v, f_v^o \rangle = 1$. Then there is a unique tempered distribution $\lambda = \otimes_v \lambda_v$ on $S(\mathbb{A})$, defined on factorizable functions by*

$$\langle \lambda, f \rangle = \prod_v \langle \lambda_v, f_v \rangle.$$

Conversely, if $\lambda \in S(\mathbb{A})'$ is nonzero, choose a factorizable f such that $\lambda(f) = 1$. Then, for any place v, we write

$$f = f_v \otimes f^v, \qquad f^v = \otimes_{w \neq v} f_w,$$

and we define a map

$$S(k_v) \longrightarrow S(\mathbb{A}), \qquad g_v \mapsto g_v \otimes f^v,$$

by just varying the v component. Thus, we obtain local distributions

$$\langle \lambda_v, g_v \rangle := \langle \lambda, g_v \otimes f^v \rangle,$$

with $\langle \lambda_v, f_v^o \rangle = 1$ for almost all v, and $\lambda = \otimes_v \lambda_v$.

The idèle group $\mathbb{A}^\times$ acts on $S(\mathbb{A})$ and $S(\mathbb{A})'$, and, for a quasicharacter ω of $\mathbb{A}^\times$, trivial on $k^\times$, we let $S'(\omega)$ be the space of ω-eigendistributions. The factorization $\omega = \otimes_v \omega_v$ of the global quasicharacter gives rise to the decomposition

$$S'(\omega) = \otimes_v S_v'(\omega_v),$$

where we write $S_v' = S(k_v)'$. The local uniqueness Theorem 3.4 then yields the global uniqueness result:

Theorem 4.2. *For all quasicharacter ω of $\mathbb{A}^\times$, the space of global ω-eigendistributions $S'(\omega)$ has dimension* 1. *For any $s \in \mathbb{C}$, the space $S'(\omega\omega_s)$ is spanned by the* standard $\omega\omega_s$-eigendistribution

$$z_o(s, \omega) := \otimes_v z_o(s, \omega_v).$$

Global zeta integrals and the completed L-function

For a character ω of $k^\times \backslash \mathbb{A}^\times$ and a function $f \in S(\mathbb{A})$, we can define a global zeta integral by

$$z(s, \omega; f) = \int_{\mathbb{A}^\times} f(x)\, \omega\omega_s(x)\, d^\times x. \tag{4.1}$$

If f is factorizable, this is a product of the corresponding local integrals

$$z(s, \omega; f) = \prod_v z(s, \omega_v; f_v). \tag{4.2}$$

A key point here is that this integral *converges* precisely when the product of the local integrals converges. But, since there is a finite set of places S such that for $v \notin S$, ω_v is unramified and $f_v = f_v^o$, we obtain convergence whenever the product

$$L^S(s,\omega) = \prod_{v \notin S} L_v(s,\omega_v)$$

is absolutely convergent, i.e., in the half plane $\mathrm{Re}(s) > 1$.

In terms of distributions, in the half plane $\mathrm{Re}(s) > 1$, we have $z(s,\omega) = \otimes_v z(s,\omega_v)$, and

$$z(s,\omega) = \Lambda(s,\omega) \cdot z_o(s,\omega), \tag{4.3}$$

where we define the *complete L-function*

$$\Lambda(s,\omega) := \prod_v L_v(s,\omega), \tag{4.4}$$

now including all local L-factors, and

$$z_o(s,\omega) = \otimes_v z_o(s,\omega_v), \tag{4.5}$$

is the standard $\omega\omega_s$-eigendistribution defined above. Thus, in the half-plane of convergence, the complete L-function can be viewed as a factor of proportionality between two "natural" global eigendistributions.

Fourier transforms and the global functional equation

We fix a global additive character ψ of $\mathbb{A}$, trivial on k. By restricting to local components, we can write

$$\psi(x) = \prod_v \psi_v(x_v).$$

For each nonarchimedean place v, the conductor $\nu_v := \nu(\psi_v)$ of ψ_v is defined. We then get an identification of $\mathbb{A}$ with its continuous character group, compatible with those defined by the components ψ_v in the local cases. We define a global Fourier transform

$$\hat{}: S(\mathbb{A}) \xrightarrow{\sim} S(\mathbb{A}),$$

compatible with the local ones, i.e., if f is factorizable, then

$$\hat{f} = \otimes_v \hat{f}_v.$$

The ψ-self dual measure dx on $\mathbb{A}$ can be written as a product of the ψ_v-self dual measures on the k_v's. We define the Fourier transform of a distribution as before.

The desired analytic continuation and functional equation of the complete L-function $\Lambda(s,\omega)$ will follow from the conjunction of (i) the proportionality (4.3), (ii) the local functional equations of the $z_o(s,\omega_v)$'s and (iii) the following global identity:

Theorem 4.3. *The distribution $z(s, \omega)$, defined by the global zeta integral* (4.1) *for* $\mathrm{Re}(s) > 1$ *has a meromorphic analytic continuation to the whole s plane and satisfies the functional equation*

$$z(\widehat{1 - s, \omega^{-1}}) = z(s, \omega).$$

Assuming Theorem 4.3, we immediately obtain the corresponding analytic continuation of $\Lambda(s, \omega)$ from (4.3), since the "normalized" global distribution $z_o(s, \omega)$ is entire. Recall that the local functional equations have the form (3.13):

$$z_o(\widehat{1 - s, \omega_v^{-1}}) = \epsilon_v(s, \omega_v, \psi_v)\, z_o(s, \omega_v),$$

where $\epsilon_v(s, \omega_v, \psi_v)$ is given explicitly in Proposition 3.7. Note that almost all of the $\epsilon_v(s, \omega_v, \psi_v)$'s are 1, so we can define the *global epsilon factor*

$$\epsilon(s, \omega) = \prod_v \epsilon_v(s, \omega_v, \psi_v), \tag{4.6}$$

and obtain the functional equation

$$z_o(\widehat{1 - s, \omega^{-1}}) = \epsilon(s, \omega)\, z_o(s, \omega), \tag{4.7}$$

by taking products of local ones. Here the additive character ψ has been omitted from the notation since, in fact, the product no longer depends on the choice made! This is due to the relations (3.29) for all places v, the fact that every global additive character has the form ψ_β, for $\beta \in k^\times$ for a given nontrivial ψ, and the product formula. Then,

$$\begin{aligned} \Lambda(1 - s, \omega^{-1}) \cdot z_o(\widehat{1 - s, \omega^{-1}}) &= z(\widehat{1 - s, \omega^{-1}}) \\ &= z(s, \omega) \\ &= \Lambda(s, \omega) \cdot z_o(s, \omega). \end{aligned} \tag{4.8}$$

Substituting relation (4.7) in the first expression in (4.8) yields

$$\Lambda(1 - s, \omega^{-1}) \cdot \epsilon(s, \omega) \cdot z_o(s, \omega) = \Lambda(s, \omega) \cdot z_o(s, \omega). \tag{4.9}$$

But, finally, the distribution $z_o(s, \omega)$ is nowhere zero by construction, and so we obtain *the functional equation*:

Corollary 4.4.

$$\Lambda(s, \omega) = \epsilon(s, \omega) \cdot \Lambda(1 - s, \omega^{-1}).$$

Finally, the proof of Theorem 4.3 comes down to Riemann's classic argument based on Poisson summation! For this very standard calculation, we refer the reader to Tate's thesis, pp. 339–341, Weil [13, pp. 121–124], or Bump [2, pp. 267–270].

5 Examples

Here we will be very brief.

For a number field k and a (unitary) character $\omega = \otimes_v \omega_v$ of $\mathbb{A}^\times / k^\times$, let

$$\mathfrak{m} = \prod_v \mathfrak{p}_v^{c_v(\omega)}, \tag{5.1}$$

where the product is taken over nonarchimedean places and $c_v(\omega) = c_v(\omega_v)$ is the conductor of ω_v. Thus, for a finite place v, $\mathfrak{p}_v \nmid \mathfrak{m}$ if and only if ω is unramified at v. We can define a character χ of $I_\mathfrak{m}$, the group of fractional ideals of k prime to $\mathfrak{m}$ by putting $\chi(\mathfrak{p}_v) = \omega_v(\varpi_v)$. Since ω_v is unramified, this value is independent of the choice of local generator ϖ_v. Let

$$k_\mathfrak{m}^\times = \{\alpha \in k^\times \mid \alpha \equiv 1 \mod (\mathfrak{m})\}, \tag{5.2}$$

where the condition $\alpha \equiv 1 \mod (\mathfrak{m})$ means that $\mathrm{ord}_v(\alpha - 1) \geq c_v(\omega)$ for all v with $c_v(\omega) > 0$ and no condition is imposed on α at other places. Let $P_\mathfrak{m}$ denote the group of principal fractional ideals (α) generated by elements $\alpha \in k_\mathfrak{m}^\times$. The fact that $\omega(\alpha) = 1$ then implies that

$$\chi((\alpha)) = \omega_\infty(\alpha)^{-1}, \tag{5.3}$$

where (α) is the principal ideal generated by α and $\omega_\infty = \prod_{v \text{ arch.}} \omega_v$ is the archimedean part of ω. Thus, a character ω determines a character χ of $I_\mathfrak{m}$ whose restriction to $P_\mathfrak{m}$ is given by evaluation of an archimedean character on a generator of the ideal.[9] Since $P_\mathfrak{m}$ has finite index in $I_\mathfrak{m}$ there are a finite number of such characters with a given conductor and a given archimedean part. For a fixed $\mathfrak{m}$, this archimedean character is constrained by the requirement that if $\epsilon \in k_\mathfrak{m}^\times \cap \mathcal{O}_k^\times$ is a unit with $\epsilon \equiv 1 \mod (\mathfrak{m})$, then

$$\omega_\infty(\epsilon) = 1. \tag{5.4}$$

Conversely, any ω_∞ satisfying this condition for some $\mathfrak{m}$ determines a character of $P_\mathfrak{m}$ which can be extended to give a finite collection of Grössencharacters χ of $I_\mathfrak{m}$.

Of course, one can recover ω from a classical Hecke character χ. Starting from such a χ with given archimedean type ω_∞, i.e., satisfying (5.3), we define unramified characters ω_v of $k_v^\times$ for $\mathfrak{p}_v \nmid \mathfrak{m}$ by setting $\omega_v(\varpi_v) = \chi(\mathfrak{p}_v)$. The remaining components ω_v of ω for v with $\mathfrak{p}_v \mid \mathfrak{m}$ are then determined uniquely by the condition that $\omega(\alpha) = 1$ for all $\alpha \in k^\times$. More precisely, the weak approximation theorem [3] implies that the image of $k^\times$ is dense in the product $\prod_{v \in S} k_v^\times$ for any finite set of places S. The component ω_v has conductor at most $\mathrm{ord}_v(\mathfrak{m})$ and conductor precisely $\mathrm{ord}_v(\mathfrak{m})$ if χ is primitive modulo $\mathfrak{m}$, and so on.

---

[9] I believe that this is what Hecke means by the term Grössencharakter, i.e., that on $(\alpha) \in P_\mathfrak{m}$, χ depends on the quantities (Grössen) $\alpha_v \in k_v$, for v archimedean.

For $\omega_\infty = 1$, we may take any $\mathfrak{m}$. For $\mathfrak{m} = \mathcal{O}_k$, we obtain everywhere unramified characters, i.e., characters of the ideal class group $H = I/P$. For more general $\mathfrak{m}$, we obtain characters of the ray class group $H_\mathfrak{m} = I_\mathfrak{m}/P_\mathfrak{m}$ mentioned in Kowalski's chapters.

The case $\omega = 1$ gives, of course, the Dedekind zeta function of the field k. This example is explained in detail in [13, Section 6, of Chapter VII] while the case of a more general ω is explained in Section 7 of that chapter.

Exercise. Let k be the real quadratic field $k = \mathbb{Q}(\sqrt{5})$. Then $\mathcal{O}_k = \mathbb{Z}+\mathbb{Z}\epsilon$, where $\epsilon = \frac{1+\sqrt{5}}{2}$ is the fundamental unit, and $\mathcal{O}_k^\times = \{\pm 1\} \times \{\epsilon^n \mid n \in \mathbb{Z}\}$.

(i) First suppose that $\mathfrak{m} = \mathcal{O}_k$. Then the character

$$\omega_\infty(x_1, x_2) = \left|\frac{x_1}{x_2}\right|^{-i\pi/\log\epsilon}$$

is trivial on $\mathcal{O}_k^\times$, and hence defines a Hecke character χ of I by

$$\chi((\alpha)) = \omega_\infty(\alpha)^{-1} = \left|\frac{\alpha}{\alpha'}\right|^{i\pi/\log\epsilon}.$$

Write down the corresponding Hecke L-series and its functional equation.

Next take $\mathfrak{m} = (\sqrt{5}) = \mathfrak{p}_5$.

(ii) Check that the group $k_\mathfrak{m}^\times \cap \mathcal{O}_k^\times$ of units $u \equiv 1 \mod (\mathfrak{m})$ is generated by $-\epsilon^2$.

(iii) Determine all characters $\omega_\infty : \mathbb{R}^\times \times \mathbb{R}^\times \to \mathbb{C}^\times$ which satisfy (5.3) for $\mathfrak{m}$. For example,

$$\omega_\infty(x_1, x_2) = \operatorname{sgn}(x_1)\left|\frac{x_1}{x_2}\right|^{-i\pi/(4\log\epsilon)} \quad \text{and} \quad \omega_\infty(x_1, x_2) = \left|\frac{x_1}{x_2}\right|^{-i\pi/(2\log\epsilon)} \tag{5.4}$$

are two possibilities.

(iv) For a given ω_∞, determine the local component ω_5 for $\mathfrak{p}_5 = \mathfrak{m}$. Note that $\mathcal{O}_5^\times/(1+\mathcal{P}_5) \simeq \mathbb{F}_5^\times$ and that the global units map onto this group. Also, $\varpi_5 = \sqrt{5}$ gives a local uniformizer. For example, for ω_∞ give by (5.4), ω_5 is the ramified character with $\omega_5(\sqrt{5}) = 1$ and with $\omega_5(\epsilon) = i$.

(v) Determine the classical Hecke character χ of $I_\mathfrak{m}$ associated to ω.

(vi) Write down the associated Hecke L-series and its functional equation.

References

[1] J. Bernstein and A. Zelevinski, Representations of the group $GL(n, F)$ where F is a nonarchimedean local field, *Russian Math. Surveys*, **3**(1976), 1–68.

[2] D. Bump, *Automorphic Forms and Representations*, Cambridge University Press, Cambridge, UK, 1997.

[3] J. W. S. Cassels and A. Fröhlich, eds., *Algebraic Number Theory*, Thompson Book Company, Washington, DC, 1967.

[4] F. G. Friedlander, *Introduction to the Theory of Distributions*, Cambridge University Press, Cambridge, UK, 1982.

[5] S. Gelbart, *Automorphic Forms on Adele Groups*, Princeton University Press, Princeton, NJ, 1975.

[6] S. Gelbart, An elementary introduction to the Langlands program, *Bull. Amer. Math. Soc.*, **10**(1984), 177–219.

[7] E. Hecke, Eine neue Art von Zetafunktionen und ihre Beziehungen zur Verteilung der Primzahlen, Erste und Zweite Mitteilung, in *Mathematische Werke*, Vandenhoeck and Ruprecht, Göttingen, The Netherlands, 1959.

[8] E. Kowalski, Elementary theory of L-functions I and II, in J. Bernstein and S. Gelbart, eds., *An Introduction to the Langlands Program*, Birkhäuser Boston, Boston, 2003, 1–37 (this volume).

[9] S. Lang, *Algebraic Number Theory*, Addison–Wesley, Reading, MA, 1970.

[10] J. Tate, *Fourier Analysis in Number Fields and Hecke's Zeta-Functions*, Ph.D. thesis, Princeton University, Princeton, NJ, 1950.

[11] R. Valenza and D. Ramakrishnan, *Fourier Analysis on Number Fields*, Graduate Texts in Mathematics 186, Springer-Verlag, New York, 1999.

[12] A. Weil, Fonction zeta et distributions, *Séminar Bourbaki*, **III**(1966); also appears in [14].

[13] A. Weil, *Basic Number Theory*, Springer-Verlag, New York, 1967.

[14] A. Weil, *Collected Papers*, Springer-Verlag, New York, 1979.

7
From Modular Forms to Automorphic Representations

Stephen S. Kudla

This chapter sketches the passage from classical holomorphic modular forms f weight k and level N to automorphic representations of $\pi = \pi(f)$ of $\mathrm{GL}_2(\mathbb{A})$.

In the much simpler case of GL_1, the analogous passage is the correspondence, described in the Tate's thesis chapter, between classical Dirichlet characters χ_N modulo N and continuous characters $\omega : \mathrm{GL}_1(\mathbb{Q})\backslash \mathrm{GL}_1(\mathbb{A}) \to \mathbb{C}^\times$. Two features of the GL_1 case will carry over to the case of GL_2. First, the character ω is a product of local characters $\omega = \otimes_v \omega_v$, and, for example, the analysis in Tate's thesis builds the global theory of the L-function of ω out of an analysis of the local components ω_v. Similarly, the automorphic representation π of $\mathrm{GL}_2(A)$ will have a factorization

$$\pi = \otimes_v \pi_v$$

into representations of the groups $\mathrm{GL}_2(\mathbb{Q}_v)$. Second, there are many classical Dirichlet characters associated to a single ω, and among them there is a unique primitive one. Similarly, there will be infinitely many f's (we allow the level N to vary arbitrarily) associated to the same π, but, among these, there will be a unique primitive form or normalized newform. Of course, a major difference is that in the $G = \mathrm{GL}_2$ case, the representation π and the local components π_v will be infinite dimensional so that the representation theory involved is much more serious (nonabelian harmonic analysis). Also, π will be realized in the "action" of the adèle group $G(\mathbb{A})$ by right translations on a certain space of functions on $G(\mathbb{Q})\backslash G(\mathbb{A})$. Later, we will discuss briefly the case of a general number field.

As with Tate's thesis, the topic of this chapter is described very well and in detail in many places. The fundamental work in this subject is the book of Jacquet and Langlands [11], which appeared in 1970, 20 years after Tate's thesis. Among many commentaries and introductions, we cite only, for example, the books of Bump [2], Gelbart [8], and Godement [10], and the article of Diamond and Im [7], any of which may be consulted for more details. For a broader perspective on the Langlands program, the reader may consult the introductory articles of Gelbart, [9] and Knapp, [12], and the surveys of Cogdell [4, 5, 6].

The input of our construction is a classical holomorphic cusp form $f \in S_k(\Gamma_0(N), \chi_N)$, where χ_N is a classical Dirichlet character modulo N. Thus, f is a holomorphic function of $\tau = u + iv \in \mathfrak{H}$ with

$$f\left(\frac{a\tau+b}{c\tau+d}\right) = \chi_N(d)\,(c\tau+d)^k\, f(\tau), \tag{1.1}$$

for all

$$\gamma \in \Gamma_0(N) = \left\{\begin{pmatrix} a & b \\ c & d \end{pmatrix} \in \mathrm{SL}_2(\mathbb{Z}) \mid c \equiv 0 \bmod(N)\right\}.$$

The q-expansion of f is

$$f(\tau) = \sum_{n=1}^{\infty} a_n\, q^n, \qquad q = e(\tau). \tag{1.2}$$

Note that the compatibility condition $\chi_N(-1) = (-1)^k$ must hold for there to be any nonzero f's.

1 Functions on $G(\mathbb{Q})\backslash G(\mathbb{A})$ associated to classical cusp forms

Let $G = \mathrm{GL}_2$, viewed as an algebraic group over $\mathbb{Q}$, and let

$$G_v = G(\mathbb{Q}_v) = \begin{cases} \mathrm{GL}_2(\mathbb{Q}_p) & \text{if } \mathbb{Q}_v = \mathbb{Q}_p, \\ \mathrm{GL}_2(\mathbb{R}) & \text{if } \mathbb{Q}_v = \mathbb{R}. \end{cases}$$

The group of adèlic points of G,

$$G(\mathbb{A}) = G(\mathbb{R})G(\mathbb{A}_f), \qquad G(\mathbb{A}_f) = \prod_p{}' G(\mathbb{Q}_p)$$

is the restricted product with respect to the subgroups $K_p = \mathrm{GL}_2(\mathbb{Z}_p) \subset \mathrm{GL}_2(\mathbb{Q}_p) = G_p$. Thus, $G(\mathbb{A})$ and $G(\mathbb{A}_f)$ are locally compact topological groups,

$$K = \prod_p K_p \subset G(\mathbb{A}_f)$$

is a compact open subgroup, and the image of $G(\mathbb{Q})$ under the diagonal embedding

$$G(\mathbb{Q}) \hookrightarrow G(\mathbb{A}), \qquad \gamma \mapsto (\dots, \gamma, \dots)$$

is a discrete subgroup of $G(\mathbb{A})$.

The analogous construction for the group SL_2 yields locally compact groups $\mathrm{SL}_2(\mathbb{A})$, $\mathrm{SL}_2(\mathbb{A}_f)$ and a discrete subgroup $\mathrm{SL}_2(\mathbb{Q}) \hookrightarrow \mathrm{SL}_2(\mathbb{A})$. The strong approximation theorem[1] asserts that the subgroup $\mathrm{SL}_2(\mathbb{Q})\,\mathrm{SL}_2(\mathbb{R})$ is dense in $\mathrm{SL}_2(\mathbb{A})$, and hence that, for any open subgroup $U \subset \mathrm{SL}_2(\mathbb{A}_f)$,

$$SL_2(\mathbb{A}) = \mathrm{SL}_2(\mathbb{Q})\,\mathrm{SL}_2(\mathbb{R})\,U, \qquad g = \gamma\, g_\infty\, u.$$

[1] See [19, p. 145] for an elementary proof in this case.

Using this result and the decomposition

$$\mathbb{A}^\times = \mathbb{Q}^\times \mathbb{R}_+^\times \widehat{\mathbb{Z}}^\times,$$

we obtain the following.

Lemma 1.1. *Suppose that $K' \subset G(\mathbb{A}_f)$ is a compact open subgroup such that* $\det(K') = \widehat{\mathbb{Z}}^\times$. *Then*

$$G(\mathbb{A}) = G(\mathbb{Q})\, G(\mathbb{R})^+\, K'.$$

In particular,

$$G(\mathbb{Q})\backslash G(\mathbb{A})/K' \simeq \Gamma'\backslash G(\mathbb{R})^+,$$

where $G(\mathbb{R})^+ = \{g \in G(\mathbb{R}) \mid \det(g) > 0\}$ and $\Gamma' = G(\mathbb{Q}) \cap G(\mathbb{R})^+ K'$ viewed as a subgroup of $G(\mathbb{R})^+$ via projection onto the archimedean component.

For $K' = K$ as above, we obtain $\Gamma' = \mathrm{SL}_2(\mathbb{Z}) \subset \mathrm{GL}_2(\mathbb{R})^+$. We could also take

$$K' = K_0(N) = \{k = \begin{pmatrix} a & b \\ c & d \end{pmatrix} \mid c \equiv 0 \bmod (N)\}, \tag{1.3}$$

in which case $\Gamma' = \Gamma_0(N) \subset \mathrm{GL}_2(\mathbb{R})^+$. Note that there is a surjective homomorphism[2]

$$K_0(N) \to (\mathbb{Z}/N\mathbb{Z})^\times, \qquad \begin{pmatrix} a & b \\ c & d \end{pmatrix} \mapsto a \bmod (N), \tag{1.4}$$

and so the classical Dirichlet character χ_N defines a character χ of $K_0(N)$.

The group $G(\mathbb{R})^+$ acts on the upper half plane by fractional linear transformations, and its center $Z_\infty = \{z \cdot 1_2 \mid z \in \mathbb{R}^\times\}$ acts trivially. For $g \in G(\mathbb{R})^+$, and $\tau \in \mathfrak{H}$, put

$$j(g, \tau) = \det(g)^{-\frac{1}{2}}\,(c\tau + d), \qquad g = \begin{pmatrix} a & b \\ c & d \end{pmatrix}. \tag{1.5}$$

Note that $j(zg, \tau) = \mathrm{sgn}(z)\, j(g, \tau)$ for $z \in Z_\infty$. Also note that if

$$k_\theta = \begin{pmatrix} \cos(\theta) & \sin(\theta) \\ -\sin(\theta) & \cos(\theta) \end{pmatrix} \in SO(2) =: K_\infty^+ \subset G(\mathbb{R})^+, \tag{1.6}$$

a maximal compact subgroup of $G(\mathbb{R})^+$, then $Z_\infty K_\infty^+$ is the stablizer of i, and

$$j(z\,k_\theta, i) = \mathrm{sgn}(z) \cdot e^{-i\theta}. \tag{1.7}$$

[2] Really this means

$$a = (\ldots, a_p, \ldots) \mapsto \prod_{p|N} (a_p \bmod (p^{n_p})) \in \prod_{p|N} (\mathbb{Z}/p^{n_p}\mathbb{Z})^\times \simeq (\mathbb{Z}/N\mathbb{Z})^\times,$$

where $n_p = \mathrm{ord}_p(N)$.

With this structural background, we return to our cusp form $f \in S_k(\Gamma_0(N), \chi_N)$, and, for $g_\infty \in G(\mathbb{R})^+$, we let

$$\phi(g_\infty) = \phi_f(g_\infty) = f(g_\infty(i))\, j(g_\infty, i)^{-k}. \tag{1.8}$$

Note that ϕ is a smooth function on $G(\mathbb{R})^+$ and that f can be recovered from it:

$$f(\tau) = f(g_\infty(i)) = \phi(g_\infty)\, j(g_\infty, i)^k, \tag{1.9}$$

for any g_∞ with $\tau = g_\infty(i)$.

Lemma 1.2.
(i) *For* $\gamma \in \Gamma_0(N)$,

$$\phi(\gamma g_\infty) = \phi(g_\infty)\, \chi_N(d), \qquad \gamma = \begin{pmatrix} a & b \\ c & d \end{pmatrix}.$$

(ii) *For* $z\, k_\theta \in Z_\infty K_\infty$,

$$\phi(g_\infty\, z\, k_\theta) = \phi(g_\infty)\ \mathrm{sgn}(z)^k\, (e^{i\theta})^k.$$

Statement (i) follows from the transformation law of f and the cocycle property $j(g_1 g_2, \tau) = j(g_1, g_2(\tau))\, j(g_2, \tau)$. Part (ii) is clear from (1.7).

Finally, we use the double coset decomposition of Lemma 1.1, with $K' = K_0(N)$, to extend $\phi = \phi_f$ to a function on $G(\mathbb{A})$ by setting

$$\phi(g) = \phi(\gamma g_\infty k) = \phi(g_\infty)\, \chi(k). \tag{1.10}$$

Note that the ambiguity in the expression $g = \gamma\, g_\infty\, k$ is just given by putting $\gamma\gamma_0^{-1}$ for γ, $\gamma_0 g_\infty$ for g_∞ and $\gamma_0 k$ for k, where $\gamma_0 \in \Gamma_0(N)$. Thus, the extension is well defined by (i) of Lemma 1.2. Note that ϕ still satisfies (ii) of that lemma and is invariant under Z_∞^+, the identity component of Z_∞.

So far, no use has been made of the holomorphy of f. This property of f is equivalent to the following property of ϕ, as a function on $G(\mathbb{R})^+$. The real Lie algebra $\mathfrak{g}_0 = \mathrm{Lie}(G(\mathbb{R})) = M_2(\mathbb{R})$ of $G(\mathbb{R})^+$ acts on smooth functions on $G(\mathbb{R})^+$ by

$$X\phi(g_\infty) = \frac{d}{dt}\{\phi(g_\infty \exp(tX))\}\Big|_{t=0}.$$

For example, if $X = \left(\begin{smallmatrix} & 1 \\ -1 & \end{smallmatrix}\right)$, then $\exp(tX) = k_t \in K_\infty^+$, and relation (ii) of Lemma 1.2, which encodes the weight of f in the transformation law (1.1), implies that, for our $\phi = \phi_f$,

$$X\phi(g_\infty) = \frac{d}{dt}\{\phi(g_\infty)(e^{it})^k\}\big|_{t=0} = ik\,\phi(g_\infty). \tag{1.11}$$

We can extend the action of $\mathfrak{g}_0$ to an action of the complexification $\mathfrak{g} = \mathfrak{g}_0 \otimes_{\mathbb{R}} \mathbb{C}$ by setting $(X + iY)\phi = X\phi + i \cdot Y\phi$. The elements

$$X_\pm = \frac{1}{2}\begin{pmatrix} 1 & \pm i \\ \pm i & -1 \end{pmatrix} \in \mathfrak{g} \tag{1.12}$$

have the property that

$$\mathrm{Ad}(k_\theta)X_\pm = k_\theta X_\pm k_\theta^{-1} = e^{\pm 2i\theta}\, X_\pm. \tag{1.13}$$

It follows that

$$X_\pm \phi(g_\infty k_\theta) = \phi(g_\infty)\,(e^{i\theta})^{k\pm 2},$$

so that $X_\pm$ raises or lowers the weight by 2. It is not difficult to express the functions $M_\pm f$ on $\mathfrak{H}$ associated to $X_\pm\phi$ via the weight $k \pm 2$ analogue of (1.9) in terms of derivatives of the original f. One finds in this way the classical Maass operators $M_\pm$ on f, and the equation $M_- f = 0$ is just the Cauchy Riemann equation; see [2, pp. 155 and 129], for example. Thus, we have the following.

Lemma 1.3. *A smooth function $f : \mathfrak{H} \to \mathbb{C}$ satisfying* (1.1) *is holomorphic if and only if the corresponding function $\phi = \phi_f$ satisfies $X_-\phi = 0$.*

We have thus explained all but the last part of the basic result:

Proposition 1.4. *The mappings* (1.8), (1.9), *and* (1.10) *determine an isomorphism*

$$S_k(\Gamma_0(N), \chi_N) \xrightarrow{\sim} \mathcal{A}_0(G)(\mathrm{hol}, k, N, \chi), \quad f \mapsto \phi_f,$$

where $\mathcal{A}_0(G)(\mathrm{hol}, k, N, \chi)$ is the space of functions ϕ on $G(\mathbb{A})$ with the following properties:

(i) *For all $\gamma \in G(\mathbb{Q})$ and $k \in K_0(N)$,*

$$\phi(\gamma g k) = \phi(g)\chi(k).$$

(ii) *For any $g_f \in G(\mathbb{A}_f)$, the function of $g_\infty \mapsto \phi(g_\infty g_f)$ is smooth on $G(\mathbb{R})^+$, invariant under Z_∞^+, and satisfies*

$$\phi(g k_\theta) = \phi(g)(e^{i\theta})^k,$$

and

$$X_-\phi = 0.$$

(iii) *The function ϕ is cuspidal, i.e.,*

$$\int_{\mathbb{Q}\backslash\mathbb{A}} \phi\left(\begin{pmatrix} 1 & x \\ & 1 \end{pmatrix} g\right) dx = 0,$$

for all $g \in G(\mathbb{A})$.

It remains to explain (iii). It is not difficult to check that, for $a \in \mathbb{R}_+^\times$,

$$\int_{\mathbb{Q}\backslash\mathbb{A}} \phi\left(\begin{pmatrix}1 & x\\ & 1\end{pmatrix}\begin{pmatrix}a & \\ & a^{-1}\end{pmatrix}\right) dx = \int_{\mathbb{Z}\backslash\mathbb{R}} \phi\left(\begin{pmatrix}1 & x\\ & 1\end{pmatrix}\begin{pmatrix}a & \\ & a^{-1}\end{pmatrix}\right) dx \tag{1.15}$$

$$= a^k \int_0^1 f(a^2 i + x)\, dx = 0,$$

since the constant term in the Fourier expansion of f at the cusp at infinity vanishes. On the other hand, the cuspidality of f asserts that the analogous constant terms vanish at *all* cusps. This is equivalent to the vanishing of the integral in (iii) for all $g \in G(\mathbb{A})$, i.e., as a function on $G(\mathbb{A})$.

Remark 1.5. The function ϕ is actually an eigenfunction of the center $Z(\mathbb{A})$. More precisely, if ω is the character of $\mathbb{A}^\times/\mathbb{Q}^\times$ associated to χ_N or χ, then

$$\phi(zg) = \omega(z)\,\phi(g). \tag{1.16}$$

Remark 1.6. Consider the function

$$\tilde{\phi}(g) = \phi(gI), \qquad I = \begin{pmatrix}1 & \\ & -1\end{pmatrix} \in O(2) =: K_\infty \subset G(\mathbb{R}). \tag{1.17}$$

Observe that

$$\tilde{\phi}(gk_\theta) = \tilde{\phi}(g)\,(e^{i\theta})^{-k}, \tag{1.18}$$

and, since $\mathrm{Ad}(I)X^{\pm} = X^{\mp}$,

$$X_+\tilde{\phi} = 0. \tag{1.19}$$

It is not difficult to check that the corresponding function on $\mathfrak{H}$ is

$$\tilde{f}(\tau) = \tilde{\phi}(g_\infty) j(g_\infty, i)^{-k} = v^k\, f(-\bar{\tau}). \tag{1.20}$$

2 The representation on $\mathcal{A}_0(G)$

To bring the representation theory of the group $G(\mathbb{A})$ into play, we enlarge the space of functions $\mathcal{A}_0(G)(\text{hol}, k, N, \chi)$ on $G(\mathbb{Q})Z_\infty^+\backslash G(\mathbb{A})$ constructed in the last section.[3]

Let ω be a character of $\mathbb{A}^\times/\mathbb{Q}^\times$ with ω_∞ trivial on $\mathbb{R}_+^\times$.

Definition 2.1. The space of *automorphic forms* $\mathcal{A}(G) = \mathcal{A}(G, \omega)$ on $G(\mathbb{A})$ with central character ω is the space of complex valued functions ϕ on $G(\mathbb{A})$ such that:

---

[3] One might initially try to take combinations of arbitrary right translates of functions in $\mathcal{A}_0(G)(\text{hol}, k, N, \chi)$ by elements $g \in G(\mathbb{A})$. It turns out to be better to insist that our functions be finite linear combinations of functions which have a weight, i.e., are $K_\infty^+ = SO(2)$-eigenfunctions. The space of such functions is not preserved under right translation by elements of $G(\mathbb{R})$, so we have to settle for an action of the complexified Lie algebra $\mathfrak{g}$.

(i) For $z \in Z(\mathbb{A}) \simeq \mathbb{A}^\times$ and $\gamma \in G(\mathbb{Q})$, $\phi(z\gamma g) = \omega(z)\,\phi(g)$.

(ii) For each $g_f \in G(\mathbb{A}_f)$, the function $g_\infty \mapsto \phi(g_\infty g_f)$ is smooth on G_∞.

(iii) The space spanned by the right translates of ϕ by elements of K_∞ is finite dimensional, i.e., ϕ is right K_∞-finite.

(iv) There is a compact open subgroup[4] K' of $G(\mathbb{A}_f)$ such that ϕ is invariant under right translation by K'.

(v) The function ϕ is $Z(\mathfrak{g})$-finite.

(vi) The function ϕ is slowly increasing.

Condition (v) has the following meaning. Because of the smoothness condition (ii), the complexified Lie algebra $\mathfrak{g}$ acts on $\phi \in \mathcal{A}_0(G)$, as explained in Section 1. Note that this action preserves conditions (i)–(iv), in particular, the K_∞-finiteness condition (iii). The action of $\mathfrak{g}$ extends to an action of the universal enveloping algebra $U(\mathfrak{g})$ of $\mathfrak{g}$. The center of this algebra $Z(\mathfrak{g})$ is spanned by $\mathrm{Lie}(Z_\infty) = \mathrm{Lie}(Z_\infty^+)$, which acts trivially by (i) and our assumption on ω, and the Casimir operator

$$C = H^2 + 2\,X_+X_- + 2X_-X_+, \quad \text{where} \quad H = -i\begin{pmatrix} & 1 \\ -1 & \end{pmatrix}, \tag{2.1}$$

and $X_\pm$ is as in (1.12); cf. [2, Section 2.2]. Condition (v) then says that there is a polynomial $R[T] \in \mathbb{C}[T]$ such that $R(C)\phi = 0$.

Finally, to express the growth condition (vi), one uses the following norm on $G(\mathbb{A})$. There is an embedding

$$G(\mathbb{A}) \hookrightarrow M_2(\mathbb{A}) \times \mathbb{A} \simeq \mathbb{A}^5, \qquad g \mapsto (g, \det(g)^{-1}). \tag{2.2}$$

For each place v, and $x = (x_i) \in \mathbb{A}^5$, put

$$||x||_v = \max_i(|x_i|_v), \tag{2.3}$$

and let

$$||x|| = \prod_v ||x||_v. \tag{2.4}$$

Note, for example, that if v is a nonarchimedean place and if g_v is in K_v, then $\det(g_v)$ is a unit, so $||g||_v = 1$. Thus almost all factors in the product are 1.

Then a function ϕ is slowly increasing or of moderate growth if there is an integer n and a constant C such that

$$|f(g)| \leq C\,||g||^n, \tag{2.5}$$

for all $g \in G(\mathbb{A})$.

---

[4]We may as well assume that $K' \subset K$. The condition of K'-invariance is equivalent to the requirement that the span of the right translates of ϕ by elements of K is finite dimensional, i.e., that ϕ is right K-finite.

Definition 2.2. The space $\mathcal{A}_0(G) = \mathcal{A}_0(G, \omega) \subset \mathcal{A}(G)$ of *cuspidal automorphic forms* with central character ω is defined by adding the following cuspidal condition:

(vii) The function ϕ is cuspidal, i.e.,

$$\int_{\mathbb{Q}\backslash\mathbb{A}} \phi\left(\begin{pmatrix} 1 & x \\ & 1 \end{pmatrix} g\right) dx = 0, \tag{2.6}$$

for all $g \in G(\mathbb{A})$.

It is easy to see that the space $\mathcal{A}_0(G)(\text{hol}, k, N, \chi)$ is contained in $\mathcal{A}_0(G)$. Indeed, (i)–(iv) and (vii) are immediate, in view of Proposition 1.4 and Remarks 1.5 and 1.6. The weight and holomorphy conditions of (ii) of Proposition 1.4 imply that[5]

$$C\phi = k(k-2)\phi. \tag{2.7}$$

The slowly increasing condition follows from the classical condition of holomorphy at the cusps [19], and the cuspidal functions are actually rapidly decreasing on $G(\mathbb{A})$, [2]. On the other hand, the construction of Section 1 can be applied to the classical (holomorphic) Eisenstein series, and the resulting functions on $G(\mathbb{A})$ lie in $\mathcal{A}(G)$, hence are slowly increasing, but are not in $\mathcal{A}_0(G)$.

Definition 2.3.

(i) A $(\mathfrak{g}, K_\infty)$-module is a complex vector V space with actions of $\mathfrak{g}$ and K_∞, such that all vectors in V are K_∞-finite and such that the two actions are compatible, i.e.,

$$k \cdot X \cdot v = \mathrm{Ad}(k)X \cdot k \cdot v \tag{2.8}$$

and, for $X \in \mathrm{Lie}(K_\infty)$,

$$X \cdot v = \frac{d}{dt}\left(\exp(tX) \cdot v\right)\Big|_{t=0}. \tag{2.9}$$

(ii) A $(\mathfrak{g}, K_\infty) \times G(\mathbb{A}_f)$-module is a $(\mathfrak{g}, K_\infty)$-module with a smooth action of $G(\mathbb{A}_f)$, commuting with the action of $(\mathfrak{g}, K_\infty)$. Here smooth means that every vector $v \in V$ is fixed by some compact open subgroup $K' \subset G(\mathbb{A}_f)$.

Note that the K_∞-finiteness of $v \in V$ is used in (2.9) to insure that the curve $\exp(tX) \cdot v$ lies in a finite-dimensional vector space.

The point of these definitions is that the spaces $\mathcal{A}(G)$ and $\mathcal{A}_0(G)$ are $(\mathfrak{g}, K_\infty) \times G(\mathbb{A}_f)$-modules and that the space $\mathcal{A}_0(G)(\text{hol}, k, N, \chi) \subset \mathcal{A}_0(G)$. In fact, there is one serious point to check, that is that the slowly increasing condition (vi) is

[5] By the weight condition $H\phi = k\phi$. In addition, $X_-\phi = 0$ and $[X_+, X_-] = H$, so that $X_-X_+\phi = (X_+X_- - H)\phi = -k\phi$.

preserved by the action of $\mathfrak{g}$. For a nice discussion of this issue, which illustrates the nontrivial role played by the finiteness conditions (iii) and (v), see Bump [2, Section 2.2—in particular, Theorem 2.9.2].

Now suppose that k is an arbitrary number field and let $G = \mathrm{GL}(2)$ as an algebraic group over k. For any place v of k, $G_v = \mathrm{GL}_2(k_v)$, and, if v is nonarchimedean, $K_v = \mathrm{GL}_2(\mathcal{O}_v)$. As before,

$$G(\mathbb{A}) = \prod_v{}' G_v$$

is the restricted product with respect to the subgroups K_v,

$$K = \prod_{v \text{ nonarch.}} K_v,$$

is a compact open subgroup of $G(\mathbb{A}_f)$, and $G(k) \hookrightarrow G(\mathbb{A})$ is a discrete subgroup, etc. We can define the space of automorphic forms $\mathcal{A}(G)$ (resp., cuspidal automorphic forms $\mathcal{A}_0(G)$) on $G(\mathbb{A})$ exactly as in Definition 2.1 (resp., Definition 2.2). Now, of course,

$$G_\infty = \prod_{v \text{ arch.}} G(k_v) \simeq \underbrace{\mathrm{GL}_2(\mathbb{R}) \times \cdots \times \mathrm{GL}_2(\mathbb{R})}_{r_1} \times \underbrace{\mathrm{GL}_2(\mathbb{C}) \times \cdots \times \mathrm{GL}_2(\mathbb{C})}_{r_2},$$

$$K_\infty = \prod_{v \text{ arch.}} K_v \simeq O(2) \times \cdots \times O(2) \times U(2) \times \cdots \times U(2),$$

and the center $Z(\mathfrak{g}) \simeq \otimes_v Z(\mathfrak{g}_v)$ is larger. Just as Tate's thesis treats Hecke characters for all number fields in a uniform way, the Jacquet–Langlands theory [11] applies to the spaces $\mathcal{A}(G) \supset \mathcal{A}_0(G)$ for any k.

To describe the representation of $(\mathfrak{g}, K_\infty) \times G(\mathbb{A}_f)$ on the space $\mathcal{A}_0(G)$, we begin with some facts from abstract representation theory.

Definition 2.4.

(i) A $(\mathfrak{g}, K_\infty) \times G(\mathbb{A}_f)$-module (π, V) is *admissible* if, for every irreducible representation σ of $K_\infty \times K$, the multiplicity of σ in V is finite.

(ii) The module (π, V) is *irreducible* if it has no proper subspaces preserved by the action of $\mathfrak{g}$, K_∞ and $G(\mathbb{A}_f)$.

There are analogous local definitions: For an archimedean place v, a $(\mathfrak{g}_v, K_v)$-module (π, V) is admissible if, for any irreducible representation σ_v of K_v, the multiplicity of σ_v in V is finite. Here $\mathfrak{g}_v = \mathrm{Lie}(G_v) \otimes \mathbb{C}$. A $(\mathfrak{g}_v, K_v)$-module is irreducible if it has no proper subspaces invariant under $\mathfrak{g}_v$ and K_v.

For a nonarchimedean place v, a representation (π, V) of G_v on a complex vector space V is *smooth* if every $\phi \in V$ is fixed by an open subgroup, admissible if, for every open subgroup K', the space of K'-fixed vectors $V^{K'}$ is finite dimensional, and irreducible if it has no proper invariant subspaces. In addition, an irreducible admissible representation (π, V) is called *unramified* or spherical if the space of K_v-invariants is nonzero. In this case, in fact, $\dim V^{K_v} = 1$.

The following factorizability theorem replaces the simple fact for GL_1 that a Hecke character ω can be written as a product of local characters, $\omega = \otimes_v \omega_v$.

Theorem 2.5.

(i) *Suppose that, for each archimedean (resp., nonarchimedean) place v of $\Bbbk$, (π_v, V_v) is an irreducible admissible $(\mathfrak{g}_v, K_v)$-module (resp., representation of G_v). Suppose, moreover, that for almost all nonarchimedean places, the representation (π_v, V_v) is unramified and a basis vector $\xi_v^0 \in V_v^{K_v}$ is given. Then the restricted tensor product*[6]

$$V = \otimes_v' V_v$$

with respect to the vectors ξ_v^0 is an irreducible admissible $(\mathfrak{g}, K_\infty) \times G(\mathbb{A}_f)$-module.

(ii) *Conversely, if (π, V) is an irreducible admissible $(\mathfrak{g}, K_\infty) \times G(\mathbb{A}_f)$-module, then there is a collection $\{(\pi_v, V_v)\}_v$ as in (i) and an isomorphism*

$$\otimes_v' V_v \xrightarrow{\sim} V$$

of $(\mathfrak{g}, K_\infty) \times G(\mathbb{A}_f)$-modules.

It is customary to write $\pi = \otimes_v \pi_v$ in both cases, although this notation omits the important isomorphism in case (ii). A detailed discussion is given in [2, Sections 3.3 and 3.4].

We now return to our $(\mathfrak{g}, K_\infty) \times G(\mathbb{A}_f)$-module of cuspidal modular forms. The main result concerning its structure is the following. Recall that the central character $\omega = \otimes_v \omega_v$ has been fixed.

Theorem 2.6.

(i) *The space $\mathcal{A}_0(G) = \mathcal{A}_0(G, \omega)$ is an algebraic direct sum of irreducible admissible $(\mathfrak{g}, K_\infty) \times G(\mathbb{A}_f)$-modules,*

$$\mathcal{A}_0(G) = \oplus_\pi \mathcal{A}_0(\pi).$$

Here $\mathcal{A}_0(\pi)$ is a summand isomorphic to π.

(ii) (multiplicity one; Jacquet–Langlands [11]) *If π_1 and π_2 are irreducible admissible $(\mathfrak{g}, K_\infty) \times G(\mathbb{A}_f)$-modules, then*

$$\pi_1 \simeq \pi_2 \implies \mathcal{A}_0(\pi_1) = \mathcal{A}_0(\pi_2).$$

(iii) (strong multiplicity one; Miyake, Casselman [8, p. 89]) *For $\pi_1 = \otimes_v \pi_{1,v}$ and $\pi_2 = \otimes_v \pi_{2,v}$ as in (ii), suppose that $\pi_{1,v} \simeq \pi_{2,v}$ for all $v \notin S$ for a finite set of places S of $\Bbbk$. Then $\pi_1 \simeq \pi_2$ and $\mathcal{A}_0(\pi_1) = \mathcal{A}_0(\pi_2)$.*

(iv) (very strong multiplicity one; Ramakrishnan [17]) *Suppose that $\pi_{1,v} \simeq \pi_{2,v}$ for all v outside of a set of places of $\Bbbk$ with density less than $\frac{1}{8}$. Then $\pi_1 \simeq \pi_2$ and $\mathcal{A}_0(\pi_1) = \mathcal{A}_0(\pi_2)$.*

---

[6] This means that V is the union, or direct limit, of the spaces $V_S = \otimes_{v \in S} V_v$ where, for $T \supset S$, the inclusion $V_S \hookrightarrow V_T$ is given by $x_S \mapsto x_S \otimes (\otimes_{v \in T \setminus S} \xi_v^0)$.

The irreducible admissible $(\mathfrak{g}, K_\infty) \times G(\mathbb{A}_f)$-modules π that occur in $\mathcal{A}_0(G)$ are the *irreducible cuspidal automorphic representations* of $G(\mathbb{A})$. Such a representation is said to be realized on the space $\mathcal{A}_0(\pi)$. Because of multiplicity one, there is no ambiguity about the space $\mathcal{A}_0(\pi)$, if it exists, once the abstract representation π is given. This is not the case for other classical groups. For example, for $G = \mathrm{SL}_n, n \geq 3$, a given abstract representation π can occur in the space of cusp forms with multiplicity more than one [1]. In this situation, it may be important to specify the space $\mathcal{A}_0(\pi)$. The proof of (i), is based on the combination of the fact that, on the one hand, the functions ϕ in the automorphic cuspidal representation π are determined by their Fourier expansions and these expansions are, in turn, determined by the global Whittaker function attached to ϕ. On the other hand, the space of such Whittaker functions associated to functions in $\pi = \otimes_v \pi_v$, is uniquely determined by the spaces of local Whittaker functions associated to the local components π_v. Details may be found in [11, 10, 8, 2].

3 Local representation theory

In this section, we very briefly review the representation theory of $G = \mathrm{GL}_2(F)$, where F is a p-adic field. At the end of the section, we add a few words about $(\mathfrak{g}, K_\infty)$-modules. We use the notation from Section 3 of the chapter on Tate's thesis. Thus F is a nonarchimedean local field with ring of integers $\mathcal{O}$. A more detailed survey can be found in [8, Chapter 4] and detailed proofs are given, for example, in [2, Chapter 4].

Let $G = \mathrm{GL}_2(F)$ and let $K = \mathrm{GL}_2(\mathcal{O})$, so that K is a maximal compact (and open) subgroup of G. Let $B = NM = MN$, where

$$M = \left\{ m = m(a) = \begin{pmatrix} a_1 & \\ & a_2 \end{pmatrix} \right\} \quad \text{and} \quad N = \left\{ n = n(x) = \begin{pmatrix} 1 & x \\ & 1 \end{pmatrix} \right\},$$

where $a_1, a_2 \in F^\times$, and $x \in F$. Let Z be the center of G.

For a pair $\mu = (\mu_1, \mu_2)$ of characters of $F^\times$, there is an induced representation

$$I(\mu) = \left\{ f : G \longrightarrow \mathbb{C} \mid f(bg) = \mu(b)\, \delta(b)^{\frac{1}{2}} f(g) \right\},$$

where f is smooth, i.e., right invariant under an open subgroup $K' \subset G$ (which can depend on f and which we may as well assume is a subgroup of K), and, for $b = nm(a)$,

$$\mu(b) = \mu_1(a_1)\mu_2(a_2) \quad \text{and} \quad \delta(b) = \left| \frac{a_1}{a_2} \right|.$$

Note that δ is the modulus for the adjoint action of B on N, i.e.,

$$d(bnb^{-1}) = \delta(b)\, dn$$

for a Haar measure dn on N. Thus, the induction is normalized so that a pair (μ_1, μ_2) of unitary characters yields a unitarizable representation $I(\mu)$. Because of the Iwasawa decomposition $G = BK$, the functions in $I(\mu)$ are determined by their restrictions to K and hence

$$I(\mu) \simeq I_{K\cap B}^{K}(\mu).$$

Also notice that the central character of $I(\mu)$ is $\omega = \mu_1\mu_2$.

Theorem 3.1.

(i) *$I(\mu)$ is irreducible if and only of $\mu_1\mu_2^{-1} \neq \omega_{\pm 1}$, where $\omega_s(x) = |x|^s$.*

(ii) *If $\mu_1\mu_2^{-1} = \omega_1$, then $I(\mu)$ has a one-dimensional quotient on which G acts by the character $\chi \circ \det$, where $\mu = (\chi\omega_{\frac{1}{2}}, \chi\omega_{-\frac{1}{2}})$, and an infinite-dimensional irreducible subrepresentation $\sigma(\mu)$.*

(iii) *If $\mu_1\mu_2^{-1} = \omega_{-1}$, then $I(\mu)$ has a one-dimensional submodule on which G acts by the character $\chi \circ \det$, where $\mu = (\chi\omega_{-\frac{1}{2}}, \chi\omega_{\frac{1}{2}})$, and an infinite-dimensional irreducible quotient $\sigma(\mu)$.*

(iv) *The only equivalences among these representations are the following. Let $\mu' = (\mu_2, \mu_1)$. If $\mu_1\mu_2^{-1} \neq \omega_{\pm 1}$, then $I(\mu) \simeq I(\mu')$. If $\mu_1\mu_2^{-1} = \omega_{\pm 1}$, then $\sigma(\mu) \simeq \sigma(\mu')$.*

Theorem 3.2. *For an irreducible admissible representation (π, V) of G, the following are equivalent:*

(i) *For all μ, $\mathrm{Hom}_G(\pi, I(\mu)) = 0$.*

(ii) *The matrix coefficients[7] $\phi_{v,v'}$ of π are compactly supported module Z, i.e., the image in G/Z of the support of $\phi_{v,v'}$ is compact.*

The representations satisfying the conditions of Theorem 3.2 are the *supercuspidal* representations of G. Together with the *irreducible principal series* representations $I(\mu)$, the *special* representations $\sigma(\mu)$ and the *one-dimensional* representations $\chi \circ \det$, they give all of the irreducible admissible representations of G, up to isomorphism.

Recall that an irreducible admissible representation (π, V) of G is unramified if $V^K \neq 0$, i.e., if there exist vectors fixed by the maximal compact subgroup K. If the characters $\mu = (\mu_1, \mu_2)$ are unramified, so that

$$\mu_j(x) = t_j^{\mathrm{ord}(x)}$$

for some pair $(t_1, t_2) \in (\mathbb{C}^\times)^2$, then $\mu|_{(B\cap K)} = 1$ and restriction to K gives an isomorphism

$$I(\mu) \simeq C^\infty(B \cap K\backslash K).$$

[7] These are defined as follows. The contragredient $\tilde{V}$ of V is the space of smooth vectors in the full dual representation $V^* = \mathrm{Hom}_{\mathbb{C}}(V, \mathbb{C})$. For $v \in V$ and $v' \in \tilde{V}$, $\phi_{v,v'}(g) = \langle \pi(g)v, v' \rangle$, where $\langle\, ,\, \rangle$ is the pairing between (π, V) and $(\tilde{\pi}, \tilde{V})$.

This isomorphism is equivariant for the action of K by right multiplication. Thus

$$\dim I(\mu)^K = 1.$$

In fact, this construction accounts for all unramified representations.

Theorem 3.3.

(i) *For every pair $(t_1, t_2) \in (\mathbb{C}^\times)^2$, there is an irreducible admissible unramified representation $\pi(t_1, t_2)$ of G.*

(ii) *Every irreducible admissible unramified representation π of G is isomorphic to one of the $\pi(t_1, t_2)$'s. The only equivalence between such representations is*

$$\pi(t_2, t_1) \simeq \pi(t_1, t_2).$$

(iii) *If $\mu_1\mu_2^{-1} \neq \omega_{\pm 1}$, i.e., if $t_1 t_2^{-1} \neq q^{\pm 1}$, then $\pi(t_1, t_2) = I(\mu)$. If $\mu_1\mu_2^{-1} = \omega_{\pm 1}$, then $\pi(t_1, t_2) = \chi \circ \det$, the one-dimensional constituent of $I(\mu)$.*

Remark. Let

$$I = \left\{ k = \begin{pmatrix} a & b \\ c & d \end{pmatrix} \in K \mid \operatorname{ord}(c) \geq 1 \right\}$$

be the Iwahori subgroup of K. Then the only irreducible admissible representations (π, V) of G with $V^I \neq 0$ are the unramified representations together with the unramified special representations, i.e., the $\sigma(\mu)$'s, where μ is unramified.

The parametrization of the unramified representations can be expressed as follows.

Corollary 3.4. *There is a bijection between isomorphism classes of irreducible admissible unramified representations of G and semisimple conjugacy classes in complex group $G^\vee := \mathrm{GL}_2(\mathbb{C})$, given by*

$$\pi = \pi(t_1, t_2) \longleftrightarrow t(\pi) := \textit{conj. class of} \begin{pmatrix} t_1 & \\ & t_2 \end{pmatrix}.$$

The semisimple conjugacy class $t(\pi)$ is called the Satake parameter of π. The analogous parametrization, proved by Satake, [18], for an arbitrary connected reductive group G involves the Langlands dual group ${}^L G$. In fact, Satake's description of the unramified representations of G, which in hindsight can be viewed as a simple case of Langlands duality, was one of the things which suggested the construction of ${}^L G$ and its fundamental importance in understanding the representation theory of G [15].

The (spherical) Hecke algebra $\mathcal{H} = \mathcal{H}(G, K)$ is the space of compactly supported functions ξ on G that are invariant under both left and right multiplication by K, i.e., $\xi(k_1 g k_2) = \xi(g)$. $\mathcal{H}$ is an algebra under convolution with identity element the characteristic function[8] of K. For any unramified representation (π, V) of G, $\mathcal{H}$ acts on the space V^K by

$$\pi(\xi)v = \int_G \xi(g)\,\pi(g)v\,dg = \lambda_\pi(\xi)\,v,$$

where $\lambda_\pi : \mathcal{H} \to \mathbb{C}$ is an algebra homomorphism. In fact, every algebra homomorphism $\mathcal{H} \to \mathbb{C}$ arises in this way, and $\mathcal{H}$ can be identified as the algebra of regular functions on the space of semisimple conjugacy classes in $G^\vee$ via the map $\xi \mapsto \lambda_\pi(\xi)$. Thus there are algebra generators ξ_1 and ξ_2 of $\mathcal{H}$ corresponding to the standard symmetric polynomials via the algebra isomorphism

$$\mathcal{H} = \mathbb{C}[\xi_1, \xi_2] \xrightarrow{\sim} \mathbb{C}[t_1 + t_2, t_1 t_2], \qquad \begin{cases} \xi_1 \mapsto t_1 + t_2 = \lambda_\pi(\xi_1), \\ \xi_2 \mapsto t_1 t_2 = \lambda_\pi(\xi_2), \end{cases}$$

where $\pi = \pi(t)$.

Finally, the following result is important for understanding the relation to the classical theory. For an integer $r \geq 0$, let

$$K_0(r) = \left\{ k = \begin{pmatrix} a & b \\ c & d \end{pmatrix} \in K \mid \operatorname{ord}(c) \geq r \right\}.$$

If ω is a character of $F^\times$ of conductor $c(\omega) \leq r$, i.e., if ω is trivial on $1 + \mathcal{P}^r$, then ω defines a character of $K_0(r)$ by

$$\omega(k) = \omega(a).$$

The following is a result of Casselman [3]; cf. [8, Theorem 4.24].

Theorem 3.5. *For any irreducible admissible representation (π, V) of G with central character ω_π, there is a minimal integer $c(\pi) \geq c(\omega_\pi)$ for which*

$$V^{K_0(r),\omega_\pi} := \{v \in V \mid \pi(k)v = \omega_\pi(k)v, \ \forall k \in K_0(r)\} \neq 0.$$

Moreover,

$$\dim V^{K_0(c(\pi)),\omega_\pi} = 1.$$

The integer $c(\pi)$ is called the *conductor* of π. For example, $c(\pi) = 0$ for unramified representations, $c(\pi) = 1$ for unramified special representations, and $c(\pi) \geq 2$ for all other infinite-dimensional irreducible admissible π's, in particular, for all supercuspidal representations.

We end this section with a brief example from the archimedean theory. Let $G = \mathrm{GL}_2(\mathbb{R})$, $\mathfrak{g}_0 = \mathrm{Lie}(G)$ and $\mathfrak{g} = \mathfrak{g}_0 \otimes \mathbb{C}$, as above. Let $K_\infty = O(2)$. For an

[8] We have fixed the Haar measure on G for which K has volume 1.

integer $k \geq 2$, there is an irreducible $(\mathfrak{g}, K_\infty)$ module $\mathrm{DS}_k = (\pi_k, V)$ defined as follows. As a complex vector space,

$$V = \bigoplus_{\substack{|\ell| \geq k \\ \ell \equiv k(2)}} \mathbb{C} v_\ell,$$

where the action of K_∞ is given by

$$\pi_k(k_\theta)\, v_\ell = e^{i\ell\theta} v_\ell \quad \text{and} \quad \pi_k\left(\begin{pmatrix} 1 & \\ & -1 \end{pmatrix}\right) v_\ell = v_{-\ell}.$$

Thus, by (2.9), $\pi_k(H)\, v_\ell = \ell\, v_\ell$. In addition, the raising and lowering operators (1.12) act via

$$\pi_k(X_\pm)\, v_\ell = \frac{1}{2}(k \pm \ell)\, v_{\ell \pm 2},$$

so that $\pi_k(X_-)v_k = 0$, $\pi_k(X_+)v_{-k} = 0$, and $\pi_k(C) = k(k-2)$. It is easy to check that these formulas define an irreducible $(\mathfrak{g}, K_\infty)$-module, an algebraic version of the discrete series representation of weight k of $\mathrm{GL}_2(\mathbb{R})$. This provides us with one possibility for the archimedean component of an irreducible admissible $(\mathfrak{g}, K_\infty) \times G(\mathbb{A}_f)$ module at a real place. For a complete list, see, for example, [2, Theorem 2.5.2, p. 219].

4 Classical newforms

We now return to the case $k = \mathbb{Q}$ and the isomorphism

$$S_k(\Gamma_0(N), \chi_N) \xrightarrow{\sim} \mathcal{A}_0(G)(\text{hol}, k, N, \chi), \qquad f \mapsto \phi_f, \tag{4.1}$$

of Proposition 1.4. If $p \nmid N$, then functions in the space on the right side are invariant under right translation by K_p, so that there is an action of the local Hecke algebra $\mathcal{H}_p = \mathcal{H}(G_p, K_p)$. When transported to $S_k(\Gamma_0(N), \chi_N)$, the action of the generators $\xi_{p,1}$ and $\xi_{p,2}$ of $\mathcal{H}_p$ are related to the classical Hecke operators T_p and R_p by

$$f_0 \mid \xi_{p,1} = p^{-\frac{k-1}{2}} f_0 \mid T_p \quad \text{and} \quad f_0 \mid \xi_{p,2} = f_0 \mid R_p, \tag{4.2}$$

[8, p. 47]. Therefore the action of the algebra

$$\mathcal{H}^N = \otimes'_{p \nmid N} \mathcal{H}_p \tag{4.3}$$

is equivalent to the action of the classical Hecke operators T_n for primes not dividing the level. This makes it possible to express the classical theory of newforms and oldforms in terms of representation theory.

First suppose that $\pi = \pi_\infty \otimes (\otimes_p \pi_p)$ is an irreducible cuspidal automorphic representation with π_∞ isomorphic to the discrete series representation DS_k described at the end of Section 3. Let $\omega = \omega_\pi$ be the central character of π and let

$$N = N(\pi) = \prod_p p^{c(\pi_p)}, \tag{4.4}$$

where $c(\pi_p)$ is the conductor of the local component π_p. Let $K_0(N)$ be as in (1.3). Then, by the results outlined in Section 3, the space

$$\mathcal{A}_0(\pi)^{\mathrm{new}}_{\mathrm{hol}} := \left\{ \phi \in \mathcal{A}_0(\pi) \;\middle|\; \begin{array}{c} \text{(i) } X_-\phi = 0, \text{ and} \\ \text{(ii) } \phi(gk) = \omega(k)\phi(g),\ \forall k \in K_0(N(\pi)) \end{array} \right\} = \mathbb{C}\,\phi_0 \tag{4.5}$$

has dimension 1 and lies in the subspace $\mathcal{A}_0(G)(\mathrm{hol}, k, N, \chi)$, where χ is the Dirichlet character associated to ω. By Proposition 1.4, there is a unique holomorphic cusp form $f_0 \in S_k(\Gamma_0(N), \chi)$ corresponding to ϕ_0.

If $p \nmid N$, the local component π_p is unramified, and the local Hecke algebra $\mathcal{H}_p = \mathcal{H}(G_p, K_p)$ acts in the space $\pi_p^{K_p}$ by the character λ_{π_p}. It follows that the cusp form f_0 is an eigenfunction of all Hecke operators for such p's:

$$f_0 \mid T_p = a_p\, f_0 = p^{\frac{k-1}{2}}(t_{p,1} + t_{p,2})\, f_0 \quad \text{and} \quad f_0 \mid R_p = \chi(p)\, f_0 = t_{p,1}t_{p,2}\, f_0, \tag{4.6}$$

where $t_{p,1}$ and $t_{p,2}$ are the components of the Satake parameter $t(\pi_p)$ of π_p. Note that, since $\omega_\pi(p) = t_{p,1}t_{p,2}$, these relations amount to

$$1 - a_p\, p^{-\frac{k-1}{2}}\, X + \chi(p) X^2 = (1 - t_{p,1}X)(1 - t_{p,2}X) = \det(1_2 - t(\pi_p)\, X), \tag{4.7}$$

where a_p is the pth coefficient in the q-expansion (1.2) of f. Thus, the Satake parameter of π_p can be read off from the classical Hecke polynomial, the left side of (4.7), and conversely.

Recall that an eigenfunction f of all T_p's and R_p's for $(p, N) = 1$ is a *newform* if f does not "come from a lower level," [7, 14, 16]. Denote the space spanned by the newforms by $S_k(\Gamma_0(N), \chi)^{\mathrm{new}}$.

Theorem 4.1.

(i) *The holomorphic cusp form* $f_0 = f_0(\pi) \in S_k(\Gamma_0(N), \chi)$ *associated to the automorphic cuspidal representation* π *is a newform.*

(ii) *Let* ω *be the Hecke character corresponding to* χ. *Then, under the isomorphism of Proposition* 1.4,

$$S_k(\Gamma_0(N), \chi)^{\mathrm{new}} \simeq \bigoplus_{\substack{\pi \\ \omega_\pi = \omega \\ \pi_\infty \simeq \mathrm{DS}_k \\ N(\pi) = N}} \mathcal{A}_0(\pi)^{\mathrm{new}}_{\mathrm{hol}},$$

so that, up to a scalar multiple, every newform is an $f_0(\pi)$.

(iii) *Moreover,*

$$S_k(\Gamma_0(N), \chi) \simeq \mathcal{A}_0(G)(\mathrm{hol}, k, N, \chi) = \bigoplus_{\substack{\pi \\ \omega_\pi = \omega \\ \pi_\infty \simeq \mathrm{DS}_k \\ N(\pi) \mid N}} \mathcal{A}_0(\pi)_{\mathrm{hol}}^{K_0(N),\omega}.$$

Here $\mathcal{A}_0(\pi)_{\mathrm{hol}}^{K_0(N),\omega}$ is defined by taking N in place of $N(\pi)$ in condition (ii) in the definition of $\mathcal{A}_0(\pi)_{\mathrm{hol}}^{\mathrm{new}}$.

Remarks.

(i) This result gives the dictionary between the classical theory of newforms and automorphic representations. In fact, the reader who is so inclined can take this result as the definition of classical newforms, oldforms, and so on.

(ii) Note that the action of the Hecke operators only involves the representation of $G(\mathbb{A}_f)$, so that it will behave in the same way if the discrete series representation DS_k is replaced by some other irreducible admissible $(\mathfrak{g}, K_\infty)$-module. This accounts for the Hecke theory of Maass forms, of a given weight, where π_∞ is a suitable principal series representation.

Finally, we return to the classical holomorphic cusp form $f \in S_k(\Gamma_0(N), \chi_N)$ which was our starting point. The associated function $\phi \in \mathcal{A}_0(G)(\mathrm{hol}, k, N, \chi)$ can be decomposed into components as in (iii) of Theorem 4.1. This amounts to the decomposition of f into a combination of newforms and oldforms in the classical theory. On the other hand, if we take the $(\mathfrak{g}, K_\infty) \times G(\mathbb{A}_f)$-submodule $\mathcal{V}(\phi)$ of $\mathcal{A}_0(G)$ generated by ϕ, then we get precisely the sum of the $\mathcal{A}_0(\pi)$'s for which the corresponding components of ϕ in (iii) are nonzero. If f is a newform, then the space $\mathcal{V}(\phi)$ is irreducible.

5 The L-functions of automorphic cuspidal representations

In this section, we describe the Langlands L-functions attached to an irreducible cuspidal automorphic representation $\pi = \otimes_v \pi_v$ for a number field k.

First recall that, for every such π, there is a finite set S of places of k, including the archimedean places, such that, for $v \notin S$, π_v is unramified. Thus, by Corollary 3.4, there is a collection $\{t(\pi_v)\}_{v \notin S}$ in $G^\vee = \mathrm{GL}_2(\mathbb{C})$ of Satake parameters of the unramified local components of π. The situation is analogous to having a finite Galois extension k'/k, where, for each finite place v of k which is unramified in k'/k, one has a conjugacy class of Frobenius elements Fr_v in $\mathrm{Gal}(k'/k)$, and hence a collection $\{\mathrm{Fr}_v\}_{v \notin S}$ of conjugacy classes in $\mathrm{Gal}(k'/k)$. By analogy with Artin L-functions, for any finite-dimensional representation r of $\mathrm{GL}_2(\mathbb{C})$, [13], and $s \in \mathbb{C}$, the associated (partial) Langlands L-function is defined as an Euler product

$$L^S(s,\pi,r) := \prod_{v\notin S} \det(1 - r(t(\pi_v))\, q_v^{-s})^{-1}.$$

This can be shown to converge in a right half plane, and one of the fundamental problems in the theory is to (i) complete the L-function by including additional factors for the places $v \in S$, and (ii) establish the analytic properties, analytic continuation and functional equation, of the completed function. For r the standard representation, a complete and beautiful theory was established by Jacquet and Langlands [11].

For example, in the case considered by Hecke, $k = \mathbb{Q}$ and π is the irreducible cuspidal automorphic representation corresponding to a holomorphic newform f of weight k and level N with q-expansion (1.2) normalized so that $a_1 = 1$. For r the standard two-dimensional representation of $\mathrm{GL}_2(\mathbb{C})$ and by (4.7),

$$\begin{aligned} L^S(s,\pi,r) &= \prod_{p\nmid N} (1 - a_p\, p^{-s-\frac{k-1}{2}} + p^{-2s})^{-1} \\ &= L^N\left(s + \frac{k-1}{2}, f\right) \\ &= \sum_{\substack{n \\ (n,N)=1}} a_n\, n^{-s}. \end{aligned}$$

For an excellent and full discussion of such L-functions, the reader may consult [2, 8, 9, 12, 7], etc.

References

[1] D. Blasius, On multiplicities for $SL(n)$, *Israel J. Math.*, **88**(1994), 1–3.

[2] D. Bump, *Automorphic Forms and Representations*, Cambridge University Press, Cambridge, UK, 1997.

[3] W. Casselman, On some results of Atkin and Lehner, *Math. Ann.*, **201**(1973), 301–314.

[4] J. Cogdell, Analytic theory of L-functions for GL_n, in J. Bernstein and S. Gelbart, eds., *An Introduction to the Langlands Program*, Birkhäuser Boston, Boston, 2003, 197–228 (this volume).

[5] J. Cogdell, Langlands conjectures for GL_n, in J. Bernstein and S. Gelbart, eds., *An Introduction to the Langlands Program*, Birkhäuser Boston, Boston, 2003, 229–249 (this volume).

[6] J. Cogdell, Dual groups and Langlands functoriality, in J. Bernstein and S. Gelbart, eds., *An Introduction to the Langlands Program*, Birkhäuser Boston, Boston, 2003, 251–269 (this volume).

[7] F. Diamond and J. Im, Modular forms and modular curves, in *Seminar on Fermat's Last Theorem*, CMS Conference Proceedings 17, American Mathematical Society, Providence, RI, 1995, 39–133.

[8] S. Gelbart, *Automorphic Forms on Adele Groups*, Princeton University Press, Princeton, NJ, 1975.

[9] S. Gelbart, An elementary introduction to the Langlands program, *Bull. Amer. Math. Soc.*, **10**(1984), 177–219.

[10] R. Godement, *Notes on the Jacquet–Langlands Theory*, Institute for Advanced Study, Princeton, NJ, 1970.

[11] H. Jacquet and R. P. Langlands, *Automorphic forms on $GL(2)$*, Lecture Notes in Mathematics 114, Springer-Verlag, New York, 1970.

[12] T. Knapp, Introduction to the Langlands program, *Proc. Sympos. Pure Math.*, **61**(1997), 245–302.

[13] E. Kowalski, Elementary theory of L-functions I and II, in J. Bernstein and S. Gelbart, eds., *An Introduction to the Langlands Program*, Birkhäuser Boston, Boston, 2003, 1–37 (this volume).

[14] S. Lang, *Introduction to modular forms*, Springer-Verlag, New York, 2001.

[15] R. P. Langlands, Problems in the theory of automorphic forms, in *Lectures in Modern Analysis and Applications*, Lecture Notes in Mathematics 170, Springer-Verlag, New York, 1970, 18–86.

[16] T. Miyake, *Modular Forms*, Springer-Verlag, New York, 1989.

[17] D. Ramakrishnan, A refinement of the strong multiplicity one theorem for $GL(2)$, appendix to R. Taylor, "ℓ-adic representations associated to modular forms over imaginary quadrtic fields II," *Invent. Math.*, **116**(1994), 645–649.

[18] I. Satake, Theory of spherical functions on reductive algebraic groups over $\mathfrak{p}$-adic fields, *Pub. IHES*, **18**(1963), 1–69.

[19] G. Shimura, *Introduction to the Arithmetic Theory of Automorphic Forms*, Princeton University Press, Princeton, NJ, 1971.

8
Spectral Theory and the Trace Formula

Daniel Bump*

We give an account of a portion of the spectral theory $\Gamma\backslash SL(2,\mathbb{R})$, particularly the Selberg trace formula, emphasizing ideas from representation theory. For simplicity, we will treat the trace formula only in the case of a compact quotient. The last section is of a different nature, intended to show a simple application of the trace formula to a lifting problem.

1 The spectral problem

The group $G = SL(2,\mathbb{R})$ acts on $\mathfrak{H} = \{x+iy \mid y > 0\}$:

$$\begin{pmatrix} a & b \\ c & d \end{pmatrix} : z \mapsto \frac{az+b}{cz+d}.$$

The stabilizer of i is $K = SO(2)$ so $\mathfrak{H} \cong G/K$. The *non-Euclidean Laplacian*

$$\Delta = -y^2 \left(\frac{\partial^2}{\partial x^2} + \frac{\partial^2}{\partial y^2} \right) \tag{1}$$

is a G-invariant differential operator. Let Γ be a discrete cocompact subgroup of $G = SL(2,\mathbb{R})$. Then $X = \Gamma\backslash\mathfrak{H}$ is a compact Riemann surface. The question at hand is to describe the spectrum of Δ on X.

Proposition 1. *The group Γ has a fundamental domain $\mathcal{F}$ on $\mathfrak{H}$ whose boundary consists of pairs of geodesic arcs α_i and $\gamma_i(\alpha_i)$, with $\gamma_i \in \Gamma$. When the boundary is traversed counterclockwise, the congruent arcs α_i and $\gamma_i(\alpha_i)$ are traversed in opposite directions.*

Sketch of proof. Choose a point $P \in \mathfrak{H}$ which is not fixed by any element of Γ except by $\pm I$. Let $\mathcal{F}$ be the set of points that are nearer to P than to $\gamma(P)$ in the

*I would like to thank Yonatan Gutman and the referee for extensive and very helpful comments on an earlier draft. Preparation of this chapter was supported in part by NSF grant DMS-9970841.

non-Euclidean metric for any $\gamma \in \Gamma$. Let $N = \{\gamma_1, \ldots, \gamma_n\}$ be the set elements of Γ such that $\gamma_i(\mathcal{F})$ is adjacent to $\mathcal{F}$.

At first we assume no $\gamma_i^2 = 1$. Evidently each $\gamma_i^{-1} \in N$, so N has an even number $2h$ of elements. We arrange it so that the γ_i so that $\gamma_{i+h} = \gamma_i^{-1}$ $(1 \leq i \leq h)$. Let α_i be the intersection of the geodesic consisting of the set of points that are equidistant from P and $\gamma_i^{-1}(P)$ with the closure of $\mathcal{F}$. Then $\gamma_i(\alpha_i) = \alpha_{i+h}$ and $\alpha_1, \ldots, \alpha_h$ satisfy our requirements.

If some $\gamma_i^2 = 1$, then the arc α_i is self-congruent by γ_i and its midpoint is fixed by γ_i. So we split such arcs in two at their midpoints and we are done. □

The Laplacian Δ acts on $C^\infty(\Gamma\backslash\mathfrak{H})$.

Proposition 2. *Δ is symmetric and positive with respect to the invariant metric $y^{-2}\,dx \wedge dy$.*

Proof. Symmetry means that

$$\langle \Delta f, g\rangle - \langle f, \Delta g\rangle = 0, \qquad f, g \in C^\infty(\Gamma\backslash\mathfrak{H}).$$

Taking a fundamental domain $\mathcal{F}$ as in Proposition 1, this equals

$$\int_{\mathcal{F}} \left(f\left(\frac{\partial^2 \overline{g}}{\partial x^2} + \frac{\partial^2 \overline{g}}{\partial y^2}\right) - \overline{g}\left(\frac{\partial^2 f}{\partial x^2} + \frac{\partial^2 f}{\partial y^2}\right)\right) dx \wedge dy = \int_{\mathcal{F}} d\omega,$$

where

$$\omega = f\frac{\partial \overline{g}}{\partial x}\,dy - f\frac{\partial \overline{g}}{\partial y}\,dx - \overline{g}\frac{\partial f}{\partial x}\,dy + \overline{g}\frac{\partial f}{\partial y}\,dx.$$

By Stokes's theorem, this is equal to the integral of ω around the boundary of $\mathcal{F}$. By Proposition 1, the contributions of the boundary arcs cancel in pairs.

Positivity means that $\langle \Delta f, f\rangle \geq 0$ with equality only if f is constant. We compute

$$\left[\left|\frac{\partial f}{\partial x}\right|^2 + \left|\frac{\partial f}{\partial y}\right|^2 + \left(\frac{\partial f^2}{\partial x^2} + \frac{\partial f^2}{\partial y^2}\right)\overline{f}\right] dx \wedge dy = d\left(\overline{f}\left(\frac{\partial f}{\partial x}\,dy - \frac{\partial f}{\partial y}\,dx\right)\right).$$

By Stokes's theorem we thus have

$$\langle \Delta f, f\rangle \geq -\int_{\partial\mathcal{F}} \overline{f}\left(\frac{\partial f}{\partial x}\,dy - \frac{\partial f}{\partial y}\,dx\right), \tag{2}$$

with equality only if $\partial f/\partial x = \partial f/\partial y = 0$ identically, that is, if f is constant. Using the Cauchy–Riemann equations and the chain rule if one may check that if $x + iy$ and $u + iv$ are related by a holomorphic mapping, such as a linear fractional transformation, then $(\partial f/\partial x)dy - (\partial f/\partial y)dx = (\partial f/\partial u)dv - (\partial f/\partial v)du$. It follows that the contributions of congruent boundary arcs cancel and (2) is zero. □

We will see later that Δ extends to a self-adjoint unbounded operator on $\mathfrak{L} = L^2(\Gamma\backslash\mathfrak{H}, y^{-2}\, dx\wedge dy)$. This means that there is a dense subspace D of $\mathfrak{L}$ containing $C^\infty(\Gamma\backslash\mathfrak{H})$ such that Δ extends to D, and that if $v \in \mathfrak{L}$ is such that $u \to \langle v, \Delta u\rangle$ is continuous on D, then $v \in D$ and $\langle \Delta v, u\rangle = \langle v, \Delta u\rangle$.

Since Δ is a self-adjoint operator it has a nice spectral theory, which we want to develop. We will accomplish this by introducing integral operators that commute with Δ. We will show that these operators are of trace class, and we will prove the Selberg trace formula for them.

Finally, we will consider the more difficult case where Γ has a cusp. In this case there are both discrete and continuous spectra, and the theory of Eisenstein series is an essential feature.

2 Rings of integral operators

The group G acts on functions $f : G \to \mathbb{C}$ by the *right regular action* $(\rho(g)f)(x) = f(xg)$. Since we will soon be discussing differential operators let us at first restrict ourselves to $f \in C^\infty(G)$.

Because $\mathfrak{H}$ is canonically identified with the homogeneous space G/K, any function f on $\mathfrak{H}$ may be regarded as a function on G that is right invariant by K. The relevance of the regular representation of G to functions on $\mathfrak{H}$ may not be immediately clear, because the property of right K invariance is not preserved under right translation. However, there *is* a relevance which we now explain.

The Lie algebra of G consists of the vector space $\mathfrak{g}$ of 2×2 real matrices of trace zero. It acts on smooth functions f on G by

$$Xf(g) = \frac{d}{dt} f(ge^{tX})\big|_{t=0}, \qquad X \in \mathfrak{g}.$$

This is a Lie algebra representation. This means that $[X, Y] = XY - YX$ (XY = matrix multiplication) has the same effect as $X \circ Y - Y \circ X$ ($X \circ Y$ = composition of operators). This representation of $\mathfrak{g}$ is the differential form of the regular representation.

The universal enveloping algebra $U(\mathfrak{g})$ is the associative $\mathbb{R}$-algebra generated by $\mathfrak{g}$ modulo the relations

$$[X, Y] - (X \cdot Y - Y \cdot X) = 0$$

($X \cdot Y$ = multiplication in $U(\mathfrak{g})$). Any Lie algebra representation extends uniquely to a representation of $U(\mathfrak{g})$. In particular the regular representation of $\mathfrak{g}$ extends to a representation of $U(\mathfrak{g})$. Thus $U(\mathfrak{g})$ is realized as a ring of differential operators on $C^\infty(G)$.

If D is an element of the *center* of $U(\mathfrak{g})$, then it commutes with the regular representation. This is intuitively reasonable and proved in Bump [6, Proposition 2.2.4]. In particular, if f is fixed under $\rho(K)$, then so is Df. Therefore D acts on $C^\infty(\mathfrak{H})$.

A particular element of the center of $U(\mathfrak{g})$ is the *Laplace–Beltrami* or *Casimir* element

$$-4\Delta = H \cdot H + 2R \cdot L + 2L \cdot R, \tag{3}$$

$$R = \begin{pmatrix} 0 & 1 \\ 0 & 0 \end{pmatrix}, \qquad L = \begin{pmatrix} 0 & 0 \\ 1 & 0 \end{pmatrix}, \qquad H = \begin{pmatrix} 1 & 0 \\ 0 & -1 \end{pmatrix}.$$

In fact, $\mathbb{C}[\Delta]$ is the center of $U(\mathfrak{g})$. We have used the same letter Δ that we previously used for the non-Euclidean Laplacian, because when this element of $U(\mathfrak{g})$ is interpreted as a differential operator on $\mathfrak{H}$, they are the same. See Bump [6, Proposition 2.2.5].

Let $\mathcal{H}$ be the convolution ring of smooth, compactly supported functions on G. Let $\mathcal{H}^\circ$ be the subring of K-biinvariant functions. These are rings without unit. We call $\mathcal{H}^\circ$ the *spherical Hecke algebra* but caution the reader that there are other natural and closely related rings which have also been called by this name. The ring $\mathcal{H}$ is noncommutative, but we have the following.

Theorem 3. *The ring $\mathcal{H}^\circ$ is commutative.*

Proof. Matrix transposition preserves K so it induces an involution ι on $\mathcal{H}^\circ$ such that $\iota(\phi * \psi) = \iota(\psi) \circ \iota(\phi)$, where $(\iota f)(g) = f(g^t)$. Every double coset in $K\backslash G/K$ has a diagonal representative. So ι is the identity map. □

This theorem of Gelfand has a representation-theoretic meaning. If (π, V) is a representation of G on a Banach space, we will denote by V^K the vector subspace of K-fixed vectors. The algebra $\mathcal{H}$ acts on V by

$$\pi(\phi)\, v = \int_G \phi(g)\, \pi(g)\, v \, dg, \qquad \phi \in \mathcal{H},\ v \in V.$$

Here $\int_G dg$ is the Haar integral. If $\phi \in \mathcal{H}^\circ$, then $\pi(\phi)\, v \in V^K$. In particular V^K is a module for $\mathcal{H}^\circ$.

Since K is compact, any irreducible representation ρ of K is finite dimensional. The *ρ-isotypic part* $V(\rho)$ of V is the direct sum of all K-invariant subspaces isomorphic to ρ as K-modules. A representation (π, V) of G is called *admissible* if $V(\rho)$ is finite dimensional for every ρ. In particular, if ρ is the trivial representation $V(\rho) = V^K$, then V^K must be finite dimensional.

Theorem 4. *If (π, V) is an irreducible admissible representation of G, then V^K is at most one dimensional.*

Proof. Since $\mathcal{H}^\circ$ is commutative, its finite-dimensional irreducible modules are one dimensional. Thus it is sufficient to show that V^K (if nonzero) is an irreducible module for $\mathcal{H}^\circ$. Suppose $L \subset V^K$ is a closed nonzero $\mathcal{H}^\circ$-invariant subspace. If $v \in V^K$ we will show that $v \in L$.

Let $\epsilon > 0$ be given. Since V is irreducible and L is nonzero, the closure of $\pi(\mathcal{H})\,L$ is V and so there exists $\phi \in \mathcal{H}$, $w \in L$ such that $\pi(\phi)\,w = v_1$, where $|v_1 - v| < \epsilon$. Let $v_2 = \int_K \pi(k)v_1\,dk$. Then v_2 is K-fixed and since v is K-fixed,

$$|v_2 - v| = \left|\int_K \pi(k)(v_1 - v)\,dk\right| \leq \int_K |\pi(k)(v_1 - v)|\,dk = \int_K |v_1 - v|\,dk < \epsilon.$$

(We normalize the Haar measure so K has volume 1.) Since w is K-fixed,

$$\pi(k)\pi(\phi)\pi(k')w = \pi(k)v_1$$

for all k, $k' \in K$. Integrating over k and k', we thus get $\pi(\phi_0)\,w = v_2$, where

$$\phi_0(g) = \int_{K\times K} \phi(kgk')\,dk\,dk'.$$

Now $\phi_0 \in \mathcal{H}^\circ$, and since L is $\mathcal{H}^\circ$-invariant, this implies that $v_2 \in L$. We can therefore approximate v arbitrarily closely by elements of L, and since $L \subset V^K$ is finite dimensional, hence closed, this implies that $v \in L$. □

More generally, if (π, V) is a representation of G and $k \in \mathbb{Z}$ we will denote by $V(k)$ the subspace of $v \in K$ satisfying

$$\pi(\kappa_\theta)\,v = e^{ik\theta}\,v, \qquad \kappa_\theta = \begin{pmatrix} \cos(\theta) & \sin(\theta) \\ -\sin(\theta) & \cos(\theta) \end{pmatrix} \in K.$$

(This is the ρ-isotypic subspace where ρ is the character $e^{ik\theta}$ of K.) If (π, V) is irreducible, then since $-I \in K$ is central, it acts by a scalar $(-1)^\epsilon$ where $\epsilon = 0$ or 1. Evidently the parity of k must be the same as ϵ for $V(k)$ to be nonzero. We will call π *even* if $\epsilon = 0$, and *odd* if $\epsilon = 1$.

Proposition 5. *If (π, V) is an irreducible admissible representation of G, then $V(k)$ is at most one dimensional.*

Sketch of proof. This may be proved along the same lines as Theorem 4. Because $K = SO(2)$ is commutative, it may be seen that the convolution ring of smooth, compactly supported functions that satisfy $f(\kappa_\theta g \kappa_{\theta'}) = e^{ik(\theta+\theta')}\,f(g)$ is commutative. See Bump [6, Proposition 2.2.8]. □

We note that while Theorem 4 generalizes directly to arbitrary reductive Lie groups, Proposition 5 does not. Thus if (π, V) is an irreducible admissible representation of a reductive Lie group, the multiplicity of the trivial representation of its maximal compact subgroup is at most one; the other irreducible representations of the maximal compact each occur with finite multiplicity (this is admissibility), but not necessarily multiplicity one. Actually irreducible representations are automatically admissible, though this fact is not needed in the theory of automorphic forms, where admissibility of automorphic representations can be proved directly.

If (π, V) is an irreducible representation of G, we will denote by V_{fin} the direct sum of the $V(k)$. This is the space of *K-finite* vectors. It is not invariant under the action of G, but it is invariant under the actions of the Lie algebra $\mathfrak{g}$ of G and of K. It is therefore a $(\mathfrak{g}, K)$-*module*. See Bump [6, p. 200] for a discussion of this concept.

If $\pi = \rho$ is the right regular representation ρ of G, we will use the notation T_ϕ instead of $\rho(\phi)$. Thus

$$T_\phi f(x) = \int_G \phi(g)\, f(xg)\, dg.$$

This is convolution with $g \to \phi(g^{-1})$. The notation is intended to suggest Hecke operators. These integral operators are important for us because they commute with Δ, yet they are easier to study than Δ. We note that $L^2(\Gamma\backslash G)$ is invariant under T_ϕ, and if $\phi \in \mathcal{H}^\circ$, then T_ϕ preserves the property of right invariance by K, so it can be regarded as an integral operator on $L^2(\Gamma\backslash\mathfrak{H})$.

Let $\phi \in \mathcal{H}$, and let

$$K_\phi(x, y) = \sum_{\gamma\in\Gamma} \phi(x^{-1}\gamma y). \tag{4}$$

At first we regard (x, y) as an element of $G \times G$. If x and y are restricted to a compact set C, then $\phi(x^{-1}gy)$ vanishes for g off a compact set C'. Therefore, only finitely many γ contribute. It follows that (4) is convergent and defines a smooth function of x and y.

A change of variables shows that $K_\phi(x, y)$ is invariant if either x or y is changed on the left by an element of γ, so we may regard this kernel as defined on either $G \times G$ or on $\Gamma\backslash G \times \Gamma\backslash G$, and it is a continuous function. If $\phi \in \mathcal{H}^\circ$, then K_ϕ may even be regarded as a function on $X \times X$. (Recall that $X = \Gamma\backslash\mathfrak{H}$.)

Proposition 6. *We have*

$$(T_\phi f)(x) = \int_{\Gamma\backslash G} K_\phi(x, y)\, f(y)\, dy. \tag{5}$$

Proof. The left side equals

$$\int_G \phi(y)\, f(xy)\, dy = \int_G \phi(x^{-1}y)\, f(y)\, dy = \sum_{\gamma\in\Gamma} \int_{\Gamma\backslash G} \phi(x^{-1}\gamma y)\, f(y)\, dy,$$

where we have used $f(\gamma y) = f(y)$. Interchanging sum and integral gives (5). □

If H is a Hilbert space, an operator $T : H \to H$ is *compact* if T maps bounded sets into compact sets.

Theorem 7. *T_ϕ is a compact operator on $L^2(\Gamma\backslash G)$.*

See Bump [6, Section II.3, particularly Theorem 2.3.2 and Proposition 2.3.1] for more complete details.

Proof. The kernel K_ϕ is continuous on the compact space $(\Gamma\backslash G) \times (\Gamma\backslash G)$, so it is certainly in $L^2\big((\Gamma\backslash G) \times (\Gamma\backslash G)\big)$. The well-known theorem of Hilbert and

Schmidt asserts that if Z is any locally compact Borel measure space such that $L^2(Z)$ is a separable Hilbert space, then integral operator

$$(Tf)(x) = \int_Z K(x, y)\, f(y)\, dy$$

with the kernel $K \in L^2(Z \times Z)$ is compact. □

If

$$\phi(g^{-1}) = \overline{\phi(g)}, \tag{6}$$

then $K_\phi(x, y) = \overline{K_\phi(y, x)}$, so T_ϕ is self-adjoint.

Theorem 8. *Let T be a compact self-adjoint operator on a separable Hilbert space H. Then H has an orthonormal basis ϕ_i ($i = 1, 2, 3, \dots$) of eigenvectors of T, so that $T\phi_i = \mu_i\phi_i$. The eigenvalues $\mu_i \to 0$ as $i \to \infty$.*

This is the Spectral Theorem for compact operators. See Bump [6, Theorem 2.3.1]. □

Thus if (6) is true, then T_ϕ is a self-adjoint compact operator whose nonzero eigenvalues $\mu_i \to 0$. The Hilbert–Schmidt property implies more: $\sum |\mu_i|^2 < \infty$. Later we will see that more is true: $\sum |\mu_i| < \infty$. This means that T_ϕ is *trace class*. This fact is important because of the Selberg trace formula.

Theorem 9. *$L^2(X)$ has a basis consisting of eigenvectors of Δ.*

Proof. The operators T_ϕ with ϕ satisfying (6) are a commuting family of self-adjoint compact operators so they can be simultaneously diagonalized. By the spectral theorem the nonzero eigenspaces are finite dimensional; there is no nonzero vector on which the operators T_ϕ are all zero, since ϕ can be chosen to be positive, of mass one and concentrated near the identity, in which case $T_\phi f$ approximates f. Therefore, the simultaneous eigenspaces of $\mathcal{H}^\circ$ are finite dimensional.

Let V be such an eigenspace. Since Δ commutes with the T_ϕ, it preserves V, and since it is symmetric, it induces a self-adjoint transformation on V. Choose an orthonormal basis for each such V consisting of eigenvectors of Δ and put these together for all V. □

Closely related to Theorem 9 is a representation-theoretic statement about $L^2(\Gamma\backslash G)$. The regular representation ρ is a unitary representation on this space.

Lemma 10. *Let H be a closed nonzero G-invariant subspace of $L^2(\Gamma\backslash G)$. Then H contains an irreducible subspace.*

Proof (Langlands). Since H is G-invariant, each T_ϕ induces an endomorphism of H. We show first that the restriction of T_ϕ to H is nonzero for suitable ϕ satisfying (6). If $0 \neq \xi \in H$, then for g near the identity $\rho(g)\xi$ is near ξ. Thus if we take ϕ

satisfying (6) such that $\phi > 0$, $\int_G \phi(g)\,dg = 1$ and such that the support of ϕ is nonzero, then $T_\phi \xi$ is near ξ so T_ϕ is nonzero on H, and T_ϕ is self-adjoint.

Let $L \subset H$ be the eigenspace of a nonzero eigenvalue of T_ϕ. It is finite dimensional by Theorem 8. Let L_0 be a nonzero subspace minimal with respect to the property of being the intersection of L with a nonzero closed invariant subspace of H. Let V be the smallest closed invariant subspace of H such that $L \cap V = L_0$. We show V is irreducible. If not, let V_1 be a proper, nonzero closed invariant subspace and let V_2 be its orthogonal complement, so $V = V_1 \oplus V_2$. Let $0 \neq f \in L_0$. Write $f = f_1 + f_2$ with $f_i \in V_i$. Since $0 = T_\phi f - \lambda f = (T_\phi f_1 - \lambda f_1) + (T_\phi f_2 - \lambda f_2)$ and $T_\phi f_i - \lambda f_i \in V_i$ we have $T_\phi f_i - \lambda f_i = 0$. Thus $f_i \in L \cap V_i$. By the minimality of L_0, $L_0 = L \cap V_i$ for some i, say $L_0 = L \cap V_1$. Now the minimality of V is contradicted. □

Theorem 11. *$L^2(\Gamma\backslash G)$ decomposes as a direct sum of closed, irreducible subspaces. Each affords an irreducible admissible representation of G.*

Proof. By Zorn's Lemma, let S be a maximal set of orthogonal closed irreducible subspaces. Let $H = \bigoplus_{V \in S} V$. If H is proper, applying Lemma 10 to its orthogonal complement contradicts the maximality of S. We leave admissibility to the reader. □

Each of these closed irreducible subspaces has at most one K-fixed vector by Theorem 4.

Theorem 11 may be thought of as a more satisfactory extension of Theorem 9. Indeed, if ϕ is an eigenfunction of Δ occurring in $L^2(\Gamma\backslash\mathfrak{H})$, then its right translates span an irreducible subspace of $L^2(\Gamma\backslash G)$. Conversely, Δ acts by a scalar on each irreducible subspace. If that subspace happens to have a K-fixed vector in it, that vector will be one of the basis elements in Theorem 9.

There will, however, be some irreducible subspaces of $L^2(\Gamma\backslash G)$ which have no K-fixed vectors. These can be constructed from *holomorphic modular forms* as follows. Let $f : \Gamma\backslash\mathfrak{H} \to \mathbb{C}$ be a holomorphic function satisfying

$$f\left(\frac{az+b}{cz+d}\right) = (cz+d)^k\, f(z), \qquad \begin{pmatrix} a & b \\ c & d \end{pmatrix} \in \Gamma.$$

If k is sufficiently large, such f will always exist, as may be shown from the Riemann–Roch theorem. Regarded as a function on G, the function f is right invariant by K, but it is not left invariant by Γ, so it is not a function on $\Gamma\backslash G$. However we may modify it, sacrificing the right invariance by K to obtain true left invariance by Γ. Define

$$F\begin{pmatrix} a & b \\ c & d \end{pmatrix} = (ci+d)^{-k}\, f\left(\frac{ai+b}{ci+d}\right).$$

Then $F(\gamma g) = F(g)$ for $\gamma \in \Gamma$, while $F(g\,\kappa_\theta) = e^{ik\theta} F(g)$. The irreducible representation spanned by F does not have a K-fixed vector. This is the weight k *holomorphic discrete series* representation.

The representation-theoretic approach has another generalization, its extension to automorphic forms on adele groups. Assume that $\Gamma = SL(2,\mathbb{Z})$. Let $\mathbb{A}$ be the adele ring of $\mathbb{Q}$. The inclusion of $SL(2,\mathbb{R}) \to GL(2,\mathbb{A})$ at the infinite place induces a homeomorphism

$$\Gamma\backslash G \to GL(2,\mathbb{Q})Z_{\mathbb{A}}\backslash GL(2,\mathbb{A})/\textstyle\prod_p K_p,$$

where $Z_{\mathbb{A}}$ is the center of $GL(2,\mathbb{A})$ and $K_p = GL(2,\mathbb{Z}_p)$ is a maximal compact subgroup of $GL(2,\mathbb{Q}_p)$. Thus functions on $\Gamma\backslash G$ may be reinterpreted as functions on $GL(2,\mathbb{Q})Z_{\mathbb{A}}\backslash GL(2,\mathbb{A})$, and in particular we may embed $L^2(\Gamma\backslash G) \to L^2\big(GL(2,\mathbb{Q})Z_{\mathbb{A}}\backslash GL(2,\mathbb{A})\big)$.

Now the study of $L^2\big(GL(2,\mathbb{Q})Z_{\mathbb{A}}\backslash GL(2,\mathbb{A})\big)$ may be carried out along exactly the same lines as we have applied above. The class of integral operators is larger now, however. In addition to the ring $\mathcal{H}$, we have its p-adic analogs, which are rings of *Hecke operators*. To see this, fix a prime p. Let $\mathcal{H}_p$ be the convolution ring of smooth (i.e., locally constant) compactly supported functions on $GL(2,\mathbb{Q}_p)$. This ring is not commutative, but the subring $\mathcal{H}_p^\circ$ of functions that are K_p-biinvariant is commutative. (Compare with Theorem 3.) For example, the characteristic function of $K_p \left(\begin{smallmatrix} p & 0 \\ 0 & 1 \end{smallmatrix}\right) K_p$ is an element of this ring, and this is the adelization of the classical Hecke operator T_p which picks off the pth Fourier coefficient of a Hecke eigenform. We will return to this point of view in the final section when we study such operators using the trace formula.

3 Green's functions and the spectral resolvent

References for this section are Hejhal [21, Section 6] and Bump [6, Chapter 2, Section 3]. We are interested in functions f on $\mathfrak{H}$ with the following.

Property S. The function f is left invariant by $K = SO(2)$, possibly singular at K's fixed point at i, and is an eigenfunction of the Laplacian.

We will see that for each eigenvalue λ of Δ there are two linearly independent such functions, one of which (the *Green's function*) has nice behavior at the boundary, the other of which (the *spherical function*) is continuous at i. Each has its uses. Since $\mathcal{H} \cong G/K$, these may be regarded as functions on $K\backslash G/K$ (possibly undefined on K) which are eigenfunctions of the Laplace Beltrami operator.

We can map the upper half plane into the unit disk $\mathfrak{D}$ by the Cayley transform $z \mapsto w = (z-i)/(z+i)$. Let $r = |w|$. Then since f as in Property S is left invariant by K, $f(z)$ depends only on r. Denote $f(z) = W(r)$. Thus a function f with Property S is determined by the function W on $(0,1)$ such that

$$W(r) = f\begin{pmatrix} y^{1/2} & \\ & y^{-1/2} \end{pmatrix},$$

where $r = (y-1)/(y+1) \in (0,1)$. The eigenvalue property amounts to the differential equation

$$W''(r) + \frac{1}{r}W'(r) + \frac{4\lambda}{(1-r^2)^2}W(r) = 0. \tag{7}$$

This differential equation has regular singular points at $(0,1)$ and there are two solutions of interest. One is nicely behaved at 0, the other at 1. (We assume familiarity with regular singular points of second order linear differential equations, particularly the indicial equation, for which see Whittaker and Watson [52, Section 10.3].) In this section we will be concerned with the solution which has nice behavior on the boundary, that is, at $r = 1$; the other one will occupy the next section.

Let $\lambda = -s$ where s is a positive real number. At $r = 1$ the roots of the indicial equation of (7) for the singularity at $r = 1$ are $\frac{1}{2}(1 \pm \sqrt{1+4s})$. Only one is positive, so there is a unique (up to multiple) solution g to (7) which vanishes near the boundary. We can use it to study the resolvent of the Laplacian.

Lemma 12. *$g_s(r)$ has a logarithmic singularity at $r = 0$.*

Proof. The roots of the indicial equation at $r = 0$ are 0 with multiplicity 2, so one solution has a logarithmic singularity, another is analytic. If g_s does not have the logarithmic singularity, then $g_s(r)$ is real and analytic on $[-1,1]$, hence has a maximum or minimum. At such a point $g_s'(r) = 0$ and since $\lambda = -s < 0$, equation (7) implies that g_s and g_s'' have the same sign, impossible at the maximum or minimum because $g_s(-1) = g_s(1) = 0$. □

Let

$$g_s(z,\zeta) = g_s\left(\left|\frac{z-\zeta}{z-\bar{\zeta}}\right|\right).$$

$z, \zeta \in \mathfrak{H}$. This is a *Green's function.*

Theorem 13.

$$\left[-y^2\left(\frac{\partial^2}{\partial x^2} + \frac{\partial^2}{\partial y^2}\right) + s\right] g(z,\zeta) = 0;$$

$$g_s(z,\zeta) \textit{ is singular on the diagonal } z = \zeta;$$

$$g_s(z,\zeta) \to 0 \textit{ as } y \to 0;$$

$$g_s(z,\zeta) = g_s(\zeta, z);$$

$$g_s\big(h(z), h(\zeta)\big) = g_s(z,\zeta), \qquad h \in SL(2,\mathbb{R}). \tag{8}$$

See Bump [6, Proposition 2.3.4, p. 181] for more complete details. Nothing here is very surprising, for example, the first property boils down to (7), the second property comes from Lemma 12, the third follows from the boundary behavior of g_s, and the last two properties follow since $\big|(z-\zeta)/(z-\bar{\zeta})\big|$ is unchanged if z and ζ are interchanged, or if an element of $SL(2,\mathbb{R})$ is applied to both. □

Since g_s has a logarithmic singularity at 0 it can be normalized so $g_s(r) - \frac{1}{2\pi}\log(r)$ is bounded as $r \to 0$. It follows that $g_s'(r) - \frac{1}{2\pi r}$ is analytic near $r = 0$.

Theorem 14. *If $f \in C_c^\infty(\mathfrak{H})$, then (writing $\zeta = \xi + i\eta$)*

$$\int_{\mathfrak{H}} g_s(z,\zeta)\left[-\eta^2\left(\frac{\partial^2}{\partial\xi^2} + \frac{\partial^2}{\partial\eta^2}\right) + s\right] f(\zeta)\,\frac{d\xi \wedge d\eta}{\eta^2} = f(z) \qquad (\zeta = \xi + i\eta).$$

Proof. See Bump [6, Proposition 2.3.4, p. 181] for more complete details. We review the proof quickly. Let $w = (z-\zeta)/(z-\overline{\zeta}) = u + iv \in \mathfrak{D}$. Let $F : \mathfrak{D} \to \mathbb{C}$ be defined by $F(w) = f(z)$. In the w coordinates we must prove

$$\int_{\mathfrak{D}} g_s(|w|)\left[-\left(\frac{\partial^2}{\partial u^2} + \frac{\partial^2}{\partial v^2}\right) + \frac{4s}{(1-|w|^2)^2}\right] F(w)\,du \wedge dv = F(0).$$

Let B_r be the disk of radius r, and let $R < 1$ be large enough that the support of F is contained in B_R. The left side equals

$$\lim_{\epsilon\to 0}\int_{B_R - B_\epsilon} g_s(|w|)\left[-\left(\frac{\partial^2}{\partial u^2} + \frac{\partial^2}{\partial v^2}\right) + \frac{4s}{(1-|w|^2)^2}\right] F(w)\,du \wedge dv.$$

Using Stokes's theorem as in Proposition 2, this is

$$\lim_{\epsilon\to 0}\int_{B_R - B_\epsilon} F(w)\left[-\left(\frac{\partial^2}{\partial u^2} + \frac{\partial^2}{\partial v^2}\right) + \frac{4s}{(1-|w|^2)^2}\right] g_s(|w|)\,du \wedge dv$$
$$+ \lim_{\epsilon\to 0}\int_{C_\epsilon} F(w)\left(\frac{\partial g_s(|w|)}{\partial u}\,dv - \frac{\partial g_s(|w|)}{\partial v}\,du\right)$$
$$- \lim_{\epsilon\to 0}\int_{C_\epsilon} g_s(|w|)\left(\frac{\partial F(w)}{\partial u}\,dv - \frac{\partial F(w)}{\partial v}\,du\right),$$

where C_ϵ is the path circling the origin counterclockwise around the circle with radius ϵ. (There would also be terms integrating around C_R, but these are zero because they lie outside the support of F.)

The first term vanishes by Theorem 13. The last term vanishes because the length of the arc shrinks faster than g_s blows up (logarithmically).

To evaluate the middle term let $w = re^{i\theta}$. By the chain rule,

$$\frac{\partial g_s(|w|)}{\partial u}\,dv - \frac{\partial g_s(|w|)}{\partial v}\,du = r\,g_s'(r)\,d\theta.$$

We obtain

$$\lim_{\epsilon\to 0}\int_0^{2\pi} F(\epsilon e^{i\theta})\,d\theta\,\epsilon\,g_s'(\epsilon) = F(0),$$

since $g_s'(\epsilon) \sim 1/(2\pi\epsilon)$ as $\epsilon \to 0$. □

Proposition 15. *The series*

$$G_s(z,\zeta) = \sum_{\gamma\in\{\pm 1\}\backslash\Gamma} g_s\big(z,\gamma(\zeta)\big) = \sum_{\gamma\in\{\pm 1\}\backslash\Gamma} g_s\big(\gamma(z),\zeta\big).$$

is absolutely convergent provided z and ζ are not Γ equivalent.

See Bump [6, Proposition 2.3.5, p. 184]. □

$G_s(z,\zeta)$ is the *automorphic Green's function*. We will see that it is an integral kernel for the resolvent of the Laplacian.

Theorem 16. *G_s is defined and real analytic for all values of (z,ζ) except where $\zeta = \gamma(z)$ for some $\gamma \in \Gamma$. We have*

$$\left[-y^2\left(\frac{\partial^2}{\partial x^2} + \frac{\partial^2}{\partial y^2}\right) + s\right] G_s(z,\zeta) = 0;$$

$$G_s\big(h(z),h(\zeta)\big) = G_s(z,\zeta), \qquad h \in G,$$

$$G_s(z,\zeta) = G_s\big(\gamma(z),\zeta\big) = G_s\big(z,\gamma(\zeta)\big), \qquad \gamma\in\Gamma,$$

$$G_s(z,\zeta) = \overline{G_s(\zeta,z)},$$

and $G_s(z,\zeta) \sim \frac{e}{2\pi}\log|z-\zeta|$ near $z=\zeta$, where e is the order of the isotropy subgroup of ζ in Γ. For $f \in C^\infty(\Gamma\backslash\mathfrak{H})$

$$\int_{\Gamma\backslash\mathfrak{H}} G_s(z,\zeta)\left[-\eta^2\left(\frac{\partial^2}{\partial \xi^2} + \frac{\partial^2}{\partial \eta^2}\right) + s\right] f(\zeta)\,\frac{d\xi\wedge d\eta}{\eta^2} = f(z) \qquad (\zeta = \xi + i\eta). \tag{9}$$

See Bump [6, Proposition 2.3.5] and Hejhal [21, Proposition 6.5, p. 33].

Proof. Most of these properties follow from the corresponding properties of g_s. We prove (9). We will need a function $F \in C_c^\infty(\mathfrak{H})$ such that $f(z) = \sum_{\gamma\in\Gamma} F(\gamma z)$. To construct F, let u be a function on $\mathfrak{H}$ which is smooth, nonnegative, and has compact support containing a fundamental domain of Γ. Then for all z, the function $\sum_{\gamma\in\Gamma} u(\gamma z)$ is positive, and for z restricted to a compact set this sum is finite. It is thus a smooth, positive valued function and we can divide by it. Now

$$F(z) = \frac{u(z)\,f(z)}{\sum_{\gamma\in\Gamma} u(\gamma z)}$$

has the required property.

Substituting this and the definition of G_s and using (8) gives

$$\sum_{\gamma,\delta\in\Gamma} \int_{\Gamma\backslash\mathfrak{H}} g_s\big(\delta(z),\gamma(\zeta)\big)\left[-\eta^2\left(\frac{\partial^2}{\partial \xi^2} + \frac{\partial^2}{\partial \eta^2}\right) + s\right] F(\gamma(\zeta))\,\frac{d\xi\wedge d\eta}{\eta^2}.$$

One of the summations may be collapsed with the integration to give

$$\sum_{\gamma\in\Gamma}\int_{\mathfrak{H}} g_s(\delta(z),\zeta)\left[-\eta^2\left(\frac{\partial^2}{\partial\xi^2}+\frac{\partial^2}{\partial\eta^2}\right)+s\right]F(\zeta)\,\frac{d\xi\wedge d\eta}{\eta^2}=\sum_{\delta}F(\delta(z))$$
$$= f(z).$$

This completes the proof. □

Let $s > 0$ be a positive constant. As is easily checked, the logarithmic singularity along the diagonal is not sufficient to cause divergence of the integral

$$\int_{\Gamma\backslash\mathfrak{H}}\int_{\Gamma\backslash\mathfrak{H}} |G_s(z,\zeta)|^2\frac{dx\wedge dy}{y^2}\,\frac{d\xi\wedge d\eta}{\eta^2}<\infty.$$

Thus the corresponding integral operator, which we shall denote $R(s,\Delta)$, is Hilbert–Schmidt. It is the *resolvent* of the Laplacian. If ϕ is an eigenfunction of Δ with eigenvalue λ, then it follows from (9) that it is also an eigenfunction of $R(s,\Delta)$ with eigenvalue $(\lambda+s)^{-1}$.

Theorem 17.

(i) *The eigenvalues* λ_i *of* Δ *on* $L^2(\Gamma\backslash\mathfrak{H})$ *tend to* ∞, *and satisfy* $\sum\lambda_i^{-2}<\infty$. *(We exclude the eigenvalue* $\lambda_0=0$ *corresponding to the constant function from this summation.)*

(ii) *The Laplacian* Δ *has an extension to a self-adjoint operator on the Hilbert space* $L^2(\Gamma\backslash\mathfrak{H})$.

Proof. Assume $\mathrm{re}(s)>0$. Since $R(s,\Delta)$ is Hilbert–Schmidt, $\sum_i(\lambda_i+s)^{-2}<\infty$, whence $\sum_i\lambda_i^{-2}<\infty$. By Theorem 9, let ϕ_i be a basis of $H=L^2(\Gamma\backslash\mathfrak{H})$ consisting of eigenvectors of Δ, with corresponding eigenvalues λ_i.

We prove (ii). Let $\mathfrak{D}_\Delta$ be the linear subspace of $\mathfrak{H}$ consisting of elements of the form $\sum a_i\phi_i$ such that $\sum\lambda_i^2|a_i|^2<\infty$; on this space, define

$$\Delta\Big(\sum a_i\phi_i\Big)=\sum\lambda_i\,a_i\,\phi_i.$$

Since the λ_i tend to infinity, and in particular are bounded away from zero, it is not hard to check that this operator is closed and in fact self-adjoint. This completes the proof. □

We arrange the eigenvalues of Δ in ascending order:

$$\lambda_0=0<\lambda_1\le\lambda_2\le\lambda_3\le\ldots.$$

The eigenvalue λ_0 corresponds to the constant function and has multiplicity exactly one. Eigenvalues in the range $(0,\frac{1}{4})$ are called *exceptional eigenvalues*. They are qualitatively different. For example, the spherical functions corresponding to exceptional eigenvalues are nontempered—they grow faster than spherical functions corresponding to $\lambda \geq \frac{1}{4}$. (See Section 4). Exceptional eigenvalues correspond to zeros of the Selberg zeta function on the real line between 0 and 1. By contrast zeros corresponding to $\lambda \geq \frac{1}{4}$ satisfy the Riemann hypothesis. (See Section 7.) Randol [39] proved that for some X, exceptional eigenvalues do occur. On the other hand, Selberg [44] conjectured that exceptional eigenvalues do not occur in the cuspidal spectrum of congruence subgroups of $SL(2, \mathbb{Z})$, and it would follow from this that they do not occur in the spectrum of compact quotients $\Gamma\backslash\mathfrak{H}$ associated to quaternion division algebras.

4 Spherical functions

As we pointed out, Property S reduces to a second order differential equation which has two independent solutions. One solution, having nice behavior at the boundary, is the Green's function. Another, having nice behavior at i, is the so-called spherical function. The substance of Lemma 12 is that these two solutions are not the same. In this section we will study the spherical function.

Definition. Let λ be a complex number. We call a function σ on $SL(2, \mathbb{R})$ a λ-*spherical function* if it is smooth, K-biinvariant, and $\Delta\sigma = \lambda\sigma$.

Here Δ is the Laplace–Beltrami operator. Before we show that such a function exists and is unique up to constant multiple, let us explain briefly how such a function fits into the representation theory.

Suppose that (π, V) is an irreducible representation of G. Let $(\hat{\pi}, \hat{V})$ be its contragredient. Thus there exists an invariant bilinear pairing $\langle\,,\rangle : V \times \hat{V} \to \mathbb{C}$.

By Theorem 4, if V^K and $\hat{V}^K$ are nonzero, they are one dimensional, and we will assume this. (Actually if one is nonzero the other is too.) Let $v^\circ \in V^K$ and $\hat{v}^\circ \in \hat{V}^K$ be nonzero K-fixed vectors. The function

$$\sigma(g) = \langle \pi(g)\, v^\circ, \hat{v}^\circ \rangle$$

is evidently K-biinvariant. Moreover, Regarding Δ as an element of the center of the universal enveloping algebra $U(\mathfrak{g})$, it acts by a scalar λ on V. and σ inherits this property. So it is a spherical function.

Without reference to this construction, we now show that spherical functions exist and are unique.

Theorem 18.

(i) *Let $\lambda \in \mathbb{C}$. Then there is a unique smooth K-biinvariant function ω_λ on $SL(2, \mathbb{R})$ such that $\Delta\omega_\lambda = \lambda\omega_\lambda$ and $\omega_\lambda(1) = 1$.*

(ii) *If $f : G \to \mathbb{C}$ is any smooth function such that $\Delta f = \lambda f$, then*

$$\int_{K\times K} f(kgk')\, dk\, dk' = f(1)\, \omega_\lambda(g). \tag{10}$$

(iii) *If f is right K-invariant and $\Delta f = \lambda f$, then*

$$\int_K f(hkg)\, dk = f(h)\, \omega_\lambda(g). \tag{11}$$

Proof. To satisfy $\Delta\omega = \lambda\omega$ we need

$$W(r) = \omega\begin{pmatrix} y^{1/2} & \\ & y^{-1/2} \end{pmatrix}, \qquad r = \frac{y-1}{y+1}$$

to satisfy (7). As we have seen, this differential equation has a regular singular point at the origin, and one solution is bounded there, whereas the other has a logarithmic singularity. Hence ω_λ, if it exists, is unique.

To show that such a function exists, let f be any continuous function on G which is an eigenfunction of Δ. The left-hand side of (10) is a K-biinvariant function which is an eigenfunction of Δ. If $f(1) = 1$, this will satisfy (i), proving existence. For example, we can take $f = f_s$ where $\lambda = s(1-s)$ and

$$f_s\left(\begin{pmatrix} y^{1/2} & * \\ & y^{-1/2} \end{pmatrix} k\right) = y^s, \qquad y > 0,\ k \in K. \tag{12}$$

Now that the existence and uniqueness of ω are established, we note that the left sides of both (10) and (11) are smooth K-biinvariant eigenfunctions of Δ, hence constant multiples of ω_λ. In both cases, the constant may be evaluated by taking $g = 1$. This proves both (ii) and (iii). □

Theorem 19. *Suppose that f is a smooth function on G which is right invariant by K and such that $\Delta f = \lambda f$. Then for $\phi \in \mathcal{H}^\circ$ we have $T_\phi f = \chi_\lambda(\phi) f$, where*

$$\chi_\lambda(\phi) = \int_G \phi(g)\,\omega_\lambda(g)\,dg. \tag{13}$$

Proof. By Theorem 18, we have

$$\int_K f(hkg)\,dk = f(h)\,\omega_\lambda(g). \tag{14}$$

We note that $T_\phi f$ is an average of right translates of f, and right translation commutes with left translation. Hence we may apply T_ϕ to both sides of (14) to obtain

$$\int_K (T_\phi f)(hkg)\,dk = f(h)\,(T_\phi\omega_\lambda)(g).$$

We take $g = 1$ in this identity. Since $T_\phi f$ is right K-invariant, the integrand on the left side becomes constant when $g = 1$ and so the left side becomes just $(T_\phi f)(h)$. On the other hand, $(T_\phi\omega_\lambda)(1)$ equals the integral (13), so $T_\phi f(h) = \chi_\lambda(\phi) f(h)$. □

Theorem 20. *If ϕ_1 and $\phi_2 \in \mathcal{H}^\circ$, then*

$$\chi_\lambda(\phi_1 * \phi_2) = \chi_\lambda(\phi_1)\,\chi_\lambda(\phi_2).$$

Proof. This follows from Theorem 19 on applying T_ϕ to any eigenfunction f, for example, f_s as in (12). □

Thus the function $\chi_\lambda : \mathcal{H}^\circ \to \mathbb{C}$ is a *character* of $\mathcal{H}^\circ$. We return to the point of view introduced at the beginning of the section to explain the meaning of $\chi(\phi)$ in terms of representations. First, we recall the construction and parametrization of a class of irreducible representations of $SL(2,\mathbb{R})$, the *principal series*.

Let s be a complex number. Let P_s^+ be the set of functions $f : G \to \mathbb{C}$ such that

$$f\left(\begin{pmatrix} y^{1/2} & xy^{-1/2} \\ & y^{-1/2} \end{pmatrix} g\right) = (-1)^{\epsilon\,\mathrm{sgn}(y)}\, y^s f(g), \tag{15}$$

where $\epsilon = 0$, and such that the restriction of f to K is square integrable. Similarly let P_s^- be the space of functions satisfying (15) with $\epsilon = 1$. G acts on these spaces by right translation. Let $(\pi_s^\pm, P_s^\pm)$ be these representations. The representation $\pi_s^\pm$ is irreducible except in the case where $2s$ is an integer and $2s$ is even for π_s^+, odd for π_s^-. We call π_s^+ and π_s^- the *odd* and *even principal series* of representations, respectively.

Suppose that $\mathrm{re}(s) > \frac{1}{2}$. An *intertwining integral* $M(s) : P_s^\pm \to P_{1-s}^\pm$ is given by

$$\big(M(s)f\big)(g) = \int_{-\infty}^{\infty} f\left(\begin{pmatrix} & -1 \\ 1 & \end{pmatrix}\begin{pmatrix} 1 & x \\ & 1 \end{pmatrix} g\right) dx. \tag{16}$$

It may be checked that the integral is absolutely convergent. Now $M(s)$ extends by analytic continuation to an intertwining map for all s such that $2s$ is not an integer congruent to ϵ mod 2. Actually we only claim that it extends to an infinitesimal equivalence, that is, an isomorphism of the underlying $(\mathfrak{g}, K)$ module, but it is a true isomorphism in the important special case $\mathrm{re}(s) = \frac{1}{2}$. Thus $\pi_{1/2+it}^\pm \cong \pi_{1/2-it}^\pm$.

Let $k \equiv \epsilon$ modulo 2. Let

$$f_{s,k}\left(\begin{pmatrix} y^{1/2} & x \\ & y^{-1/2} \end{pmatrix} \kappa_\theta\right) = \mathrm{sgn}(y)^\epsilon\, y^s\, e^{ik\theta}. \tag{17}$$

This is a K-finite vector in $P_s^\pm$ where $\pm$ is $(-1)^\epsilon$. Let $\tilde{f}_{1-s,k} = M(s)\, f_{s,k}$. One may compute

$$\tilde{f}_{k,1-s} = (-i)^k \sqrt{\pi}\, \frac{\Gamma(s)\,\Gamma\left(s-\frac{1}{2}\right)}{\Gamma\left(s+\frac{k}{2}\right)\Gamma\left(s-\frac{k}{2}\right)}\, f_{k,1-s}. \tag{18}$$

See Bump [6, Proposition 2.6.3].

The representation $\pi_{1/2+it}^\pm$ is unitary if t is real. These representations are called the *unitary principal series* representations. Also π_s^+ is unitary if s is a real number between 0 and 1. These are the *complementary series* representations. They correspond to exceptional eigenvalues of the Laplacian in the automorphic spectrum. There are a few other irreducible representations, namely the discrete series (related to holomorphic modular forms) and the trivial representations. But only the even principal series π_s^+ have K-fixed vectors. All of these facts are proved in Bump [6, Chapter 2].

The K-fixed vector in P_s^+ is precisely the function f_s in (12). Now if $\phi \in \mathcal{H}^\circ$, then since ϕ is K-biinvariant, the operator $\pi^+(\phi)$ maps all of P_s^+ onto the unique K-fixed vector f_s. Thus the trace of the rank one operator $\pi_s^+(\phi)$ is just its eigenvalue on this vector f_s, so

$$\operatorname{tr} \pi_s^+(\phi) = \chi_\lambda(\phi). \tag{19}$$

5 The Plancherel formula

Let $\phi \in \mathcal{H}^\circ$. Define

$$g(u) = e^{u/2} \int_{-\infty}^{\infty} \phi\left(\begin{pmatrix} e^{u/2} & \\ & e^{-u/2} \end{pmatrix}\begin{pmatrix} 1 & x \\ & 1 \end{pmatrix}\right) dx, \tag{20}$$

and let

$$h(t) = \int g(u)\, e^{iut}\, dt \tag{21}$$

be its Fourier transform.

Theorem 21. *The functions g and h are even, and g is compactly supported. If $\lambda = \frac{1}{4} + t^2$, then*

$$\chi_\lambda(\phi) = h(t). \tag{22}$$

Proof. Let f_s be as in (12) with $s = \frac{1}{2} + it$ so that $\lambda = \frac{1}{4} + t^2 = s(1-s)$. By Theorem 18,

$$\int_K f_s(kg)\, dk = \omega_\lambda(g),$$

and

$$\chi_\lambda(\phi) = \int_G \int_K \phi(g)\, f_s(gk)\, dk\, dg.$$

Interchanging the order of integration and making the variable change $g \to gk^{-1}$, since ϕ is right k invariant, we obtain

$$\chi_\lambda(\phi) = \int_G \phi(g)\, f_s(g)\, dg.$$

Now we use the coordinates

$$g = \begin{pmatrix} e^{u/2} & \\ & e^{-u/2} \end{pmatrix}\begin{pmatrix} 1 & x \\ & 1 \end{pmatrix} \kappa_\theta, \qquad dg = \frac{1}{2\pi}\, du\, dx\, d\theta. \tag{23}$$

Noting that in these coordinates $f_s(g) = e^{u/2}\, e^{iut}$, we obtain

$$\chi_\lambda(\phi) = \int_{\infty}^{\infty} g(u)\, e^{itu}\, du,$$

proving (22). We note that ω_λ, and therefore the character $\chi_\lambda(\phi)$ is unchanged if $s \to 1 - s$, that is, if $t \to -t$. Hence (22) implies that h is an even function. By Fourier inversion, so is g. □

Theorem 22.
(i) *We have*

$$\phi(1) = \frac{1}{2\pi} \int_0^\infty t\, h(t) \tanh(\pi t)\, dt.$$

(ii) *If* ϕ_1, ϕ_2 *are* K*-invariant and compactly supported, then*

$$\int_G \phi_1(g)\, \overline{\phi_2(g)}\, dg = \frac{1}{2\pi} \int_0^\infty t\, h_{\phi_1}(t)\, \overline{h_{\phi_2}(t)} \tanh(\pi t)\, dt.$$

Proofs may be found in Knapp [27, Chapter 11], Gelfand, Graev, and Piatetski-Shapiro [16, Chapter 2, Section 6], and Varadarajan [47, Theorem 39, p. 205]. We will give a proof (after some Lemmas) which uses no Lie theory. A portion of the argument parallels of Hejhal [20, Proposition 4.1, p. 15], and we have made our notation consistent with his.

Theorem 22 (i) is the Fourier inversion formula on the noncommutative group $SL(2,\mathbb{R})$. It is sometimes called the *Plancherel formula* because it implies (ii), which is the true Plancherel formula. The measure $\frac{1}{2\pi} t \tanh(\pi t)\, dt$ is called the *Plancherel measure* on the even unitary principal series. (Since we have only considered $\phi \in \mathcal{H}^\circ$ we do not need the other irreducible unitary representations.) The Plancherel measure is closely related to the intertwining integrals (16). Indeed, $M(\frac{1}{2} - it) \circ M(\frac{1}{2} + it)$ is a G-equivariant endomorphism of the irreducible space $P^+_{1/2+it}$, so by Schur's Lemma it is a scalar. One may check using (18) that the reciprocal of this scalar is $\frac{1}{\pi} t \tanh(\pi t)$. See Knapp and Stein [28].

Lemma 23. *Let* $\eta \geq y > 1$. *Then*

$$\phi\left(\begin{pmatrix} 1 & x \\ & 1 \end{pmatrix} \begin{pmatrix} y^{1/2} & \\ & y^{-1/2} \end{pmatrix}\right) = \phi\begin{pmatrix} \eta^{1/2} & \\ & \eta^{-1/2} \end{pmatrix},$$

where

$$x = \pm\sqrt{y(\eta + \eta^{-1} - y - y^{-1})}.$$

Proof. Identifying $K\backslash G/K$ with $K\backslash\mathcal{H}$, we want to find x such that the images ηi and $z = x + iy$ of the two matrices in $\mathcal{H}$ are in the same $SO(2)$ orbit. Mapping $\mathcal{H}$ to the unit disk by the Cayley transform $z \to (z - i)/(z + i)$, their images must therefore be equidistant from the origin. That is,

$$\frac{\eta - 1}{\eta + 1} = \left|\frac{z - i}{z + i}\right| = \sqrt{\frac{a - 2y}{a + 2y}}, \qquad a = x^2 + y^2 + 1.$$

Applying the map $t \to (1 + t^2)/(1 - t^2)$ to both sides of this equation, $\eta + \eta^{-1} = a/y$. This equation can now be solved for x. □

Define a function Φ on $\mathbb{R}^+$ by

$$\Phi(e^v + e^{-v} - 2) = \phi\begin{pmatrix} e^{v/2} & \\ & e^{-v/2} \end{pmatrix}. \tag{24}$$

Lemma 24. *If $U = e^u + e^{-u} - 2$, then $g(u) = Q(U)$, where*

$$Q(U) = \int_U^\infty \frac{\Phi(V)\,dV}{\sqrt{V-U}}. \tag{25}$$

Proof. Let $V = e^v + e^{-v} - 2$. With $y = e^u$, $\eta = e^v$ and x as in Lemma 23, think of x as function of $V \geq U$. Then $dx = \frac{1}{2}\,e^u\,(V-U)^{-1/2}\,dV$. We integrate V from U to infinity and double the result to account for both positive and negative x. □

Lemma 25. *We have*

$$\Phi(U) = -\frac{1}{\pi}\int_U^\infty \frac{Q'(V)\,dV}{\sqrt{V-U}}. \tag{26}$$

Proof. We would like to integrate under the integral sign in (25) but since the left endpoint depends on U we must be careful. Integrate (25) by parts to obtain

$$Q(U) = -2\int_U^\infty \Phi'(V)\,\sqrt{V-U}\,dV.$$

The integrand now vanishes at the left endpoint, so we may differentiate under the integral sign, then integrate by parts again to obtain

$$Q'(U) = \int_U^\infty \frac{\Phi'(V)}{\sqrt{V-U}}\,dV = -2\int_U^\infty \Phi''(V)\,\sqrt{V-U}\,dV.$$

We substitute this into the right side of (26), then switch the order of integration:

$$\begin{aligned} -\frac{1}{\pi}\int_U^\infty \frac{Q'(V)\,dV}{\sqrt{V-U}} &= \frac{2}{\pi}\int_U^\infty\int_V^\infty \sqrt{\frac{W-V}{V-U}}\,\Phi''(W)\,dW\,dV \\ &= \frac{2}{\pi}\int_U^\infty\int_U^W \sqrt{\frac{W-V}{V-U}}\,dV\,\Phi''(W)\,dW. \end{aligned} \tag{27}$$

To evaluate the inner integral, let $V = U + (W-U)\,v$. We have

$$\int_U^W \sqrt{\frac{W-V}{V-U}}\,dV = (W-U)\int_0^1 \sqrt{\frac{1-v}{v}}\,dv = \frac{\pi}{2}(W-U)$$

since $\int_0^1 v^{-1/2}(1-v)^{1/2}\,dv = B(\frac{1}{2},\frac{3}{2}) = \frac{\pi}{2}$. Thus (27) equals

$$\int_U^\infty (U-W)\,\Phi''(W)\,dW = -\int_U^\infty \Phi'(W)\,dW = \Phi(U),$$

where we have integrated twice by parts. □

Proof of Theorem 22. Take $U = 0$ and write Lemma 25 in the form

$$\phi(1) = \Phi(0) = -\frac{1}{\pi}\int_0^\infty \frac{g'(u)\,du}{e^{u/2} - e^{-u/2}}. \tag{28}$$

Since g is even and g' is odd, we have the Fourier inversion formula

$$g(u) = \frac{1}{\pi}\int_0^\infty h(t)\,e^{-itu}\,dt, \qquad g'(u) = \frac{1}{i\pi}\int_0^\infty t\,h(t)\,e^{-itu}\,dt,$$

and we may change the limits in (28) to $(-\infty, \infty)$, dividing by 2, then interchange the order of integration to obtain

$$\phi(1) = \frac{1}{2\pi}\int_0^\infty t\,h(t)\,\frac{i}{\pi}\int_{-\infty}^\infty \frac{e^{-iut}\,du}{e^{u/2} - e^{-u/2}}\,dt.$$

The inner integral passes through the pole at $u = 0$ of the integrand and is interpreted as the principal value. The Plancherel formula now follows from

$$\int_{-\infty}^\infty \frac{e^{-iut}\,du}{e^{u/2} - e^{-u/2}} = -i\pi\tanh(\pi t), \tag{29}$$

which we may prove as follows. Since $t > 0$, the numerator e^{-iut} is small for u in the lower half plane, and we may move the path of integration downwards. The left side of (29) $-2\pi i$ times the sum of the residues at $u = -2\pi i n$ of the integrand in the lower half plane. The residue at $u = 0$ is only counted half since the path of integration passes through this point. We get

$$-2\pi i\left(\frac{1}{2} + \sum_{k=1}^\infty (-1)^k e^{-2\pi k t}\right) = -\pi i\,\tanh(\pi t)$$

proving (i).

To prove (ii) define $\phi_2'(g) = \overline{\phi_2(g^{-1})}$. Since ϕ_2 is K-biinvariant, and since the K-double cosets are stable under $g \to g^{-1}$ this equals $\overline{\phi_2(g)}$. Now apply (i) to $\phi = \phi_1 * \phi_2'$, then $\phi(0) = \langle\phi_1, \phi_2\rangle_2$. On the other hand, $h_\phi(t) = h_{\phi_1}(t)\,\overline{h_{\phi_2}(t)}$ by Theorem 20, so (ii) follows. □

Proposition 26. *Let g be an even, compactly supported function on $\mathbb{R}$. There exists $\phi \in \mathcal{H}^\circ$ such that (20) is true.*

Proof. Let $Q : \mathbb{R}^+ \to \mathbb{C}$ be defined by $Q(e^u + e^{-u} - 2) = g(u)$. We define $\Phi : \mathbb{R}^+ \to \mathbb{C}$ by (26) or, integrating by parts and substituting $W = V - U$,

$$\Phi(U) = \frac{2}{\pi}\int_0^\infty Q''(W + U)\sqrt{W}\,dW.$$

We may differentiate under the integral sign arbitrarily many times, so Φ is smooth, even at $U = 0$. Now $u \to \Phi(e^u + e^{-u} - 2)$ is a smooth, even, compactly supported function, and it follows that there is a unique smooth, compactly supported K-biinvariant ϕ satisfying (24). Now the statement follows from Lemma 24. □

6 The Selberg trace formula

We have already shown that the integral operators T_ϕ are Hilbert–Schmidt, hence compact. More is true: they are *trace class*. A compact operator is *trace class* if it can be factored as the composite of two Hilbert–Schmidt operators. If it is self-adjoint, and has eigenvalues λ_i, it is easy to see that this is equivalent to $\sum |\lambda_i| < \infty$. Lang's [30] contains much useful material about trace class operators.

Let f_i be a basis of $L^2(\Gamma\backslash\mathfrak{H})$ consisting of eigenfunctions of T_ϕ that are also eigenfunctions of Δ. Assuming that $\phi \in \mathcal{H}^\circ$ satisfies (6), let μ_i be the eigenvalues of T_ϕ with respect to this basis. Making a Fourier expansion we have

$$K_\phi(z, w) = \sum \mu_i f_i(z) \overline{f_i(w)}. \tag{30}$$

Theorem 27. *If $\phi \in \mathcal{H}^\circ$, then T_ϕ is trace class.*

Proof. A linear combination of trace class operators is trace class. Hence it is sufficient to prove this with ϕ replaced by $\frac{1}{2}\big(\phi(g) + \overline{\phi(g^{-1})}\big)$ and by $\frac{1}{2i}\big(\phi(g) - \overline{\phi(g^{-1})}\big)$. We may thus assume that ϕ satisfies (6) and so T_ϕ is self-adjoint. Let μ_i be its (nonzero) eigenvalues. Let λ_i be the corresponding eigenvalues of Δ. Thus $\sum \lambda_i^{-2} < \infty$.

Applying Δ to $K_\phi(z, w)$ in the first variable gives a new kernel $\Delta_z K_\phi$. We will show that

$$(\Delta_z K_\phi)(z, w) = \sum \mu_i \lambda_i f_i(x) \overline{f_i(y)}. \tag{31}$$

Formally, this follows from (30) by termwise differentiation. However, this must be justified since the convergence in (30) is in the L^2 sense, and Δ is an unbounded operator. Note that $\Delta_x K_\phi(x, y)$ is continuous, hence has an expansion

$$(\Delta_z K_\phi)(z, w) = \sum \nu_i f_i(x) \overline{f_i(y)}.$$

Let $s > 0$. Consider

$$\int_{\Gamma\backslash\mathfrak{H}} G_s(z, \zeta) \left[(\Delta_\zeta K_\phi)(\zeta, w) + s K_\phi(\zeta, w)\right] \frac{d\xi \wedge d\eta}{\eta^2}.$$

On the one hand, by (9) this is just (30). On the other hand, the term in square brackets is $\sum_i (\nu_i + s\mu_i) f_i(\zeta) \overline{f_i(w)}$, and since the resolvent is compact, we may apply it term by term. Using $R(s, \Delta) f_i = (s + \lambda_i)^{-1} f_i$ we get $(\nu_i + s\mu_i)(\lambda_i + s)^{-1} = \mu_i$, so $\nu_i = \lambda_i \mu_i$ proving (31).

Since this function $\Delta_z K(z, w)$ is continuous, it is Hilbert–Schmidt, and so we obtain the bound

$$\sum |\mu_i \lambda_i|^2 < \infty. \tag{32}$$

Now $\sum |\mu_i| < \infty$ follows from $\sum |\lambda_i|^{-2} < \infty$ and (32) by Cauchy–Schwarz. □

The trace $\operatorname{tr} T$ of a self-adjoint trace class operator T is by definition the sum of its eigenvalues.

Theorem 28. *If $\phi \in \mathcal{H}^\circ$ satisfies (6), and if μ_i are the eigenvalues of T_ϕ, the trace*

$$\operatorname{tr} T_\phi = \int_{\Gamma\backslash\mathfrak{H}} K_\phi(z,z)\,\frac{dx\wedge dy}{y^2}. \tag{33}$$

Proof. This follows from orthonormality on integrating (30). □

The Selberg trace formula is a more explicit formula for its trace. Let $\{\gamma\}$ denote a set of representatives for the conjugacy classes of Γ. Let $Z_\Gamma(\gamma)$ denote the centralizer in Γ of γ.

Theorem 29. *We have*

$$\operatorname{tr} T_\phi = \sum_{\{\gamma\}} \int_{Z_\Gamma(\gamma)\backslash G} \phi(g^{-1}\gamma g)\,dg. \tag{34}$$

Proof. We rewrite the right side of (33) as

$$\sum_{\gamma\in\Gamma}\int_G \phi(g^{-1}\gamma g)\,dg = \sum_{\{\gamma\}}\sum_{\delta\in Z_\Gamma\backslash\Gamma}\int_G \phi(g^{-1}\delta^{-1}\gamma\,\delta\,g)\,dg.$$

Combining the integral and the summation gives (34). □

This is a primitive form of the trace formula. To make this more explicit, we must study more explicitly the *orbital integrals* on the right side. An element $1 \neq \gamma \in \Gamma$ is *hyperbolic* if its eigenvalues are real, *elliptic* if complex of absolute value 1. Γ is *hyperbolic* if each $1 \neq \gamma \in \Gamma$ is hyperbolic. For example, let X be a compact Riemann surface of genus ≥ 2. Its universal cover is $\mathfrak{H}$ and $\Gamma = \pi_1(X)$ acts with quotient X. These examples are precisely the hyperbolic groups.

We assume now that Γ is a hyperbolic group. If $1 \neq \gamma \in \Gamma$, define $N = N(\gamma)$ by asking that γ be conjugate to

$$\begin{pmatrix} N^{1/2} & \\ & N^{-1/2}\end{pmatrix} \tag{35}$$

for some N. Let $N_0 = N_0(\gamma)$ be such that

$$\begin{pmatrix} N_0^{1/2} & \\ & N_0^{-1/2}\end{pmatrix}$$

is conjugate to a generator of $Z_\Gamma(\gamma)$. We may obviously assume that N and N_0 are > 1. We note that $Z_G(\gamma)$ is conjugate to the diagonal subgroup. Its image in X is a closed geodesic. So the numbers $N(\gamma)$ are thus the lengths of closed geodesics, and the numbers $N_0(\gamma)$ are the lengths of prime geodesics.

Theorem 30. *With g the function in (20),*

$$\int_{Z_\Gamma(\gamma)\backslash G} \phi(g^{-1}\gamma g)\,dg = \frac{\log N_0}{N^{1/2}-N^{-1/2}}\,g(\log N). \tag{36}$$

Proof. We may assume that γ equals (35). Using Iwasawa coordinates (23) the integral is

$$\int_0^{\log N_0} du \int_{-\infty}^{\infty} \phi\left(\begin{pmatrix}1 & -x\\ & 1\end{pmatrix}\begin{pmatrix}N^{1/2} & \\ & N^{-1/2}\end{pmatrix}\begin{pmatrix}1 & x\\ & 1\end{pmatrix}\right) dx =$$
$$|\log(N_0)| \int_{-\infty}^{\infty} \phi\left(\begin{pmatrix}N^{1/2} & \\ & N^{-1/2}\end{pmatrix}\begin{pmatrix}1 & (1-N^{-1})x\\ & 1\end{pmatrix}\right) dx,$$

and a change of variables proves (36). □

Theorem 31. *Let g be a smooth, even, compactly supported function and let h be its Fourier transform, defined by (21). If* $\frac{1}{4}+t_i^2$ *are the eigenvalues of* Δ *on* $\mathfrak{H}$, *and if* $\log(N)$ *runs through the lengths of closed geodesics of* Γ, *where for each N we let* N_0 *be the length of the corresponding prime geodesic, we have*

$$\sum h(t_i) = \frac{\operatorname{vol}(\Gamma\backslash\mathfrak{H})}{4\pi}\int_{-\infty}^{\infty} t\,h(t)\,\tanh(\pi t)\,dt + \sum_N \frac{\log N_0}{N^{1/2}-N^{-1/2}}\,g(\log N). \tag{37}$$

This is the *Selberg trace formula.*

Proof. We choose ϕ as in Proposition 26. By Theorem 19 and Theorem 21, the $h(t_i)$ are the eigenvalues of T_ϕ on the eigenfunctions of the Laplacian, so the left side of (37) is the trace of T_ϕ. By Theorem 22 and Theorem 30, the right side of (37) is the sum of the orbital integrals. Thus the identity follows from Theorem 29. □

7 The Selberg zeta function

In order to get useful applications the class of functions g and h in the trace formula must be expanded.

Theorem 32. *The trace formula Theorem* 31 *remains true, provided h is an even function analytic in the strip* $\operatorname{im}(z) \le \frac{1}{2}+\delta$, *such that* $h(r) = O(1+|r|)^{-2-\delta}$ *in this strip. This assumption implies that the Fourier transform* $g(u) = O(e^{-(\frac{1}{2}+\delta)|u|})$.

See Hejhal [20, Chapter 1, Section 7] for the proof by an approximation argument that Theorem 31 implies this stronger statement. □

Theorem 33 (Weyl's law). *The number of j with* $t_j \le x$ *is asymptotically* $\frac{1}{4\pi}\operatorname{vol}(\Gamma\backslash\mathfrak{H})\,x^2$.

This must be mentioned as the first significant application of the trace formula. Noting that $\lambda_j = \frac{1}{4}+t_j^2$, this means that the number of $\lambda_j \le x$ is asymptotically $\frac{1}{4\pi}\operatorname{vol}(\Gamma\backslash\mathfrak{H})\,x$.

We will not prove Weyl's law. Briefly, taking $h(t) = e^{-t^2T}$, the hyperbolic contributions in Theorem 31 are of smaller magnitude and the first term in (37) predominates. One obtains $\sum e^{-\lambda_j T} \cong \frac{1}{4\pi T}\operatorname{vol}(\Gamma\backslash\mathfrak{H})$, and Weyl's law follows from a Tauberian theorem. See Hejhal [20, Chapter 2, Section 2]. □

The trace formula may be regarded as a duality between the length spectrum of $\Gamma\backslash\mathfrak{H}$ (that is, the set of lengths N of closed geodesics) and the numbers t_i such that $\frac{1}{4}+t_i^2$ are the eigenvalues of the Laplacian. A similar duality, which we next discuss, pertains between the set of prime numbers and the zeros of the Riemann zeta function.

Theorem 34. *Let g be compactly supported, smooth and even, and let $h(t) = \int_{-\infty}^{\infty} e^{itu}\, g(u)\, du$. Let $\frac{1}{2}+it_i$ denote the zeros of ζ in the cricital strip. Then*

$$h\left(-\frac{i}{2}\right) - \sum h(t_i) + h\left(\frac{i}{2}\right) = -\frac{1}{2\pi}\int_{-\infty}^{\infty} h(t)\frac{\Gamma'\left(\frac{1+2it}{4}\right)}{\Gamma\left(\frac{1+2it}{4}\right)}\, dt$$
$$+ g(0)\log(\pi) + 2\sum \frac{\log(p)}{\sqrt{p^n}}\, g\big(\log(p^n)\big). \tag{38}$$

Since h is assumed even, $h(-i/2) = h(i/2)$. However there are good reasons for writing the formula this way. (See Remark 4 below.) Special cases ("explicit formulae") were found by Riemann, von Mangoldt, Hadamard, de la Vallee Poussin, and Ingham. These are discussed in Ingham [24]. Weil [50] formalized the duality.

Proof. To prove this, let $\xi(s) = \Gamma_{\mathbb{R}}(s)\,\zeta(s)$, where $\Gamma_{\mathbb{R}}(s) = \pi^{-s/2}\Gamma\left(\frac{s}{2}\right)$. The function $h(t)$ is entire and we write $h(t) = H(\frac{1}{2}+it)$, $H(s) = h\big(-i(s-1/2)\big)$. Consider, for $\delta > 0$

$$\frac{1}{2\pi i}\int_{1+\delta-i\infty}^{1+\delta+i\infty} \frac{\xi'(s)}{\xi(s)}\, H(s)\, ds.$$

Moving the path of integration to $\operatorname{re}(s) = -\delta$ and using the functional equation $\xi(s) = \xi(1-s)$, we obtain the negative of this integral plus the sum of the residues; so

$$\frac{1}{\pi i}\int_{1+\delta-i\infty}^{1+\delta+i\infty} \frac{\xi'(s)}{\xi(s)}\, H(s)\, ds = \sum h(t_i) - h\left(\frac{i}{2}\right) - h\left(-\frac{i}{2}\right).$$

We have $\xi'/\xi = \Gamma_{\mathbb{R}}'/\Gamma_{\mathbb{R}} + \zeta'/\zeta$, and the integral over $\Gamma_{\mathbb{R}}$ can be moved left to $\operatorname{re}(s) = \frac{1}{2}$. We have

$$\frac{1}{\pi i}\int_{\frac{1}{2}-i\infty}^{\frac{1}{2}+i\infty} \frac{\Gamma_{\mathbb{R}}'(s)}{\Gamma_{\mathbb{R}}(s)}\, h\big(-i(s-\tfrac{1}{2})\big)\, ds$$
$$= -g(0)\,\log(\pi) + \frac{1}{2\pi}\int_{-\infty}^{\infty} h(t)\,\frac{\Gamma'\left(\frac{1+2it}{4}\right)}{\Gamma\left(\frac{1+2it}{4}\right)}\, dt.$$

On the other hand, we have $-\zeta'(s)/\zeta(s) = \sum \Lambda(n)\,n^{-s}$, where $\Lambda(p^k) = \log(p)$, p prime, while $\Lambda(n) = 0$ if n is not a prime power. We have

$$\frac{1}{2\pi i}\int_{2-i\infty}^{2+i\infty} H(s)\,n^{-s}\,ds = \frac{g\big(\log(n)\big)}{\sqrt{n}},$$

and assembling the pieces we get (38). □

Remark 1. Moving the line of integration requires knowing that a path from $2+iT$ to $-1+iT$ can be found for arbitrarily large T where ζ'/ζ is not too large. This can be accomplished by choosing the path to lie about half way between a pair of zeros that are not too close together. See Ingham [24, Theorem 2, p. 71], where he proves that one can always find t near T such that ζ/ζ' is $O(\log^2(T))$ on the line from $2+it$ to $-1+it$.

Remark 2. In practice the condition that g be compactly supported is too strong. It is sufficient that g and h be as in Theorem 32.

Remark 3. If H is not too big in the left half plane we have

$$\frac{1}{2\pi}\int_{-\infty}^{\infty} h(t)\frac{\Gamma'\left(\frac{1+2it}{4}\right)}{\Gamma\left(\frac{1+2it}{4}\right)}\,dt = \frac{1}{2\pi i}\int_{\frac{1}{2}-i\infty}^{\frac{1}{2}+i\infty} H(s)\frac{\Gamma'(\frac{s}{2})}{\Gamma(\frac{s}{2})}\,ds = 2\sum_{n=0}^{\infty} H(-2n),$$

where the sum is over the residues at the poles of $\Gamma(s/2)$.

Remark 4. Let X be a nonsingular complete curve of genus g over the finite field $k = \mathbb{F}_q$. If P is a prime divisor of degree $d(P)$, we denote $N(P) = q^{d(P)}$. The zeta function of X is

$$\prod(1 - N(P)^{-s})^{-1} = \frac{\prod_{j=1}^{2g}(1-\alpha_j q^{-s})^{-1}}{(1-q^{-s})(1-q^{1-s})},$$

where α_i are the eigenvalues of the Frobenius map in $H^1(X, \mathbb{Q}_l)$, for any prime $l \neq p$. The Riemann hypothesis is that $|\alpha_j| = \sqrt{q}$. We write $\alpha_j = q^{\frac{1}{2}+i\theta_j}$. Let $g(n) : \mathbb{Z} \to \mathbb{C}$ be a sequence which is even and nonzero for only finitely many n, and let

$$h(t) = \sum_n g(n)\,e^{itn\log(q)}.$$

Thus h is entire, even, and periodic with period $2\pi/\log(q)$. Then

$$h\left(-\frac{i}{2}\right) - \sum h(\theta_j) + h\left(\frac{i}{2}\right) = -(2g-2)\,g(0) + \sum_P\sum_{m=1}^{\infty}\frac{d(P)\,g\big(m\,d(P)\big)}{\sqrt{q^{md(P)}}}.$$

This function field analog of Theorem 34 may be proved by considering

$$\frac{1}{2\pi i}\int_2^{2+2\pi it/\log(q)}\left(\frac{Z'(s)}{Z(s)} + (g-1)\log(q)\right)H(s)\,ds,$$

where $h(t) = H(\frac{1}{2} + it)$. It may be regarded as an application of the Lefschetz–Grothendieck fixed point formula applied to a correspondence on X, and the three contributions $h(-i/2)$, $\sum h(\theta_j)$ and $h(i/2)$ are the traces of the correspondence applied to $H^0(X)$, $H^1(X)$ and $H^2(X)$. See Patterson [38, Chapter 5] and the first part of Connes [10].

The analogy between the Selberg trace formula and the explicit formulae has been the source of much speculation. Whether this analogy is misleading or not remains to be seen. One major difference between the explicit formulae and the Selberg trace formula is that in the trace formula, $\sum h(t_i)$ appears with a positive sign, while in (38), it appears with a negative sign. As Connes [10] points out, this difference may be very significant.

One tangible fruit of the analogy is Selberg's discovery of a zeta function which bears a relationship to the trace formula similar to that of the Riemann zeta function to the explicit formulae. With notation as in the previous section, Selberg considered

$$Z(s) = \prod_{\{N_0\}} \prod_{k=0}^{\infty} (1 - N_0^{-s-k}),$$

where $\log(N_0)$ runs through the lengths of prime geodesics.

Theorem 35. *$Z(s)$ has analytic continuation to all s, with zeros at the negative integers and at $\frac{1}{2} \pm it_k$, where $\frac{1}{4} + t_k^2$ are the eigenvalues of the Laplacian on $\Gamma\backslash\mathfrak{H}$.*

Proof and discussion. To motivate the introduction of the Selberg zeta function, we would like to take

$$g(u) = e^{-|u|(s-1/2)}$$

in the trace formula. Unfortunately we cannot use this function, but if we could, the geometric side of the trace formula would be the logarithmic derivative of $Z(s)$.

Lemma 36.

$$\sum_{\{N\}} \frac{\log(N_0)\, N^{-(s-1/2)}}{N^{1/2} - N^{-1/2}} = \frac{Z'(s)}{Z(s)}.$$

Proof. Writing $N = N_0^m$

$$\sum_{\{N_0\}} \sum_{m=1}^{\infty} \frac{\log(N_0)\, N_0^{-m(s-1/2)}}{N_0^{m/2} - N_0^{-m/2}} = \sum_{\{N_0\}} \sum_{m=1}^{\infty} \frac{\log(N_0)\, N_0^{-ms}}{1 - N_0^{-m}}.$$

Substituting $(1 - N_0^{-m})^{-1} = \sum_{k=0}^{\infty} N_0^{-ks}$ and interchanging the order of summation, this equals

$$\sum_{\{N_0\}} \frac{\log(N_0)\, N_0^{-(s+k)}}{1 - N_0^{-(s+k)}} = \frac{Z'(s)}{Z(s)}.$$

This completes the proof. □

Although we cannot use $g(u) = e^{-|u|(s-1/2)}$, we may use

$$g(u) = \frac{e^{-|u|(s-1/2)}}{2s-1} - \frac{e^{-|u|(\sigma-1/2)}}{2\sigma-1},$$

where s and σ are distinct. This device of subtraction eliminates the discontinuity in g' at $u = 0$. We will hold σ fixed and vary s. Initially, they are both assumed to have large real part. The Fourier transform $h(t) = \int_{-\infty}^{\infty} g(u)\, e^{itu}\, du$ is

$$h(t) = \frac{1}{(s-\frac{1}{2})^2 + t^2} - \frac{1}{(\sigma-\frac{1}{2})^2 + t^2}.$$

Next we prove, assuming that s and σ have real part $\geq \frac{1}{2}$, that

$$\int_{-\infty}^{\infty} h(t)\, t \tanh(\pi t)\, dt = 2\sum_{k=0}^{\infty}\left[\frac{1}{s+k} - \frac{1}{\sigma+k}\right]. \tag{39}$$

To prove this, use the partial fraction decomposition

$$\frac{t}{(s-\frac{1}{2})^2 + t^2} = \frac{1}{2i}\left[\frac{1}{(s-\frac{1}{2}) - it} - \frac{1}{(s-\frac{1}{2}) + it}\right].$$

The left side of (39) equals

$$\frac{1}{2i}\int_{-\infty}^{\infty}\left[\frac{1}{(s-\frac{1}{2}) - it} - \frac{1}{(\sigma-\frac{1}{2}) - it}\right]\tanh(\pi t)\, dt$$
$$-\frac{1}{2i}\int_{-\infty}^{\infty}\left[\frac{1}{(s-\frac{1}{2}) + it} - \frac{1}{(\sigma-\frac{1}{2}) + it}\right]\tanh(\pi t)\, dt.$$

Since tanh is odd, the two contributions are equal, and the first integral may be evaluated by moving the path of integration down into the lower half plane, where the only poles are at the poles $-i(k+\frac{1}{2})$ of $\tanh(\pi t)$, $k = 0, 1, 2, \ldots$. The residue of $\tanh(\pi t)$ at these points is π^{-1}, and we obtain (39).

We now use the Selberg Trace Formula. We note that the volume of the fundamental domain is $4\pi(g-1)$. So for these functions g and h we get

$$\frac{1}{2s-1}\frac{Z'(s)}{Z(s)} - \frac{1}{2\sigma-1}\frac{Z'(\sigma)}{Z(\sigma)}$$
$$= -(2g-2)\sum_{k=0}^{\infty}\left[\frac{1}{s+k} - \frac{1}{\sigma+k}\right] + \sum\left[\frac{1}{(s-\frac{1}{2})^2 + t_i^2} - \frac{1}{(\sigma-\frac{1}{2})^2 + t_i^2}\right].$$

Fixing σ and varying s, the right side has meromorphic continuation to all s. Multiplying by $2s-1$ see that the poles of Z'/Z are simple and have integer residues. Hence

$$Z(s) = \exp\int_{\infty}^{s}\frac{Z'(s)}{Z(s)}\, ds$$

has analytic continuation to all s. □

8 Groups with one cusp

If the group Γ is of cofinite volume but has a cusp, the spectral theory is complicated by a continuous spectrum, coming from the Eisenstein series. However the cuspidal spectrum is discrete. For simplicity assume that ∞ is the only cusp of Γ, and the stabilizer of ∞ in Γ is

$$\Gamma_\infty = \left\{ \begin{pmatrix} 1 & n \\ & 1 \end{pmatrix} \Big| n \in \mathbb{Z} \right\}.$$

We will also denote $G_\infty = \left\{ \begin{pmatrix} 1 & x \\ & 1 \end{pmatrix} \mid x \in \mathbb{R} \right\}$. Let $L_0^2(\Gamma\backslash G)$ be the subspace of "cusp forms" $f \in L^2(\Gamma\backslash G)$ satisfying

$$\int_0^1 f\left(\begin{pmatrix} 1 & x \\ & 1 \end{pmatrix} g \right) dx = 0. \tag{40}$$

Also let $L_0^2(\Gamma\backslash\mathfrak{H})$ be the right K-invariant elements of $L_0^2(\Gamma\backslash G)$, which may be regarded as functions on $\mathfrak{H}$.

In contrast with Theorem 7, the operators T_ϕ are no longer compact. However we still have:

Theorem 37 (Gelfand, Graev, and Piatetski-Shapiro). *If $\phi \in \mathcal{H}$, the restriction of T_ϕ to $L_0^2(\Gamma\backslash G)$ is a compact operator.*

Proof. We refer to Bump [6, Section 3.2] for an exposition of Godement's proof. □

Theorem 38.
(i) *$L_0^2(\Gamma\backslash\mathfrak{H})$ has a basis consisting of eigenfunctions of Δ.*
(ii) *$L_0^2(\Gamma\backslash G)$ decomposes as a direct sum of irreducible invariant subspaces. Any K-finite element of one of these spaces is of rapid decay.*

Sketch of proof. The proofs Theorem 9 and Theorem 11 are easily adapted. For the second assertion in (ii), it is sufficient to show that if $V \subset L_0^2(\Gamma\backslash G)$ is an irreducible subspace and $f \in V(k)$, then f is of rapid decay. We may find $\phi \in \mathcal{H}$ satisfying $\phi(\kappa_\theta g \kappa_\sigma) = e^{ik(\theta+\sigma)}\phi(g)$, and such that

$$y \to \phi\begin{pmatrix} y^{1/2} & \\ & y^{-1/2} \end{pmatrix}$$

is a positive function of mass 1 concentrated near $y = 1$. Then $T_\phi f$ is near f, therefore nonzero, and it is in $V(k)$, which is one dimensional by Proposition 5, so it is proportional to f. It follows from the rapid decay of the kernel $K_0(g, h)$, defined in Bump [6, Proposition 3.2.3] that $T_\phi f$ is of rapid decay, thus so is f. □

As in the case of compact quotient, the Laplacian acts by scalars on each irreducible one-dimensional subspace. The cuspidal spectrum behaves much as the entire spectrum in the compact case. On the other hand, the orthogonal complement of $L_0^2(\Gamma\backslash G)$ contains a continuous spectrum. The eigenfunctions of the Laplacian relevant to the spectral theorem are themselves not square integrable.

To understand how this can be, and to get some intuition as to the nature of the continuous spectrum, consider the following example. The group $\mathbb{R}$ acts on itself by translation, and the Laplacian $-d^2/dx^2$ is an invariant differential operator. It has eigenfunctions $f_a(x) = e^{2\pi i a x}$ with eigenvalues a^2. Any L^2 function has a Fourier expansion

$$\phi(x) = \int_{-\infty}^{\infty} \hat{\phi}(a)\, f_a(x)\, da,$$

but f_a is itself not L^2. If $T \subset \mathbb{R}$ is measurable, the Fourier transforms of L^2 functions supported on T form an invariant subspace. There are no minimal invariant subspaces, so $L^2(\mathbb{R})$ does not decompose as a direct sum of irreducible representations.

In order to formulate the basic theorem about the Eisenstein series which belong to both the odd and the even principal series representations, let us slightly generalize our setup. We assume that the discrete Γ contains $-I$, which acts trivially on $\mathfrak{H}$. Fix a unitary character χ of Γ such that $\chi(-I) \equiv (-1)^{\epsilon}$ mod 2 and such that $\chi | \Gamma_\infty$ is trivial, where $\epsilon = 0$ or 1. Let $k \equiv \epsilon$ modulo 2 and let $\pm = (-1)^{\epsilon}$. Recall that $f_{s,k}$ was defined by (15),

$$E(g, f_{s,k}, \chi) = \sum_{\Gamma_\infty\backslash\Gamma} \overline{\chi(\gamma)}\, f_{s,k}(\gamma g). \tag{41}$$

Proposition 39. *The series (41) converges absolutely if* $\mathrm{re}(s) > 1$.

Sketch of proof. Let σ be the real part of s. What we need to show is that if $\sigma > 1$, then

$$\sum_{\Gamma_\infty\backslash\Gamma} f_{\sigma,0}(\gamma g) < \infty. \tag{42}$$

We define a measure μ_σ on the upper half plane by $\mu_\sigma = y^{\sigma-2}\, dx \wedge dy$. Let B be a small neighborhood around $z \in \mathfrak{H}$. The Jacobian of the map $z \to \gamma(z)$, where $\gamma = \left(\begin{smallmatrix} a & b \\ c & d \end{smallmatrix}\right) \in SL(2, \mathbb{R})$ is $|cz + d|^{-4}$. Hence the μ_σ-volume of $\gamma(B)$ is (approximately) $|cz + d|^{-2\sigma}$ vol(B).

We may choose the representatives $\gamma \in \Gamma_\infty\backslash\Gamma$ so that the images $\gamma(B)$ all lie within the rectangle $0 \le x \le 1, 0 < y \le C$ for some constant C. (Actually this is not *quite* true. If one $\gamma(B)$ happens to lie on the left or right edge of this region, cut it into two pieces along this edge and move one piece by ± 1 back into the region.) This rectangle has finite volume, so $\sum_{\Gamma_\infty\backslash\Gamma} |cz + d|^{-2\sigma} < \infty$, which implies (42). □

The analytic continuation of the Eisenstein series is closely connected with the analytic continuation of its *constant term*

$$E_0(g, f_{s,k}, \chi) = \int_0^1 E\left(\begin{pmatrix} 1 & t \\ & 1 \end{pmatrix} g, f_{s,k}, \chi\right) dt. \tag{43}$$

We recall that an intertwining operator $M(s) : P_s^{\pm} \to P_{1-s}^{\pm}$ defined by (16).

Proposition 40. *Assume* $\mathrm{re}(s) > 1$. *There exists an analytic function* $c(s)$ *independent of k which is bounded on vertical strips (to the right of* 1) *such that*

$$E_0(g, f_{s,k}, \chi) = f_{s,k}(g) + c(s)\, M(s)\, f_{s,k}. \tag{44}$$

Proof. Substitute the definition of $E(g, f_{s,k}, \chi)$ into (43). The coset Γ_∞ in $\Gamma_\infty\backslash\Gamma$ contributes $f_{s,k}$. The remaining terms contribute

$$\int_0^1 \sum_{\substack{\gamma \in \Gamma_\infty\backslash\Gamma/\Gamma_\infty \\ \gamma \notin \Gamma_\infty}} \sum_{\delta\in\Gamma_\infty} \overline{\chi(\gamma\delta)}\, f_{s,k}\left(\gamma\delta \begin{pmatrix} 1 & x \\ & 1 \end{pmatrix} g\right) dx$$

$$= \sum_{\substack{\gamma \in \Gamma_\infty\backslash\Gamma/\Gamma_\infty \\ \gamma \notin \Gamma_\infty}} \overline{\chi(\gamma)} \int_{-\infty}^{\infty} f_{s,k}\left(\gamma \begin{pmatrix} 1 & x \\ & 1 \end{pmatrix} g\right) dx.$$

If $\gamma = \left(\begin{smallmatrix} a & b \\ c & d \end{smallmatrix}\right) \notin \Gamma_\infty$, then $c \neq 0$ and

$$\gamma = \begin{pmatrix} c^{-1} & a \\ & c \end{pmatrix} \begin{pmatrix} & -1 \\ 1 & \end{pmatrix} \begin{pmatrix} 1 & d/c \\ & 1 \end{pmatrix},$$

so the variable change $x \to x - d/c$ shows that

$$\int_{-\infty}^{\infty} f_{s,k}\left(\gamma \begin{pmatrix} 1 & x \\ & 1 \end{pmatrix} g\right) dx = |c|^{-2s}\, M(s)\, f_{s,k}.$$

Thus (44) is satisfied, where

$$c(s) = \sum_{\substack{\gamma \in \Gamma_\infty\backslash\Gamma/\Gamma_\infty \\ \gamma \notin \Gamma_\infty}} \chi(\gamma)\, |c|^{-2s}. \tag{45}$$

Since we are within the region of absolute convergence of the Eisenstein series this Dirichlet series is convergent if $\mathrm{re}(s) > 1$, and (45) shows that it is an analytic function bounded in vertical strips. □

We may rewrite this in the form

$$E_0(g, f_{s,k}, \chi) = f_{s,k}(g) + A_k(s)\, f_{1-s,k}, \tag{46}$$

where A_k is c times the constant on the right side in (18). It follows from Stirling's formula and the boundedness of $c(\sigma + it)$ that

$$A(\sigma + it) = O(|t|^{-1/2})$$

when $\sigma > 1$.

The analytic continuation of $E(g, f_{s,k}, \chi)$ and $c(s)$ are closely connected with each other. Selberg proved:

Theorem 41. *$A_k(s)$ and $E(g, f_{s,k}, \chi)$ have meromorphic continuations and the same poles. They are analytic for* $\mathrm{re}(s) > \frac{1}{2}$ *except possibly for a finite number of poles on the real axis in* $(\frac{1}{2}, 1]$. *On the line* $\mathrm{re}(s) = \frac{1}{2}$ *they are holomorphic. The Eisenstein series satisfies the functional equation*

$$E(g, f_{s,k}, \chi) = A_k(s)\, E(g, f_{1-s,k}, \chi).$$

We will only prove this in the special case where $\Gamma = SL(2, \mathbb{Z})$ (Theorem 44). For the general case, there are various accounts in the literature. See Borel [5], Cohen and Sarnak [9], Colin de Verdier [8], Efrat [11], Elstrodt [12], Fadeev [13], Harish-Chandra [19], Hejhal [21, Chapter 6 and Appendix F, with discussion of the literature on p. 225], Kubota [29], Jacquet [25], Langlands [32], Lax and Phillips [34], Moeglin and Waldspurger [36], Osborne and Warner [37], Venkov [48], and Wong [53]. The most general treatments are Langlands' historically important work [32] and the careful modern treatise of Moeglin and Waldspurger [36].

This body of literature all owes something to Selberg, who (according to Hejhal [21]) found three proofs. Generally speaking, one shows the analytic continuation of $A_k(s)$ and $E(g, f_{s,k}, \chi)$ in parallel. The basic principle is that the resolvent of an operator has analytic continuation to the complement of its spectrum. Applied as, for example, in Kubota [29], Venkov [48], or Langlands [32, Appendix IV], this gives the analytic continuation of $A_k(s)$ and $E(g, f_{s,k}, \chi)$ to the region $\mathrm{re}(s) > \frac{1}{2}$, $s \notin (\frac{1}{2}, 1]$. In this approach, then, obtaining the meromorphic continuation to the entire plane presents some difficulties.

Selberg and Bernstein realized independently that these difficulties could be avoided by combining the resolvent principle with another one, the insight that if a system of inhomogeneous linear equations having analytic coefficients has a *unique* solution, then the solution has meromorphic continuation to wherever the coefficients in the linear equations do. Selberg's published comments on this idea are in his introduction (written in 1988) to his Göttingen lectures [43]. In addition to the analytic continuation of Eisenstein series Bernstein gave other applications, such as to the analytic continuation of the intertwining integrals $M(s)$ and their p-adic analogs [2, 3], and his method has become an standard technique in the representation theory of p-adic groups.

Assuming the analytic continuation, we note the following two propositions.

Proposition 42. *The functions $E(g, s, f_{s,k})$ for all $k \equiv \epsilon$ modulo 2 comprise a $(\mathfrak{g}, K)$-module isomorphic to $P_s^{\pm}$.*

Proof. First suppose that $s > 1$. Then $E(g, s, f_{s,k})$ is constructed by averaging left translates of $f_{s,k}$. The representation of the group and its Lie algebra on functions is via right translation. The two operations commute with each other, so the statement is clear in this case. The identities implied are preserved under analytic continuation. □

Proposition 43. *If s is not a pole of $E(g, f_{s,k}, \chi)$, then $E - E_0$ is of rapid decay.*

Sketch of proof. Let $\phi \in \mathcal{H}$. The function

$$\tilde{E}(g) = E(g, f_{s,k}) - E_0(g, f_{s,k})$$

is not automorphic with respect to Γ, but modifying the proof of Proposition 3.2.3 of Bump [6], where the kernel K_0 is defined, we have

$$T_\phi \tilde{E}(g) = \int_{\Gamma_\infty \backslash G} K_0(g, h)\, \tilde{E}(h, f_{s,k}, \chi)\, dh.$$

By the same reasoning as in Theorem 38 (ii) we may choose ϕ so that $T_\phi \tilde{f}_{s,k}$ is a nonzero multiple of $f_{s,k}$. Now $\tilde{E}$ lies in $V(k)$, where (π, V) is a space of functions isomorphic to $\pi_s^{\pm}$, so $T_\phi \tilde{E}$ is the same constant multiple of $\tilde{E}$. The rapid decay now follows from the corresponding property of the kernel $K_0(g, h)$. □

The asymptotic behavior of $E(g, s, f_s)$ near the cusp is therefore determined by its constant term. The constant term, we see, consists of two parts, one of the order y^s and the other of order y^{1-s}. The smallest growth evidently occurs on the line $\mathrm{re}(s) = \frac{1}{2}$, where it is of order $\sqrt{y}$. On this line, the Eisenstein series is *almost* but *not quite* L^2.

There is a substantial difference between the arithmetic case (such as $\Gamma = SL(2, \mathbb{Z})$, $\chi = 1$ and the nonarithmetic case). In the arithmetic case it is possible to normalize the Eisenstein series. This means there is a multiple of the Eisenstein series which has a simple functional equation, and only a finite number of poles.

For example, suppose that if $\Gamma = SL(2, \mathbb{Z})$, $\chi = 1$ and $k = 0$. Then the Eisenstein series

$$E(z, s) = \sum_{\gamma \in \Gamma_\infty \backslash SL(2,\mathbb{Z})} f_{s,0}(\gamma z)$$

and the *normalized Eisenstein series* is

$$E^*(z, s) = \pi^s\, \Gamma(s)\, \zeta(2s)\, E(z, s) \tag{47}$$

Theorem 44. *$E^*(z, s)$ has analytic continuation to all s except $s = 0, 1$, where it has simple poles; the residue at $s = 1$ is the constant function $\frac{1}{2}$. It satisfies the functional equation $E^*(z, s) = E^*(z, 1 - s)$.*

Sketch of proof. If $z = x + iy \in \mathcal{H}$ and $t > 0$, let

$$\Theta(t) = \sum_{(m,n) \in \mathbb{Z}^2} e^{-\pi |mz+n|^2 t/y}.$$

It follows from Euler's integral for Γ that

$$E^*(z, s) = \frac{1}{2} \int_0^\infty \left(\Theta(t) - 1\right) t^s \frac{dt}{t}.$$

The Poisson summation formula implies that $\Theta(t) = t^{-1}\,\Theta(t^{-1})$. From this one gets

$$E^*(z,s) = \tfrac{1}{2}\int_0^\infty \Theta(t)\,(t^s + t^{1-s})\,\frac{dt}{t} - \frac{1}{2s} - \frac{1}{2-2s}.$$

This expression gives the analytic continuation and functional equation. □

Theorem 44 is a special case of Theorem 41. It is typical of the arithmetic case, but not of the general case.

9 Spectral expansion

The spectral expansion was obtained by Roelcke [40], modulo the analytic continuation of the Eisenstein series. Numerous accounts are in the literature, of which Godement [17, 18] is a good and influential one. In this section we discuss the spectral expansion for $L^2(\Gamma\backslash\mathfrak{H})$, where as in Section 6 the group has a single cusp. We will assume the analytic continuation of the Eisenstein series (Theorem 41).

Let ϕ be a K-finite element of $C_c^\infty(G_\infty\backslash G)$. The *incomplete theta series*

$$\theta_\phi(g) = \sum_{\Gamma_\infty\backslash\Gamma} \phi(\gamma g) \tag{48}$$

is something like an Eisenstein series but it is not a Δ-eigenfunction. (Godement's term "incomplete theta series" seems something of a misnomer.) If $f \in C^\infty(\Gamma\backslash G)$, let

$$f_0(g) = \int_{\Gamma_\infty\backslash G_\infty} f(ug)\,du$$

be its constant term, which is in $C^\infty(G_\infty\backslash G)$.

Proposition 45. *The incomplete theta series and constant term maps are adjoints; that is,*

$$\int_{\Gamma\backslash G} \theta_\phi(g)\,\overline{f(g)}\,dg = \int_{G_\infty\backslash G} \phi(g)\,\overline{f_0(g)}\,dg.$$

Proof. The left side is

$$\int_{\Gamma\backslash G} \sum_{\gamma\in\Gamma_\infty\backslash\Gamma} \phi(\gamma g)\,\overline{f(g)}\,dg = \int_{\Gamma_\infty\backslash G} \phi(g)\,\overline{f(g)}\,dg$$

$$= \int_{G_\infty\backslash G}\int_{\Gamma_\infty\backslash G_\infty} \phi(ug)\,\overline{f(ug)}\,du\,dg = \int_{G_\infty\backslash G} \phi(g) \int_{\Gamma_\infty\backslash G_\infty} \overline{f(ug)}\,du\,dg$$

which equals the right side. □

The incomplete theta series are compactly supported modulo Γ, hence are square-integrable. For this reason they are easier to work with than the Eisenstein series.

Proposition 46. *$L_0^2(\Gamma\backslash G)$ is the orthogonal complement in $L^2(\Gamma\backslash G)$ of the closed subspace spanned by the incomplete theta series.*

Proof. Immediate from Proposition 45, since the cuspidal spectrum is characterized by vanishing of its constant terms. □

For the remainder of this section, we make the simplifying assumption that the character χ be trivial and we only consider functions on G which are right invariant by K, that is, which may be regarded as functions on $\mathfrak{H}$. We will therefore denote $E(g, f_{s,0}, 1)$ as just $E(g, s)$, or as $E(z, s)$ where $z = g(i)$. We will also denote $f_s = f_{s,0}$. We will write the constant term in the form

$$E_0(g, s) = f_s(g) + A(s)\, f_{1-s}(g). \tag{49}$$

(See Proposition 40.)

Using Theorem 38, let ξ_i be a basis of $L_0^2(\Gamma\backslash G)$. Also, note that by Theorem 41 there may be a finite number of poles of the Eisenstein series $E(z, s)$ at locations $s_j \in (\frac{1}{2}, 1]$ $(j = 1, \dots, N)$. These are also poles of $A(s)$. Let α_j be the residue of $A(s)$ at $s = s_j$, and let

$$\eta_j = \frac{1}{\sqrt{|\alpha_j|}} \operatorname{res}_{s=s_j} E(g, s). \tag{50}$$

Proposition 47. *If ϕ is a cusp form, then*

$$\int_{\Gamma\backslash G} \phi(g)\, E(g, s)\, dg = 0 \tag{51}$$

Proof. Assume first $\mathrm{re}(s) > 1$. Then (51) is

$$\begin{aligned}\int_{\Gamma\backslash G} \sum_{\gamma\in\Gamma_\infty\backslash\Gamma} \phi(g)\, f_s(\gamma g)\, dg &= \int_{\Gamma_\infty\backslash G} \phi(g)\, f_s(g)\, dg \\ &= \int_{G_\infty\backslash G} \int_{\Gamma_\infty\backslash G_\infty} \phi(ug)\, f_s(ug)\, du\, dg \\ &= \int_{G_\infty\backslash G} f_s(g) \int_{\Gamma_\infty\backslash G_\infty} \phi(ug)\, du\, dg = 0\end{aligned}$$

since ϕ is a cusp form. The general case follows by analytic continuation. □

Proposition 48.

(i) *The constant term of η_j is $\sqrt{|\alpha_j|}\, f_{1-s_j}$.*

(ii) *The functions η_j are real valued, square integrable and orthogonal to the cusp forms.*

Proof. Part (i) is immediate from (49) since the first term on the right has no pole but the second one does. The η_j are real valued since the s_j are real. It follows from Proposition 43 that η_i is asymptotic to its constant term near the cusp so using (i) and the fact that $s_j > \frac{1}{2}$, we find that η_i is square integrable. Taking the residue in Proposition 47 shows that the η_j are orthogonal to the cusp forms. □

Proposition 49. *Let $\phi \in C_c^\infty(G_\infty \backslash G/K)$. Define*

$$\hat{\phi}(s) = \int_{G_\infty \backslash G} \phi(g)\, f_{1-s}(g)\, dg. \tag{52}$$

This is an entire function of s and $t \to \hat{\phi}(\sigma + it)$ is of Schwartz class for all real σ. For real σ we have

$$\phi(g) = \frac{1}{2\pi i} \int_{\sigma - i\infty}^{\sigma + i\infty} \hat{\phi}(s)\, f_s(g)\, ds. \tag{53}$$

(The contour integral is over the vertical line with real part σ.)

Proof. We have

$$\hat{\phi}(s) = \int_0^\infty \phi\begin{pmatrix} y^{1/2} & \\ & y^{-1/2} \end{pmatrix} y^{-s} \frac{dy}{y}.$$

Since ϕ is compactly supported and smooth, $t \to \hat{\phi}(\sigma + it)$ is the Fourier transform of a Schwartz function, hence is Schwartz itself. By the Mellin inversion formula

$$\phi\begin{pmatrix} y^{1/2} & \\ & y^{-1/2} \end{pmatrix} = \frac{1}{2\pi i} \int_{\sigma - i\infty}^{\sigma + i\infty} \hat{\phi}(s)\, y^s\, ds. \tag{54}$$

This verifies (53) on the diagonal, and since both sides are left invariant by G_∞ and right invariant by K, the general case follows. □

Proposition 50. *We have*

$$\langle \theta_\phi, \eta_j \rangle = \sqrt{|\alpha_j|}\, \hat{\phi}(s_j).$$

Proof. This follows from Proposition 45, Proposition 48 and from the definition of $\hat{\phi}$. □

Theorem 51. *Let $\psi \in L^2(\Gamma \backslash \mathfrak{H})$. Assume also that $\psi(g)\, f_{1/2}(g)$ is integrable over the Siegel set $\mathcal{S}$. Then*

$$\psi(g) = \sum_{j=0}^{N} \langle \psi, \eta_j \rangle \eta_j(g) + \sum_{i=0}^{\infty} \langle \psi, \xi_i \rangle \xi_i(g)$$
$$+ \frac{1}{4\pi} \int_{-\infty}^{\infty} \langle \psi, E(-, \tfrac{1}{2} + it) \rangle E(g, \tfrac{1}{2} + it)\, dt. \tag{55}$$

The integrability assumption on that $\psi(g)\, f_{1/2}(g)$ guarantees that the inner products occurring in the expansion are convergent. Correctly interpreted, the expansion is valid for all $f \in L^2(\Gamma\backslash\mathfrak{H})$. This is like the Fourier inversion formula for $\mathbb{R}$, which is valid for all $L^2(\mathbb{R})$ though the usual definition of the Fourier transform as an integral is only strictly correct on $L^2(\mathbb{R}) \cap L^1(\mathbb{R})$.

Sketch of proof. By Proposition 48 and Proposition 47 this is true for cusp forms. By Proposition 46 it is therefore sufficient to prove it for incomplete theta series; that is,

$$\theta_\phi(g) = \sum_{j=1}^{N} \langle \theta_\phi, \eta_j \rangle \eta_j(g) + \frac{1}{4\pi} \int_0^\infty \langle \theta_\phi, E(\text{-}, \tfrac{1}{2} + it) \rangle E(g, \tfrac{1}{2} + it)\, dt \qquad (56)$$

when $\phi \in C_c^\infty(G_\infty\backslash\mathfrak{H})$.

Indeed, both sides are left invariant by G_∞, right invariant by K, and agree on the diagonal by (54), so this follows by the Iwasawa decomposition. We take $\sigma > 1$ and sum over γg with $\gamma \in \Gamma_\infty\backslash\Gamma$ to obtain

$$\theta_\phi(g) = \frac{1}{2\pi i} \int_{\sigma - i\infty}^{\sigma + i\infty} \hat{\phi}(s)\, E(g, s)\, ds. \qquad (57)$$

Now we move the path of integration to the left. The pole at $s = s_j$ contributes a residue which by (50) and Proposition 50 equals $\langle \theta_\phi, \eta_j \rangle\, \eta_j$. Thus

$$\theta_\phi(g) = \sum_{j=1}^{N} \langle \theta_\phi, \eta_j \rangle \eta_j(g) + \frac{1}{2\pi} \int_{-\infty}^{\infty} \hat{\phi}(\tfrac{1}{2} + it)\, E(g, \tfrac{1}{2} + it)\, dt \qquad (58)$$

Now let us look at the integral in (56). Using Proposition 45, (49) and (52) the integrand is

$$\begin{aligned} &\frac{1}{4\pi} \int_{\Gamma\backslash G} \theta_\phi(h)\, E(h, \tfrac{1}{2} - it)\, dh \cdot E(g, \tfrac{1}{2} + it) \\ &= \frac{1}{4\pi} \left(\hat{\phi}(\tfrac{1}{2} + it)\, E(g, \tfrac{1}{2} + it) + \hat{\phi}(\tfrac{1}{2} - it)\, A(\tfrac{1}{2} - it)\, E(g, \tfrac{1}{2} + it) \right) \\ &= \frac{1}{4\pi} \left(\hat{\phi}(\tfrac{1}{2} + it)\, E(g, \tfrac{1}{2} + it) + \hat{\phi}(\tfrac{1}{2} - it)\, E(g, \tfrac{1}{2} - it) \right). \end{aligned}$$

Integrating with respect to t, combining the duplicate contributions and applying (58) now proves (56). □

10 Liftings and the trace formula

One of the most interesting applications of the trace formula is to liftings of automorphic forms. The method involves comparison of two different trace formulae, on different groups, leading to the conclusion that automorphic forms on one group can be lifted to automorphic forms on the other.

Jacquet and Langlands [26] gave an early application in a lifting from automorphic forms on a division algebra to $GL(2)$. A variation of this theme noted by Gelbart and Jacquet [15] is probably the simplest example of this type, since in this case neither trace formula involves a continuous spectrum. It will be convenient to switch to an adelic point of view, but we think the reader who has read Section 5 will not have trouble making the transition.

If D_1 and D_2 are central division algebras over a field F, then $D_1 \otimes D_2 \cong \mathrm{Mat}_k(D_3)$ for some D_3 and k, and $D_1, D_2 \to D_3$ is an associative multiplication on the set $B(F)$ of isomorphism classes of central division algebras. Thus $B(F)$ becomes a group, called the *Brauer group*.

If D is a central division algebra over F, then the dimension of D is a square d^2, and if E/F is any field extension of degree d which can be embedded in D, then $E \otimes D \cong \mathrm{Mat}_d(E)$. Thus a division ring is a Galois twisted form of a matrix ring. The composite map

$$D \to E \otimes D \cong \mathrm{Mat}_d(E) \to E,$$

the last map being either the trace or determinant, takes values in F, and gives us the *reduced trace* or *reduced norm*.

The Brauer group of a local or global field F admits a simple and beautiful description related to the reciprocity laws of class field theory. See Section 1 of Chapter 6 ("Local Class Field Theory" by J.-P. Serre) and Section 9 of Chapter 7 ("Global Class Field Theory" by J. Tate) in Cassels and Fröhlich [7].

Let F be a global field, $\mathbb{A}$ its adele ring, and D a central division algebra of degree p^2 over F, where p is a prime. Let $Z \cong F^\times$ be the center of $D^\times$. Let S be the finite set of places where D_v is a division ring. If $v \notin S$ we identify $D_v = \mathrm{Mat}_p(F_v)$.

Let $\mathcal{H}$ be the set of functions on $D_\mathbb{A}^\times = \prod_v D_v^\times$ which are finite linear combinations of functions of the form $\prod_v \phi_v$, where for each v, $\phi_v : D_v^\times \to \mathbb{C}$ is smooth and compactly supported modulo Z_v, satisfies $\phi_v(z_v g_v) = \phi_v(g_v)$ when $z_v \in Z_v$, and agrees with the characteristic function of $Z_v \mathrm{Mat}_p(\mathfrak{o}_v)$ for almost all places v of F. The ring $\mathcal{H}$ contains the classical Hecke operators as well as the integral operators introduced in Section 7 above.

$Z_\mathbb{A} D_F^\times \backslash D_\mathbb{A}^\times$ is compact. As with $SL(2, \mathbb{R})$, $L^2(Z_\mathbb{A} D_F^\times \backslash D_\mathbb{A}^\times)$ admits integral operators T_ϕ for $\phi \in \mathcal{H}$:

$$(T_\phi f)(g) = \int_{Z_\mathbb{A} \backslash D_\mathbb{A}^\times} \phi(h)\, f(gh)\, dh.$$

Let $\{\gamma\}$ be a set of representatives for the conjugacy classes of $D_F^\times$. We denote by C_γ the centralizer of γ in $D_F^\times$. It is an algebraic group, so $C_\gamma(\mathbb{A}) \subset D_\mathbb{A}^\times$ will denote its points in $\mathbb{A}$.

Theorem 52 (Selberg trace formula).

$$\operatorname{tr} T_\phi = \sum_{\{\gamma\}} \operatorname{vol}(C_\gamma \backslash C_\gamma(\mathbb{A})) \int_{C_\gamma(\mathbb{A}) Z_\mathbb{A} \backslash D_\mathbb{A}^\times} \phi(g^{-1}\gamma g)\, dg. \tag{59}$$

Proof. The proof of Proposition 6 goes through without change, so

$$(T_\phi f)(g) = \int_{Z_\mathbb{A} D_F^\times \backslash D_\mathbb{A}^\times} K_\phi(g,h)\, f(h)\, dh,$$

$$K_\phi(g,h) = \sum_{\gamma \in D_F^\times / Z_F^\times} \phi(g^{-1}\gamma h).$$

As with $SL(2,\mathbb{R})$, the operator T_ϕ is thus Hilbert–Schmidt, and with more work may be shown to be trace class. As in Theorem 11,

$$\operatorname{tr} T_\phi = \int_{Z_\mathbb{A} D_F^\times \backslash D_\mathbb{A}^\times} K_\phi(g,g)\, dg$$

Now (59) follows as in Theorem 28. □

The conjugacy classes of $D_F^\times$ are easily described.

Proposition 53. *If $\alpha \in D_F^\times - Z_F$, then $F(\alpha)$ is a field extension of F of degree p. Elements α and β are conjugate in $D_F^\times$ if and only if there is a field isomorphism $F(\alpha) \to F(\beta)$ such that $\alpha \mapsto \beta$. If $F(\alpha)$ is a field extension of degree p, then $F(\alpha)$ may be embedded in D_F if and only if $[F_v(\alpha) : F_v] = p$ for all $v \in S$.*

Proof. The conjugacy of α and β follows from the *Skolem–Noether Theorem* (see Herstein [22, p. 99]). The last statement follows from (i) $\leftrightarrow$ (ii) in Weil [51, Proposition VIII.5, p. 253]. □

The trace formula can be used to prove functorial liftings in many cases.

Let E be another division algebra of degree p^2, and assume that the set of places where E_v is a division ring agrees with the set S of places where D_v is. If $p = 2$, this implies that D and E are isomorphic, but not in general. (This follows from the computation of the Brauer group in the global class field theory. See [7, Chapters 6 and 7].) Thus we want $p > 2$.

We will show that two spaces of automorphic forms on D and on E are isomorphic.

Suppose that $\pi = \otimes_v \pi_v$ is an irreducible constituent of $L^2(Z_\mathbb{A} D_F^\times \backslash D_\mathbb{A}^\times)$. Since $Z_v \backslash D_v^\times$ is compact for $v \in S$, π_v is finite dimensional at these places. We assume that π_v is trivial when $v \in S$.

If $v \notin S$, then $D_v \cong E_v \cong \operatorname{Mat}_p(F_v)$. We may therefore identify π_v with an irreducible representation π_v' of E_v when $v \notin S$, and if $v \in S$ we take $\pi_v' = 1$. Let $\pi' = \otimes \pi_v'$. It is an irreducible representation of $E_\mathbb{A}^\times$.

Theorem 54. *π' occurs in $L^2(Z_\mathbb{A} E_F^\times \backslash E_\mathbb{A}^\times)$.*

The correspondence $\pi \to \pi'$ is a functorial lift of automorphic forms in the sense of Langlands (see Langlands [33], Borel [4]).

Sketch of proof. If $v \in S$, then $Z_v \backslash D_v^\times$ is compact, so the constant function $\phi_v^\circ(g_v) = 1$ is in $C_c^\infty(F_v)$. Let $\mathcal{H}_S$ be the subalgebra of $\mathcal{H}$ spanned by functions $\prod \phi_v$ such that $\phi_v = \phi_v^\circ$ for $v \in S$. It is isomorphic to the corresponding Hecke ring on E. Let $\phi \to \phi'$ denote this isomorphism.

By Proposition 53, noncentral conjugacy classes in $D_F^\times$ and $E_F^\times$ are both in bijection with the set of Galois equivalence classes of elements α of field extensions $[F(\alpha) : F] = p$ such that $[F_v(\alpha) : F_v] = p$ for all $p \in S$. This intrinsic characterization shows that we may identify the conjugacy classes of D_F and E_F, and compare trace formulae to get

$$\operatorname{tr} T_\phi = \operatorname{tr} T'_\phi. \tag{60}$$

This is almost but not quite as easy as we have made it sound, because one must show that the volumes on the right side of (59) are the same for the two trace formulae.

It follows from (60) that the representations of $\mathcal{H}_S$ on the spaces

$$L^2\left(Z_{\mathbb{A}} D_F^\times \prod_{v \in S} D_v^\times \backslash D_{\mathbb{A}}^\times\right) \quad \text{and} \quad L^2\left(Z_{\mathbb{A}} E_F^\times \prod_{v \in S} E_v^\times \backslash E_{\mathbb{A}}^\times\right)$$

are isomorphic, and the theorem follows. □

Underlying the final step is the fact that two representations of rings are characterized by their traces. For example, if R is an algebra over a field of characteristic zero and if M_1, M_2 are finite-dimensional semisimple R-modules, and if for every $\alpha \in R$ the induced endomorphisms of M_1 and M_2 have the same trace, then the modules are isomorphic (see Lang [31, Corollary 3.8, p. 650]). This statement is not directly applicable here but it gives the flavor.

For the remainder we take $p = 2$, and review the *Jacquet–Langlands correspondence*. Let D be as before. The Jacquet–Langlands correspondence is a lifting of automorphic representations from $D^\times$ to $GL(2, F)$.

There is a local correspondence for $v \in S$. $D_v^\times$ is compact modulo its center, so its irreducible representations are finite dimensional. These lift to irreducible representations of $GL(2, F_v)$ having the same central character. The lifting was constructed by Jacquet and Langlands by use of the theta correspondence. Indeed, $Z_v \backslash D_v^\times$ is a quotient of the orthogonal group $GO(4)$ while $GL(2)$ is the same as $GSp(2)$, and the theta correspondence $GO(4) \leftrightarrow GSp(2)$ gives the Jacquet–Langlands correspondence. Its image is the square integrable representations (that is, the supercuspidals and the Steinberg representation).

Jacquet and Langlands constructed a global correspondence from automorphic forms on $D^\times$ to automorphic forms on $GL(2)$ first using the converse theorem in Section 14 of their book. To prove functional equations of L-functions on $D^\times$, they use the Godement–Jacquet construction, because the Hecke integral is not available in this context.

Finally, they reconsidered the lifting from the point of view of the trace formula. This allowed them to characterize the image of the lift. They sketched a proof (and later Gelbart and Jacquet completed) of:

Theorem 55. *An automorphic representation π of $GL(2, \mathbb{A})$ is the lift of an automorphic representation of $D_{\mathbb{A}}^{\times}$ if and only if π_v is square integrable for every $v \in S$.*

The remarkable fact is that their proof of this fact uses so many different techniques which have proved important in the subsequent 30 years: the Hecke and Godement–Jacquet integral constructions of L-functions, the Weil representation and the trace formula.

The trace formula on $GL(2)$ is harder than on the division ring because of the presence of the continuous spectrum. We have avoided this problem by proving Theorem 54 instead of Theorem 55.

References

[1] J. Arthur, The trace formula and Hecke operators, in K. E. Aubert, E. Bombieri and D. Goldfeld, eds., *Number Theory, Trace Formulas and Discrete Groups: Symposium in Honor of Atle Selberg Oslo, Norway, July* 14–21, 1987, Academic Press, New York, 1989, pp. 11–27.

[2] I. Bernstein, *Letter to Piatetski-Shapiro*, unpublished manuscript, 1986.

[3] I. Bernstein, *Meromorphic Continuation of Eisenstein Series*, unpublished manuscript, 1987.

[4] A. Borel, Automorphic L-functions, in A. Borel and W. Casselman, eds., *Automorphic L-functions: Automorphic Forms, Representations and L-Functions, Part* 2, Proceedings of Symposia in Pure Mathematics 33, AMS, Providence, RI, 1979, pp. 27–62.

[5] A. Borel, *Automorphic Forms on* $SL(2, \mathbb{R})$, Cambridge University Press, Cambridge, UK, 1997.

[6] D. Bump, Automorphic Forms and Representations, Cambridge University Press, Cambridge, UK, 1997.

[7] J. Cassels and A. Fröhlich, *Algebraic Number Theory*, Academic Press, New York, 1967.

[8] Y. Colin de Verdier, Une nouvelle démonstration du prolongement méromorphe des séries d'Eisenstein, *C. R. Acad. Sci. Paris Sér.* I *Math.*, **293**(1981), 361–363.

[9] P. Cohen and P. Sarnak, *Discontinuous Groups and Harmonic Analysis*, unpublished lecture notes.

[10] A. Connes, Noncommutative geometry and the Riemann zeta function, in V. I. Arnold, ed., *Mathematics: Frontiers and Perspectives*, AMS, Providence, RI, 2002, 35–54.

[11] I. Efrat, The Selberg trace formula for $PSL_2(\mathbb{R})^n$, *Mem. Amer. Math. Soc.*, **65**(1987), 359.

[12] J. Elstrodt, Die Resolvente zum Eigenwertproblem der automorphen Formen in der hyperbolischen Ebene I, II, III, *Math. Ann.*, **203**(1973), 295–300; *Math. Z.*, **132**(1973), 99–134; *Math. Ann.*, **208**(1974), 99–132.

[13] L. Faddeev, The eigenfunction expansion of Laplace's operator on the fundamental domain of a discrete group on the Lobačevskiĭ, plane, Trudy Moskov. Mat. Obšč. **17**(1967) 323–350. *Trans. Moscow Math. Soc.*, **17**(1967), 357–386.

[14] S. Gelbart, *Lectures on the Arthur–Selberg Trace Formula*, AMS, Providence, RI, 1996.

[15] S. Gelbart and H. Jacquet, Forms of $GL(2)$ from an analytic point of view, in A. Borel and W. Casselman, eds., *Automorphic Forms, Representations, and L-functions, Part* 1, Proceedings of Symposia in Pure Mathematics 33, AMS, Providence, RI, 1979, pp. 213–254.

[16] I. Gelfand, M. Graev, and I. Piatetskii-Shapiro, *Representation Theory and Automorphic Functions*, Saunders, San Francisco, 1969; reprinted by Academic Press, New York, 1990.

[17] R. Godement, The decomposition of $L^2(G/\Gamma)$ for $\Gamma = SL(2, \mathbb{Z})$, in *Algebraic Groups and Discontinuous Subgroups*, in A. Borel and G. Mostow, eds., *Algebraic Groups and Discontinuous Subgroups*, Proceedings of Symposia in Pure Mathematics 9, AMS, Providence, RI, 1966, 225–234.

[18] R. Godement, The spectral decomposition of cusp forms, in A. Borel and G. Mostow, eds., *Algebraic Groups and Discontinuous Subgroups*, Proceedings of Symposia in Pure Mathematics 9 AMS, Providence, RI, 1966, 225–234.

[19] H. Harish-Chandra, *Automorphic Forms on Semisimple Lie Groups*, Lecture Notes in Mathematics 62, Springer-Verlag, New York, 1968.

[20] D. Hejhal, *The Selberg Trace Formula for $PSL(2, R)$*, Vol. 1, Lecture Notes in Mathematics 548, Springer-Verlag, New York, 1976.

[21] D. Hejhal, *The Selberg Trace Formula for $PSL(2, R)$*, Vol. 2, Lecture Notes in Mathematics 1001, Springer-Verlag, New York, 1983.

[22] I. Herstein, *Noncommutative Rings*, Carus Monograph 15, MAA, Washington, DC, 1968.

[23] H. Iwaniec, *Introduction to the Spectral Theory of Automorphic Forms*, Biblioteca de la Revista Matematica Iberoamericana, Madrid, 1995.

[24] A. E. Ingham, *Distribution of Prime Numbers*, Cambridge University Press, Cambridge, UK, 1932.

[25] H. Jacquet, Note on the analytic continuation of Eisenstein series: An appendix to A. Knapp, "Theoretical aspects of the trace formula for $GL(2)$," in *Representation Theory and Automorphic Forms* (*Edinburgh*, 1996), Proceedings of Symposia in Pure Mathematics 61, AMS, Providence, RI, 355–405.

[26] H. Jacquet and R. Langlands, *Automorphic Forms on $GL(2)$*, Lecture Notes in Mathematics 114, Springer-Verlag, New York, 1970.

[27] A. Knapp, *Representation Theory of Semisimple Groups: An Overview Based on Examples*, Princeton University Press, Princeton, NJ, 1986.

[28] A. Knapp and E. Stein, Intertwining operators for semisimple groups, *Ann. Math.* (2), **93**(1971), 489–578.

[29] T. Kubota, *Elementary Theory of Eisenstein Series*, John Wiley, New York, 1973.

[30] S. Lang, $SL(2, \mathbb{R})$, Addison–Wesley, Reading, MA, 1975.

[31] S. Lang, *Algebra*, 3rd ed., Addison–Wesley, Reading, MA, 1993.

[32] R. Langlands, *On the Functional Equations Satisfied by Eisenstein Series*, Lecture Notes in Mathematics 544, Springer-Verlag, New York, 1976.

[33] R. Langlands, Problems in the theory of automorphic forms, in *Lectures in Modern Analysis and Applications* III, Lecture Notes in Mathematics 170, Springer-Verlag, New York, 1970.

[34] P. Lax and R. Phillips, *Scattering Theory for Automorphic Functions*, Annals of Mathematics Studies 87, Princeton University Press, Princeton, NJ, 1976.

[35] H. Maass, Über eine neue Art von nichtanalytischen automorphen Funktionen und die Bestimmung Dirichletscher Reihen durch Funktionalgleichungen, *Math. Ann.*, **121**(1949), 141–183.

[36] C. Moeglin and J.-L. Waldspurger, *Décomposition Spectrale et Séries d'Eisenstein*, Birkhäuser, Basel, 1993 (in French); *Spectral Decomposition and Eisenstein Series*, Cambridge University Press, Cambridge, UK, 1995 (in English).

[37] M. S. Osborne and G. Warner, *The Theory of Eisenstein Systems*, Academic Press, New York, 1981.

[38] S. Patterson, *An Introduction to the Theory of the Riemann Zeta-Function*, Cambridge University Press, Cambridge, UK, 1988.

[39] B. Randol, Small eigenvalues of the Laplace operator on compact Riemann surfaces, *Bull. Amer. Math. Soc.*, **80**(1974), 996–1000.

[40] W. Roelcke, Uber die Wellengleichung bei Grenzkreisgruppen erster Art, *S.-B. Heidelberger Akad. Wiss. Math.-Nat. Kl.* 1953/1955 (1956), 159–267.

[41] A. Selberg, Harmonic analysis and discontinuous groups in weakly symmetric Riemannian spaces with applications to Dirichlet series, *J. Indian Math. Soc. (N.S.)*, **20**(1956), 47–87.

[42] A. Selberg, Discontinuous groups and harmonic analysis, in *Proceedings of the International Congress of Mathematicians* (*Stockholm*, 1962), 1963, pp. 177–189.

[43] A. Selberg, Harmonic analysis, in A. Selberg, *Collected Works*, Springer-Verlag, New York, 1989, article 39; this is the annotated surviving portion of the 1955 Göttingen lecture notes.

[44] A. Selberg, Atle, On the estimation of Fourier coefficients of modular form, in A. Whiteman, ed., *Theory of Numbers*, Proceedings of Symposia in Pure Mathematics 8, AMS, Providence, RI, 1965, 1–15.

[45] T. Tamagawa, On Selberg's trace formula, *J. Fac. Sci. Univ. Tokyo Sec.* I, **8**(1960), 363–386.

[46] A. Terras, *Harmonic Analysis on Symmetric Spaces and Applications* I, Springer-Verlag, New York, 1985.

[47] V. Varadarajan, *An Introduction to Harmonic Analysis on Semisimple Lie Groups*, Cambridge University Press, Cambridge, UK, 1999.

[48] A. Venkov, Spectral theory of automorphic functions, the Selberg zeta function and some problems of analytic number theory and mathematical physics, *Uspekhi Mat. Nauk*, **34**-3 (1979), 69–135; *Russian Math. Surveys*, **34**(1979), 79–153.

[49] A. Venkov, *Spectral Theory of Automorphic Functions and Its Applications*, Kluwer, Dordrecht, The Netherlands, 1990.

[50] A. Weil, Sur les "formules explicites" de la théorie des nombres premiers, *Comm. Sém. Math. Univ. Lund.*, volume dedicated to Marcel Riesz, 1952, 252–265.

[51] A. Weil, *Basic Number Theory*, Springer-Verlag, New York, 1967.

[52] E. Whittaker and G. Watson, *A Course of Modern Analysis*, 4th ed., Cambridge University Press, Cambridge, UK, 1927.

[53] S.-T. Wong, The meromorphic continuation and functional equations of cuspidal Eisenstein series for maximal cuspidal groups, *Mem. Amer. Math. Soc.*, **83**(1990), 423.

9
Analytic Theory of L-Functions for GL_n

J.W. Cogdell

The purpose of this chapter is to describe the analytic theory of L-functions for cuspidal automorphic representations of GL_n over a global field. There are two approaches to L-functions of GL_n: via integral representations or through analysis of Fourier coefficients of Eisenstein series. In this chapter we will discuss the theory via integral representations.

The theory of L-functions of automorphic forms (or modular forms) via integral representations has its origin in the paper of Riemann on the ζ-function. However the theory was really developed in the classical context of L-functions of modular forms for congruence subgroups of $\mathrm{SL}_2(\mathbb{Z})$ by Hecke and his school. Much of our current theory is a direct outgrowth of Hecke's. L-functions of automorphic representations were first developed by Jacquet and Langlands for GL_2. Their approach followed Hecke combined with the local-global techniques of Tate's thesis. The theory for GL_n was then developed along the same lines in a long series of papers by various combinations of Jacquet, Piatetski-Shapiro, and Shalika. In addition to associating an L-function to an automorphic form, Hecke also gave a criterion for a Dirichlet series to come from a modular form, the so-called converse theorem of Hecke. In the context of automorphic representations, the converse theorem for GL_2 was developed by Jacquet and Langlands, extended and significantly strengthened to GL_3 by Jacquet, Piatetski-Shapiro, and Shalika, and then extended to GL_n with Piatetski-Shapiro. What we have attempted to present here is a synopsis of this work. An expanded version of this chapter can be found in [1].

There is another body of work on integral representations of L-functions for GL_n which developed out of the classical work on zeta functions of algebras. This is the theory of principal L-functions for GL_n, as developed by Godement and Jacquet [15, 20]. This approach is related to the one pursued here, but we have not attempted to present it.

The other approach to these L-functions is via the Fourier coefficients of Eisenstein series. This approach also has a classical history. In the context of automorphic representations, and in a broader context than GL_n, this approach was originally laid out by Langlands [29] but then most fruitfully pursued by Shahidi. Some of the major papers of Shahidi on this subject are [35, 36, 37, 38, 39, 40, 41]. In particular, in [38] he shows that the two approaches give the same L-functions for GL_n. We will not pursue this approach here.

1 Fourier expansions

In this section we let k denote a global field, $\mathbb{A}$, its ring of adeles, and ψ will denote a continuous additive character of $\mathbb{A}$ which is trivial on k.

We begin with a cuspidal automorphic representation (π, V_π) of $\mathrm{GL}_n(\mathbb{A})$. For us, automorphic forms are assumed to be smooth (of uniform moderate growth) but not necessarily K_∞-finite at the archimedean places. This is most suitable for the analytic theory. For simplicity, we assume the central character ω_π of π is unitary. Then V_π is the space of smooth vectors in an irreducible unitary representation of $\mathrm{GL}_n(\mathbb{A})$. We will always use cuspidal in this sense: the smooth vectors in an irreducible unitary cuspidal automorphic representation. (Any other smooth cuspidal representation π of $\mathrm{GL}_n(\mathbb{A})$ is necessarily of the form $\pi = \pi^\circ \otimes |\det|^t$ with π° unitary and t real, so there is really no loss of generality in the unitarity assumption. It merely provides us with a convenient normalization.) By a cusp form on $\mathrm{GL}_n(\mathbb{A})$ we will mean a function lying in a cuspidal representation. By a cuspidal function we will simply mean a smooth function φ on $\mathrm{GL}_n(k)\backslash \mathrm{GL}_n(\mathbb{A})$ satisfying $\int_{\mathrm{U}(k)\backslash \mathrm{U}(\mathbb{A})} \varphi(ug)\, du \equiv 0$ for every unipotent radical U of standard parabolic subgroups of GL_n.

The basic references for this section are the papers of Piatetski-Shapiro [31, 32] and Shalika [42].

1.1 *The Fourier expansions*

If $f(\tau)$ is a holomorphic cusp form on the upper half plane $\mathfrak{H}$, say with respect to $\mathrm{SL}_2(\mathbb{Z})$, then f is invariant under integral translations, $f(\tau+1) = f(\tau)$ and thus has a Fourier expansion of the form

$$f(\tau) = \sum_{n=1}^{\infty} a_n e^{2\pi i n \tau}.$$

If $\varphi(g)$ is a smooth cusp form on $\mathrm{GL}_2(\mathbb{A})$, then the translations correspond to the maximal unipotent subgroup $\mathrm{N}_2 = \{n = \left(\begin{smallmatrix} 1 & x \\ 0 & 1 \end{smallmatrix}\right)\}$ and $\varphi(ng) = \varphi(g)$ for $n \in \mathrm{N}_2(k)$. So, if ψ is any continuous character of $k\backslash\mathbb{A}$ we can define the ψ-Fourier coefficient or ψ-Whittaker function by

$$W_{\varphi,\psi}(g) = \int_{k\backslash\mathbb{A}} \varphi\left(\begin{pmatrix} 1 & x \\ 0 & 1 \end{pmatrix} g\right) \psi^{-1}(x)\, dx.$$

We have the corresponding Fourier expansion

$$\varphi(g) = \sum_{\psi} W_{\varphi,\psi}(g).$$

(Actually from abelian Fourier theory, one has

$$\varphi\left(\begin{pmatrix} 1 & x \\ 0 & 1 \end{pmatrix} g\right) = \sum_{\psi} W_{\varphi,\psi}(g)\psi(x)$$

as a periodic function of $x \in \mathbb{A}$. Now set $x = 0$.)

If we fix a single nontrivial character ψ of $k\backslash\mathbb{A}$, then the additive characters of the compact group $k\backslash\mathbb{A}$ are isomorphic to k via the map $\gamma \in k \mapsto \psi_\gamma$ where ψ_γ is the character $\psi_\gamma(x) = \psi(\gamma x)$. An elementary calculation shows that $W_{\varphi,\psi_\gamma}(g) = W_{\varphi,\psi}\left(\begin{pmatrix} \gamma & \\ & 1 \end{pmatrix} g\right)$ if $\gamma \neq 0$. If we set $W_\varphi = W_{\varphi,\psi}$ for our fixed ψ, then the Fourier expansion of φ becomes

$$\varphi(g) = W_{\varphi,\psi_0}(g) + \sum_{\gamma \in k^\times} W_\varphi\left(\begin{pmatrix} \gamma & \\ & 1 \end{pmatrix} g\right).$$

Since φ is cuspidal

$$W_{\varphi,\psi_0}(g) = \int_{k\backslash\mathbb{A}} \varphi\left(\begin{pmatrix} 1 & x \\ 0 & 1 \end{pmatrix} g\right) dx \equiv 0$$

and the Fourier expansion for a cusp form φ becomes simply

$$\varphi(g) = \sum_{\gamma \in k^\times} W_\varphi\left(\begin{pmatrix} \gamma & \\ & 1 \end{pmatrix} g\right).$$

We will need a similar expansion for cusp forms φ on $\mathrm{GL}_n(\mathbb{A})$. The translations still correspond to the maximal unipotent subgroup

$$\mathrm{N}_n = \left\{ n = \begin{pmatrix} 1 & x_{1,2} & & * \\ & 1 & \ddots & \\ & & \ddots & \ddots \\ & & 1 & x_{n-1,n} \\ 0 & & & 1 \end{pmatrix} \right\},$$

but now this is nonabelian. This difficulty was solved independently by Piatetski-Shapiro [31] and Shalika [42]. We fix our nontrivial continuous character ψ of $k\backslash\mathbb{A}$ as above. Extend it to a character of N_n by setting $\psi(n) = \psi(x_{1,2} + \cdots + x_{n-1,n})$ and define the associated Fourier coefficient or Whittaker function by

$$W_\varphi(g) = W_{\varphi,\psi}(g) = \int_{\mathrm{N}_n(k)\backslash \mathrm{N}_n(\mathbb{A})} \varphi(ng)\psi^{-1}(n)\, dn.$$

Since φ is continuous and the integration is over a compact set, this integral is absolutely convergent, uniformly on compact sets. The Fourier expansion takes the following form.

Theorem 1.1. *Let $\varphi \in V_\pi$ be a cusp form on $\mathrm{GL}_n(\mathbb{A})$ and W_φ its associated ψ-Whittaker function. Then*

$$\varphi(g) = \sum_{\gamma \in \mathrm{N}_{n-1}(k)\backslash GL_{n-1}(k)} W_\varphi\left(\begin{pmatrix} \gamma & \\ & 1 \end{pmatrix} g\right)$$

with convergence absolute and uniform on compact subsets.

The proof of this fact is an induction. It utilizes the *mirabolic subgroup* P_n of GL_n which seems to be ubiquitous in the study of automorphic forms on GL_n. Abstractly, a mirabolic subgroup of GL_n is simply the stabilizer of a nonzero vector in (either) standard representation of GL_n on k^n. We denote by P_n the stabilizer of the row vector $e_n = (0, \ldots, 0, 1) \in k^n$. So

$$\mathrm{P}_n = \left\{ p = \begin{pmatrix} h & y \\ & 1 \end{pmatrix} \middle| h \in \mathrm{GL}_{n-1}, y \in k^{n-1} \right\} \simeq \mathrm{GL}_{n-1} \ltimes \mathrm{Y}_n$$

where

$$\mathrm{Y}_n = \left\{ y = \begin{pmatrix} I_{n-1} & y \\ & 1 \end{pmatrix} \middle| y \in k^{n-1} \right\} \simeq k^{n-1}.$$

Simply by restriction of functions, a cusp form on $\mathrm{GL}_n(\mathbb{A})$ restricts to a smooth cuspidal function on $\mathrm{P}_n(\mathbb{A})$ which remains left invariant under $\mathrm{P}_n(k)$. (A smooth function φ on $\mathrm{P}_n(\mathbb{A})$ which is left invariant under $\mathrm{P}_n(k)$ is called cuspidal if $\int_{\mathrm{U}(k)\backslash \mathrm{U}(\mathbb{A})} \varphi(up)\, du \equiv 0$ for every standard unipotent subgroup $\mathrm{U} \subset \mathrm{P}_n$.) Since $\mathrm{P}_n \supset \mathrm{N}_n$ we may define a Whittaker function attached to a cuspidal function φ on $\mathrm{P}_n(\mathbb{A})$ by the same integral as on $\mathrm{GL}_n(\mathbb{A})$, namely

$$W_\varphi(p) = \int_{\mathrm{N}_n(k)\backslash \mathrm{N}_n(\mathbb{A})} \varphi(np)\psi^{-1}(n)\, dn.$$

One proves by induction on n that for a cuspidal function φ on $\mathrm{P}_n(\mathbb{A})$ we have

$$\varphi(p) = \sum_{\gamma \in \mathrm{N}_{n-1}(k)\backslash \mathrm{GL}_{n-1}(k)} W_\varphi\left(\begin{pmatrix} \gamma & 0 \\ 0 & 1 \end{pmatrix} p\right)$$

with convergence absolute and uniform on compact subsets.

To obtain the Fourier expansion on GL_n from this, if φ is a cusp form on $\mathrm{GL}_n(\mathbb{A})$, then for $g \in \Omega$, a compact subset, the functions $\varphi_g(p) = \varphi(pg)$ form a compact family of cuspidal functions on $\mathrm{P}_n(\mathbb{A})$. So we have

$$\varphi_g(1) = \sum_{\gamma \in \mathrm{N}_{n-1}(k)\backslash \mathrm{GL}_{n-1}(k)} W_{\varphi_g}\left(\begin{pmatrix} \gamma & 0 \\ 0 & 1 \end{pmatrix}\right)$$

with convergence absolute and uniform. Hence

$$\varphi(g) = \sum_{\gamma \in \mathrm{N}_{n-1}(k)\backslash \mathrm{GL}_{n-1}(k)} W_\varphi\left(\begin{pmatrix} \gamma & 0 \\ 0 & 1 \end{pmatrix} g\right)$$

again with absolute convergence, uniform for $g \in \Omega$.

1.2 *Whittaker models and multiplicity one*

Consider now the functions W_φ appearing in the Fourier expansion of a cusp form φ. These are all smooth functions $W(g)$ on $\mathrm{GL}_n(\mathbb{A})$ which satisfy $W(ng) = \psi(n)W(g)$ for $n \in N_n(\mathbb{A})$. If we let $\mathcal{W}(\pi, \psi) = \{W_\varphi \mid \varphi \in V_\pi\}$, then $\mathrm{GL}_n(\mathbb{A})$ acts on this space by right translation and the map $\varphi \mapsto W_\varphi$ intertwines V_π with $\mathcal{W}(\pi, \psi)$. $\mathcal{W}(\pi, \psi)$ is called the *Whittaker model* of π.

The notion of a Whittaker model of a representation makes perfect sense over a local field. Let k_v be a local field (a completion of k for example) and let (π_v, V_{π_v}) be an irreducible admissible smooth representation of $\mathrm{GL}_n(k_v)$. Fix a nontrivial continuous additive character ψ_v of k_v. Let $\mathcal{W}(\psi_v)$ be the space of all smooth functions $W(g)$ on $\mathrm{GL}_n(k_v)$ satisfying $W(ng) = \psi_v(n)W(g)$ for all $n \in \mathrm{N}_k(k_v)$, that is, the space of all smooth Whittaker functions on $\mathrm{GL}_n(k_v)$ with respect to ψ_v. This is also the space of the smooth induced representation $\mathrm{Ind}_{\mathrm{N}_v}^{\mathrm{GL}_n}(\psi_v)$. $\mathrm{GL}_n(k_v)$ acts on this by right translation. If we have a nontrivial continuous intertwining $V_{\pi_v} \to \mathcal{W}(\psi_v)$, we will denote its image by $\mathcal{W}(\pi_v, \psi_v)$ and call it a Whittaker model of π_v.

Whittaker models for a representation (π_v, V_{π_v}) are equivalent to continuous Whittaker functionals on V_{π_v}, that is, continuous functionals Λ_v satisfying $\Lambda_v(\pi_v(n)\xi_v) = \psi_v(n)\Lambda_v(\xi_v)$ for all $n \in \mathrm{N}_n(k_v)$. To obtain a Whittaker functional from a model, set $\Lambda_v(\xi_v) = W_{\xi_v}(e)$, and to obtain a model from a functional, set $W_{\xi_v}(g) = \Lambda_v(\pi_v(g)\xi_v)$. This is a form of Frobenius reciprocity, which in this context is the isomorphism between $\mathrm{Hom}_{\mathrm{N}_n}(V_{\pi_v}, \mathbb{C}_{\psi_v})$ and $\mathrm{Hom}_{\mathrm{GL}_n}(V_{\pi_v}, \mathrm{Ind}_{\mathrm{N}_n}^{\mathrm{GL}_n}(\psi_v))$ constructed above.

The fundamental theorem on the existence and uniqueness of Whittaker functionals and models is the following.

Theorem 1.2. *Let (π_v, V_{π_v}) be a smooth irreducible admissible representation of $\mathrm{GL}_n(k_v)$. Let ψ_v be a nontrivial continuous additive character of k_v. Then the space of continuous ψ_v-Whittaker functionals on V_{π_v} is at most one dimensional. That is, Whittaker models, if they exist, are unique.*

This was first proved for nonarchimedean fields by Gelfand and Kazhdan [14] and their results were later extended to archimedean local fields by Shalika [42].

A smooth irreducible admissible representation (π_v, V_{π_v}) of $\mathrm{GL}_n(k_v)$ which possesses a Whittaker model is called *generic* or *nondegenerate*. Gelfand and Kazhdan in addition show that π_v is generic iff its contragredient $\widetilde{\pi}_v$ is generic, in fact that $\widetilde{\pi} \simeq \pi^\iota$ where ι is the outer automorphism $g^\iota = {}^t g^{-1}$, and in this case the Whittaker model for $\widetilde{\pi}_v$ can be obtained as $\mathcal{W}(\widetilde{\pi}_v, \psi_v^{-1}) = \{\widetilde{W}(g) = W(w_n\, {}^t g^{-1}) \mid W \in \mathcal{W}(\pi, \psi_v)\}$.

As a consequence of the local uniqueness of the Whittaker model we can conclude a global uniqueness. If (π, V_π) is an irreducible smooth admissible representation of $\mathrm{GL}_n(\mathbb{A})$, then π factors as a restricted tensor product of local representations $\pi \simeq \otimes' \pi_v$ taken over all places v of k [10, 13]. Consequently we have a continuous embedding $V_{\pi_v} \hookrightarrow V_\pi$ for each local component. Hence any Whittaker functional Λ on V_π determines a family of local Whittaker functionals

Λ_v on each V_{π_v} and conversely such that $\Lambda = \otimes' \Lambda_v$. Hence global uniqueness follows from the local uniqueness. Moreover, once we fix the isomorphism of V_π with $\otimes' V_{\pi_v}$ and define global and local Whittaker functions via Λ and the corresponding family Λ_v we have a factorization of global Whittaker functions

$$W_\xi(g) = \prod_v W_{\xi_v}(g_v)$$

for $\xi \in V_\pi$ which are factorizable in the sense that $\xi = \otimes' \xi_v$ corresponds to a pure tensor. As we will see, this factorization, which is a direct consequence of the uniqueness of the Whittaker model, plays a most important role in the development of Eulerian integrals for GL_n.

Now let us see what this means for our cuspidal representations (π, V_π) of $\mathrm{GL}_n(\mathbb{A})$. We have seen that for any smooth cusp form $\varphi \in V_\pi$ we have the Fourier expansion

$$\varphi(g) = \sum_{\gamma \in \mathrm{N}_{n-1}(k) \backslash GL_{n-1}(k)} W_\varphi\left(\begin{pmatrix} \gamma & \\ & 1 \end{pmatrix} g\right).$$

We can thus conclude that $\mathcal{W}(\pi, \psi) \neq 0$ and that π is (globally) generic with Whittaker functional

$$\Lambda(\varphi) = W_\varphi(e) = \int \varphi(ng)\psi^{-1}(n)\, dn.$$

Thus φ is completely determined by its associated Whittaker function W_φ. From the uniqueness of the global Whittaker model we can derive the Multiplicity One Theorem of Piatetski-Shapiro [32] and Shalika [42].

Multiplicity one. *Let (π, V_π) be an irreducible smooth admissible representation of $\mathrm{GL}_n(\mathbb{A})$. Then the multiplicity of π in the space of cusp forms on $\mathrm{GL}_n(\mathbb{A})$ is at most one.*

2 Eulerian integral representations

Let $f(\tau)$ again be a holomorphic cusp form of weight k on $\mathfrak{H}$ for the full modular group with Fourier expansion

$$f(\tau) = \sum a_n e^{2\pi i n \tau}.$$

Then Hecke [18] associated to f an L-function

$$L(s, f) = \sum a_n n^{-s}$$

and analyzed its analytic properties, namely continuation, order of growth, and functional equation, by writing it as the Mellin transform of f

$$\Lambda(s, f) = (2\pi)^{-s}\Gamma(s)L(s, f) = \int_0^\infty f(iy)y^s d^\times y.$$

An application of the modular transformation law for $f(\tau)$ under the transformation $\tau \mapsto -1/\tau$ gives the functional equation

$$\Lambda(s, f) = (-1)^{k/2}\Lambda(k-s, f).$$

Moreover, if f is an eigenfunction of all Hecke operators, then $L(s, f)$ has an Euler product expansion

$$L(s, f) = \prod_p (1 - a_p p^{-s} + p^{k-1-2s})^{-1}.$$

There is a similar theory for cuspidal automorphic representations (π, V_π) of $\mathrm{GL}_n(\mathbb{A})$. For applications to the Langlands conjectures and to functoriality via the Converse Theorem, we will need not only the standard L-functions $L(s, \pi)$ but the twisted L-functions $L(s, \pi \times \pi')$ for $(\pi', V_{\pi'})$ a cuspidal automorphic representation of $\mathrm{GL}_m(\mathbb{A})$ for $m < n$ as well.

The basic references for this section are Jacquet–Langlands [21], Jacquet, Piatetski-Shapiro, and Shalika [22], and Jacquet and Shalika [25].

2.1 *Integral representations for* GL_2

Let us first consider the L-functions for cuspidal automorphic representations (π, V_π) of $\mathrm{GL}_2(\mathbb{A})$ with twists by an idele class character χ, or, which is the same, a (cuspidal) automorphic representation of $\mathrm{GL}_1(\mathbb{A})$, as in Jacquet–Langlands [21].

Following Jacquet and Langlands, who were following Hecke, for each $\varphi \in V_\pi$ we consider the integral

$$I(s; \varphi, \chi) = \int_{k^\times\backslash\mathbb{A}^\times} \varphi\begin{pmatrix} a & \\ & 1 \end{pmatrix} \chi(a)|a|^{s-1/2}\, d^\times a.$$

Since a cusp form on $\mathrm{GL}_2(\mathbb{A})$ is rapidly decreasing upon restriction to $\mathbb{A}^\times$ as in the integral, it follows that the integral is absolutely convergent for all s, uniformly for $\mathrm{Re}(s)$ in an interval. Thus $I(s; \varphi, \chi)$ is an entire function of s, bounded in any vertical strip $a \leq \mathrm{Re}(s) \leq b$. Moreover, if we let $\widetilde{\varphi}(g) = \varphi({}^t g^{-1}) = \varphi(w_n\, {}^t g^{-1})$, then $\widetilde{\varphi} \in V_{\widetilde{\pi}}$ and the simple change of variables $a \mapsto a^{-1}$ in the integral shows that each integral satisfies a functional equation of the form

$$I(s; \varphi, \chi) = I(1-s; \widetilde{\varphi}, \chi^{-1}).$$

So these integrals individually enjoy rather nice analytic properties.

If we replace φ by its Fourier expansion from Section 1 and unfold, we find

$$\begin{aligned} I(s; \varphi, \chi) &= \int_{k^\times\backslash\mathbb{A}^\times} \sum_{\gamma \in k^\times} W_\varphi\begin{pmatrix} \gamma a & \\ & 1 \end{pmatrix} \chi(a)|a|^{s-1/2}\, d^\times a \\ &= \int_{\mathbb{A}^\times} W_\varphi\begin{pmatrix} a & \\ & 1 \end{pmatrix} \chi(a)|a|^{s-1/2}\, d^\times a \end{aligned}$$

where we have used the fact that the function $\chi(a)|a|^{s-1/2}$ is invariant under $k^\times$. By standard gauge estimates on Whittaker functions [22] this converges for $\mathrm{Re}(s) \gg 0$ after the unfolding. As we have seen in Section 1, if $W_\varphi \in \mathcal{W}(\pi, \psi)$ corresponds to a decomposable vector $\varphi \in V_\pi \simeq \otimes' V_{\pi_v}$, then the Whittaker function factors into a product of local Whittaker functions

$$W_\varphi(g) = \prod_v W_{\varphi_v}(g_v).$$

Since the character χ and the adelic absolute value factor into local components and the domain of integration $\mathbb{A}^\times$ also factors, we find that our global integral naturally factors into a product of local integrals

$$\int_{\mathbb{A}^\times} W_\varphi\begin{pmatrix} a & \\ & 1\end{pmatrix}\chi(a)|a|^{s-1/2}\,d^\times a = \prod_v \int_{k_v^\times} W_{\varphi_v}\begin{pmatrix} a_v & \\ & 1\end{pmatrix}\chi_v(a_v)|a_v|^{s-1/2}\,d^\times a_v,$$

with the infinite product still convergent for $\mathrm{Re}(s) \gg 0$, or

$$I(s;\varphi,\chi) = \prod_v \Psi_v(s; W_{\varphi_v}, \chi_v)$$

with the obvious definition of the local integrals

$$\Psi_v(s; W_{\varphi_v}, \chi_v) = \int_{k_v^\times} W_{\varphi_v}\begin{pmatrix} a_v & \\ & 1\end{pmatrix}\chi_v(a_v)|a_v|^{s-1/2}\,d^\times a_v.$$

Thus each of our global integrals is Eulerian.

In this way, to π and χ we have associated a family of global Eulerian integrals with nice analytic properties as well as for each place v a family of local integrals convergent for $\mathrm{Re}(s) \gg 0$.

2.2 *Integral representations for* $\mathrm{GL}_n \times \mathrm{GL}_m$ *with* $m < n$

Now let (π, V_π) be a cuspidal representation of $\mathrm{GL}_n(\mathbb{A})$ and $(\pi', V_{\pi'})$ a cuspidal representation of $\mathrm{GL}_m(\mathbb{A})$ with $m < n$. Take $\varphi \in V_\pi$ and $\varphi' \in V_{\pi'}$. At first glance, a natural analogue of the integrals we considered for GL_2 with GL_1 twists would be

$$\int_{\mathrm{GL}_m(k)\backslash \mathrm{GL}_m(\mathbb{A})} \varphi\begin{pmatrix} h & \\ & I_{n-m}\end{pmatrix}\varphi'(h)|\det(h)|^{s-(n-m)/2}\,dh.$$

This family of integrals would have all the nice analytic properties as before (entire functions of finite order satisfying a functional equation), but they would not be Eulerian except in the case $m = n-1$, which proceeds exactly as in the GL_2 case. The problem is that the restriction of the form φ to GL_m is too brutal to allow a nice unfolding when the Fourier expansion of φ is inserted. Instead we will introduce projection operators from cusp forms on $\mathrm{GL}_n(\mathbb{A})$ to cuspidal functions on on $P_{m+1}(\mathbb{A})$ which are given by part of the unipotent integration through which the Whittaker function is defined.

In GL_n, let $\mathrm{Y}_{n,m}$ be the unipotent radical of the standard parabolic subgroup attached to the partition $(m+1, 1, \ldots, 1)$. If ψ is our standard additive character of $k\backslash\mathbb{A}$, then ψ defines a character of $\mathrm{Y}_{n,m}(\mathbb{A})$ trivial on $\mathrm{Y}_{n,m}(k)$ since $\mathrm{Y}_{n,m} \subset \mathrm{N}_n$. The group $\mathrm{Y}_{n,m}$ is normalized by $\mathrm{GL}_{m+1} \subset \mathrm{GL}_n$ and the mirabolic subgroup $\mathrm{P}_{m+1} \subset \mathrm{GL}_{m+1}$ is the stabilizer in GL_{m+1} of the character ψ.

If $\varphi(g)$ is a cusp form on $\mathrm{GL}_n(\mathbb{A})$ define the projection operator $\mathbb{P}^n_m$ from cusp forms on $\mathrm{GL}_n(\mathbb{A})$ to cuspidal functions on $\mathrm{P}_{m+1}(\mathbb{A})$ by

$$\mathbb{P}^n_m\varphi(p) = |\det(p)|^{-\left(\frac{n-m-1}{2}\right)} \int_{\mathrm{Y}_{n,m}(k)\backslash \mathrm{Y}_{n,m}(\mathbb{A})} \varphi\left(y\begin{pmatrix} p & \\ & I_{n-m-1}\end{pmatrix}\right)\psi^{-1}(y)\,dy$$

for $p \in \mathrm{P}_{m+1}(\mathbb{A})$. As the integration is over a compact domain, the integral is absolutely convergent. One can easily check that $\mathbb{P}^n_m\varphi(p)$ is indeed cuspidal on $\mathrm{P}_{m+1}(\mathbb{A})$. From Section 1, we know that cuspidal functions on $\mathrm{P}_{m+1}(\mathbb{A})$ have a Fourier expansion summed over $\mathrm{N}_m(k)\backslash \mathrm{GL}_m(k)$. Applying this expansion to our projected cusp form on $\mathrm{GL}_n(\mathbb{A})$ we find that for $h \in \mathrm{GL}_m(\mathbb{A})$, $\mathbb{P}^n_m\varphi\begin{pmatrix} h & \\ & 1\end{pmatrix}$ has the Fourier expansion

$$\mathbb{P}^n_m\varphi\begin{pmatrix} h & \\ & 1\end{pmatrix} = |\det(h)|^{-\left(\frac{n-m-1}{2}\right)} \sum_{\gamma\in \mathrm{N}_m(k)\backslash \mathrm{GL}_m(k)} W_\varphi\left(\begin{pmatrix} \gamma & 0 \\ 0 & I_{n-m}\end{pmatrix}\begin{pmatrix} h & \\ & I_{n-m}\end{pmatrix}\right)$$

with convergence absolute and uniform on compact subsets.

We now have the prerequisites for writing down a family of Eulerian integrals for cusp forms φ on GL_n twisted by automorphic forms on GL_m for $m < n$. Let $\varphi \in V_\pi$ be a cusp form on $\mathrm{GL}_n(\mathbb{A})$ and $\varphi' \in V_{\pi'}$ a cusp form on $\mathrm{GL}_m(\mathbb{A})$. (Actually, we could take φ' to be an arbitrary automorphic form on $\mathrm{GL}_m(\mathbb{A})$.) Consider the integrals

$$I(s;\varphi,\varphi') = \int_{\mathrm{GL}_m(k)\backslash \mathrm{GL}_m(\mathbb{A})} \mathbb{P}^n_m\varphi\begin{pmatrix} h & 0 \\ 0 & 1\end{pmatrix}\varphi'(h)|\det(h)|^{s-1/2}\,dh.$$

The integral $I(s;\varphi,\varphi')$ is absolutely convergent for all values of the complex parameter s, uniformly in compact subsets, since the cusp forms are rapidly decreasing. Hence it is entire and bounded in any vertical strip as before.

Let us now investigate the Eulerian properties of these integrals. We first replace $\mathbb{P}^n_m\varphi$ by its Fourier expansion to obtain

$$I(s;\varphi,\varphi')$$
$$= \int_{\mathrm{GL}_m(k)\backslash \mathrm{GL}_m(\mathbb{A})} \sum_{\gamma\in \mathrm{N}_m(k)\backslash \mathrm{GL}_m(k)} W_\varphi\begin{pmatrix} \gamma h & 0 \\ 0 & I_{n-m}\end{pmatrix}\varphi'(h)|\det(h)|^{s-(n-m)/2}\,dh.$$

Since $\varphi'(h)$ is automorphic on $\mathrm{GL}_m(\mathbb{A})$ and $|\det(\gamma)| = 1$ for $\gamma \in \mathrm{GL}_m(k)$ we may interchange the order of summation and integration for $\mathrm{Re}(s) \gg 0$ and then recombine to obtain

$$I(s;\varphi,\varphi') = \int_{\mathrm{N}_m(k)\backslash \mathrm{GL}_m(\mathbb{A})} W_\varphi\begin{pmatrix} h & 0 \\ 0 & I_{n-m}\end{pmatrix}\varphi'(h)|\det(h)|^{s-(n-m)/2}\,dh.$$

This integral is absolutely convergent for $\mathrm{Re}(s) \gg 0$ by the gauge estimates of [22, Section 13] and this justifies the interchange. Let us now integrate first over $\mathrm{N}_m(k)\backslash \mathrm{N}_m(\mathbb{A})$. Recall that for $n \in \mathrm{N}_m(\mathbb{A}) \subset \mathrm{N}_n(\mathbb{A})$ we have $W_\varphi(ng) = \psi(n) W_\varphi(g)$. Hence we obtain

$$\begin{aligned} I(s;\varphi,\varphi') &= \int_{\mathrm{N}_m(\mathbb{A})\backslash \mathrm{GL}_m(\mathbb{A})} W_\varphi \begin{pmatrix} h & 0 \\ 0 & I_{n-m} \end{pmatrix} W'_{\varphi'}(h) |\det(h)|^{s-(n-m)/2}\, dh \\ &= \Psi(s; W_\varphi, W'_{\varphi'}), \end{aligned}$$

where $W'_{\varphi'}(h)$ is the ψ^{-1}-Whittaker function on $\mathrm{GL}_m(\mathbb{A})$ associated to φ', i.e.,

$$W'_{\varphi'}(h) = \int_{\mathrm{N}_m(k)\backslash \mathrm{N}_m(\mathbb{A})} \varphi'(nh)\psi(n)\, dn,$$

and we retain absolute convergence for $\mathrm{Re}(s) \gg 0$.

From this point, the fact that the integrals are Eulerian is a consequence of the uniqueness of the Whittaker model for GL_n. Take φ a smooth cusp form in a cuspidal representation π of $\mathrm{GL}_n(\mathbb{A})$. Assume in addition that φ is factorizable, i.e., in the decomposition $\pi = \otimes' \pi_v$ of π into a restricted tensor product of local representations, $\varphi = \otimes \varphi_v$ is a pure tensor. Then as we have seen there is a choice of local Whittaker models so that $W_\varphi(g) = \prod W_{\varphi_v}(g_v)$. Similarly for decomposable φ' we have the factorization $W'_{\varphi'}(h) = \prod W'_{\varphi'_v}(h_v)$. If we substitute these factorizations into our integral expression, then since the domain of integration factors $\mathrm{N}_m(\mathbb{A})\backslash \mathrm{GL}_m(\mathbb{A}) = \prod \mathrm{N}_m(k_v)\backslash \mathrm{GL}_m(k_v)$ we see that our integral factors into a product of local integrals as

$$I(s;\varphi,\varphi') = \Psi(s; W_\varphi, W'_{\varphi'}) = \prod_v \Psi_v(s; W_{\varphi_v}, W'_{\varphi'_v})$$

where the local integrals are given by

$$\begin{aligned} &\Psi_v(s; W_{\varphi_v}, W'_{\varphi'_v}) \\ &\quad = \int_{\mathrm{N}_m(k_v)\backslash \mathrm{GL}_m(k_v)} W_{\varphi_v} \begin{pmatrix} h_v & 0 \\ 0 & I_{n-m} \end{pmatrix} W'_{\varphi'_v}(h_v) |\det(h_v)|_v^{s-(n-m)/2}\, dh_v. \end{aligned}$$

The individual local integrals converge for $\mathrm{Re}(s) \gg 0$ by the gauge estimate of [22, Proposition 2.3.6]. We now see that we now have constructed a family of Eulerian integrals.

Now let us return to the question of a functional equation. As in the case of GL_2, the functional equation is essentially a consequence of the existence of the outer automorphism $g \mapsto \iota(g) = g^\iota = {}^t g^{-1}$ of GL_n. If we define the action of this automorphism on automorphic forms by setting $\widetilde{\varphi}(g) = \varphi(g^\iota) = \varphi(w_n g^\iota)$ and let $\widetilde{\mathbb{P}}^n_m = \iota \circ \mathbb{P}^n_m \circ \iota$, then our integrals naturally satisfy the functional equation

$$I(s;\varphi,\varphi') = \widetilde{I}(1-s; \widetilde{\varphi}, \widetilde{\varphi}')$$

where

$$\widetilde{I}(s;\varphi,\varphi') = \int_{\mathrm{GL}_m(k)\backslash \mathrm{GL}_m(\mathbb{A})} \widetilde{\mathbb{P}}^n_m\varphi\begin{pmatrix} h & \\ & 1\end{pmatrix}\varphi'(h)|\det(h)|^{s-1/2}\,dh.$$

We have established the following result.

Theorem 2.1. *Let $\varphi \in V_\pi$ be a cusp form on $\mathrm{GL}_n(\mathbb{A})$ and $\varphi' \in V_{\pi'}$ a cusp form on $\mathrm{GL}_m(\mathbb{A})$ with $m < n$. Then the family of integrals $I(s;\varphi,\varphi')$ define entire functions of s, bounded in vertical strips, and satisfy the functional equation*

$$I(s;\varphi,\varphi') = \widetilde{I}(1-s;\widetilde{\varphi},\widetilde{\varphi}').$$

Moreover the integrals are Eulerian and if φ and φ' are factorizable, we have

$$I(s;\varphi,\varphi') = \prod_v \Psi_v(s;W_{\varphi_v},W'_{\varphi'_v})$$

with convergence absolute and uniform for $\mathrm{Re}(s) \gg 0$.

The integrals occurring in the right-hand side of our functional equation are again Eulerian. One can unfold the definitions to find first that

$$\widetilde{I}(1-s;\widetilde{\varphi},\widetilde{\varphi}') = \widetilde{\Psi}(1-s;\rho(w_{n,m})\widetilde{W}_\varphi,\widetilde{W}'_{\varphi'})$$

where the unfolded global integral is

$$\widetilde{\Psi}(s;W,W') = \int\int W\begin{pmatrix} h & & \\ x & I_{n-m-1} & \\ & & 1\end{pmatrix} dx\, W'(h)|\det(h)|^{s-(n-m)/2}\,dh$$

with the h integral over $\mathrm{N}_m(\mathbb{A})\backslash \mathrm{GL}_m(\mathbb{A})$ and the x integral over $M_{n-m-1,m}(\mathbb{A})$, the space of $(n-m-1)\times m$ matrices, ρ denoting right translation, and $w_{n,m}$ the Weyl element $w_{n,m} = \begin{pmatrix} I_m & \\ & w_{n-m}\end{pmatrix}$ with $w_{n-m} = \begin{pmatrix} & & 1\\ & \cdot^{\cdot^{\cdot}} & \\ 1 & & \end{pmatrix}$ the standard long Weyl element in GL_{n-m}. Also, for $W \in \mathcal{W}(\pi,\psi)$, we set $\widetilde{W}(g) = W(w_n g^\iota) \in \mathcal{W}(\widetilde{\pi},\psi^{-1})$. The extra unipotent integration is the remnant of $\widetilde{\mathbb{P}}^n_m$. As before, $\widetilde{\Psi}(s;W,W')$ is absolutely convergent for $\mathrm{Re}(s) \gg 0$. For φ and φ' factorizable as before, these integrals $\widetilde{\Psi}(s;W_\varphi,W'_{\varphi'})$ will factor as well. Hence we have

$$\widetilde{\Psi}(s;W_\varphi,W'_{\varphi'}) = \prod_v \widetilde{\Psi}_v(s;W_{\varphi_v},W'_{\varphi'_v})$$

where

$$\widetilde{\Psi}_v(s;W_v,W'_v) = \int\int W_v\begin{pmatrix} h_v & & \\ x_v & I_{n-m-1} & \\ & & 1\end{pmatrix} dx_v\, W'_v(h_v)|\det(h_v)|^{s-(n-m)/2}\,dh_v,$$

where now with the h_v the integral is over $\mathrm{N}_m(k_v)\backslash \mathrm{GL}_m(k_v)$ and the x_v integral is over the matrix space $M_{n-m-1,m}(k_v)$. Thus, coming back to our functional equation, we find that the right-hand side is Eulerian and factors as

$$\widetilde{I}(1-s;\widetilde{\varphi},\widetilde{\varphi}') = \widetilde{\Psi}(1-s;\rho(w_{n,m})\widetilde{W}_\varphi,\widetilde{W}'_{\varphi'}) = \prod_v \widetilde{\Psi}_v(1-s;\rho(w_{n,m})\widetilde{W}_{\varphi_v},\widetilde{W}'_{\varphi'_v}).$$

2.3 *Integral representations for* $\mathrm{GL}_n \times \mathrm{GL}_n$

The paradigm for integral representations of L-functions for $\mathrm{GL}_n \times \mathrm{GL}_n$ is not Hecke but rather the classical papers of Rankin [33] and Selberg [34]. These were first interpreted in the framework of automorphic representations by Jacquet for $\mathrm{GL}_2 \times \mathrm{GL}_2$ [20] and then Jacquet and Shalika in general [25].

Let (π, V_π) and $(\pi', V_{\pi'})$ be two cuspidal representations of $\mathrm{GL}_n(\mathbb{A})$. Let $\varphi \in V_\pi$ and $\varphi' \in V_{\pi'}$ be two cusp forms. The analogue of the construction above would be simply

$$\int_{\mathrm{GL}_n(k)\backslash \mathrm{GL}_n(\mathbb{A})} \varphi(g)\varphi'(g)|\det(g)|^s \, dg.$$

This integral is essentially the L^2-inner product of φ and φ' and is not suitable for defining an L-function, although it will occur as a residue of our integral at a pole. Instead, following Rankin and Selberg, we use an integral representation that involves a third function: an Eisenstein series on $\mathrm{GL}_n(\mathbb{A})$. This family of Eisenstein series is constructed using the mirabolic subgroup once again.

To construct our Eisenstein series we return to the observation that $\mathrm{P}_n \backslash \mathrm{GL}_n \simeq k^n - \{0\}$. If we let $\mathcal{S}(\mathbb{A}^n)$ denote the Schwartz–Bruhat functions on $\mathbb{A}^n$, then each $\Phi \in \mathcal{S}$ defines a smooth function on $\mathrm{GL}_n(\mathbb{A})$, left invariant by $\mathrm{P}_n(\mathbb{A})$, by $g \mapsto \Phi((0, \dots, 0, 1)g) = \Phi(e_n g)$. Let η be a unitary idele class character. (For our application η will be determined by the central characters of π and π'.) Consider the function

$$F(g, \Phi; s, \eta) = |\det(g)|^s \int_{\mathbb{A}^\times} \Phi(a e_n g)|a|^{ns}\eta(a) \, d^\times a.$$

If we let $\mathrm{P}'_n = \mathrm{Z}_n \mathrm{P}_n$ be the parabolic of GL_n associated to the partition $(n-1, 1)$ and extend η to a character of P'_n by $\eta(p') = \eta(d)$ for $p' = \left(\begin{smallmatrix} h & y \\ 0 & d \end{smallmatrix}\right) \in \mathrm{P}'_n(\mathbb{A})$ with $h \in \mathrm{GL}_{n-1}(\mathbb{A})$ and $d \in \mathbb{A}^\times$, we have that $F(g, \Phi; s, \eta)$ is a smooth section of the normalized induced representation $\mathrm{Ind}_{\mathrm{P}'_n(\mathbb{A})}^{\mathrm{GL}_n(\mathbb{A})}(\delta_{\mathrm{P}'_n}^{s-1/2}\eta)$. Since the inducing character $\delta_{\mathrm{P}'_n}^{s-1/2}\eta$ of $\mathrm{P}'_n(\mathbb{A})$ is invariant under $\mathrm{P}'_n(k)$ we may form Eisenstein series from this family of sections by

$$E(g, \Phi; s, \eta) = \sum_{\gamma \in \mathrm{P}'_n(k)\backslash \mathrm{GL}_n(k)} F(\gamma g, \Phi; s, \eta).$$

This is absolutely convergent for $\mathrm{Re}(s) > 1$ [25].

If we replace F in this sum by its definition and unfold, we can rewrite this Eisenstein series as

$$E(g, \Phi; s, \eta) = |\det(g)|^s \int_{k^\times \backslash \mathbb{A}^\times} \Theta'_\Phi(a, g)|a|^{ns}\eta(a) \, d^\times a.$$

This second expression essentially gives the Eisenstein series as the Mellin transform of the Theta series

$$\Theta_\Phi(a, g) = \sum_{\xi \in k^n} \Phi(a\xi g),$$

where in the above we have written

$$\Theta'_\Phi(a,g) = \sum_{\xi \in k^n - \{0\}} \Phi(a\xi g) = \Theta_\Phi(a,g) - \Phi(0).$$

This allows us to obtain the analytic properties of the Eisenstein series from the Poisson summation formula for Θ_Φ. Poisson summation gives

$$\begin{aligned} E(g,\Phi,s,\eta) = {} & |\det(g)|^s \int_{|a|\geq 1} \Theta'_\Phi(a,g)|a|^{ns}\eta(a)\,d^\times a \\ & + |\det(g)|^{s-1} \int_{|a|\geq 1} \Theta'_{\hat\Phi}(a,{}^tg^{-1})|a|^{n(1-s)}\eta^{-1}(a)\,d^\times a + \delta(s), \end{aligned}$$

where the Fourier transform $\hat\Phi$ on $\mathcal{S}(\mathbb{A}^n)$ is defined by

$$\hat\Phi(x) = \int_{\mathbb{A}^\times} \Phi(y)\psi(y\,{}^tx)\,dy$$

and

$$\delta(s) = \begin{cases} 0 & \text{if } \eta \text{ is ramified} \\ -c\Phi(0)\frac{|\det(g)|^s}{s+i\sigma} + c\hat\Phi(0)\frac{|\det(g)|^{s-1}}{s-1+i\sigma} & \text{if } \eta(a) = |a|^{in\sigma} \text{ with } \sigma \in \mathbb{R} \end{cases}$$

with c a nonzero constant.

From this we derive easily the basic properties of our Eisenstein series [25, Section 4]. The Eisenstein series $E(g,\Phi;s,\eta)$ has a meromorphic continuation to all of $\mathbb{C}$ with at most simple poles at $s = -i\sigma, 1 - i\sigma$ when η is unramified of the form $\eta(a) = |a|^{in\sigma}$. As a function of g it is smooth of moderate growth, and as a function of s it is bounded in vertical strips (away from the possible poles), uniformly for g in compact sets. Moreover, we have the functional equation

$$E(g,\Phi;s,\eta) = E(g^\iota,\hat\Phi;1-s,\eta^{-1}),$$

where $g^\iota = {}^tg^{-1}$. Note that under the center the Eisenstein series transforms by the central character η^{-1}.

Now let us return to our Eulerian integrals. Let π and π' be our irreducible cuspidal representations. Let their central characters be ω and ω'. Set $\eta = \omega\omega'$. Then for each pair of cusp forms $\varphi \in V_\pi$ and $\varphi' \in V_{\pi'}$ and each Schwartz–Bruhat function $\Phi \in \mathcal{S}(\mathbb{A}^n)$, set

$$I(s;\varphi,\varphi',\Phi) = \int_{\mathrm{Z}_n(\mathbb{A})\,\mathrm{GL}_n(k)\backslash \mathrm{GL}_n(\mathbb{A})} \varphi(g)\varphi'(g)E(g,\Phi;s,\eta)\,dg.$$

Since the two cusp forms are rapidly decreasing on $\mathrm{Z}_n(\mathbb{A})\,\mathrm{GL}_n(k)\backslash \mathrm{GL}_n(\mathbb{A})$ and the Eisenstein is only of moderate growth, we see that the integral converges absolutely for all s away from the poles of the Eisenstein series and is hence meromorphic. It will be bounded in vertical strips away from the poles and satisfies the functional equation

$$I(s;\varphi,\varphi',\Phi)=I(1-s;\widetilde{\varphi},\widetilde{\varphi}',\hat{\Phi}),$$

coming from the functional equation of the Eisenstein series, where we still have $\widetilde{\varphi}(g)=\varphi(g^\iota)=\varphi(w_n g^\iota)\in V_{\widetilde{\pi}}$ and similarly for $\widetilde{\varphi}'$.

These integrals will be entire unless we have $\eta(a)=\omega(a)\omega'(a)=|a|^{in\sigma}$ is unramified. In that case, the residue at $s=-i\sigma$ will be

$$\operatorname*{Res}_{s=-i\sigma} I(s;\varphi,\varphi',\Phi)=-c\Phi(0)\int_{Z_n(\mathbb{A})\,\mathrm{GL}_n(\mathbb{A})\backslash \mathrm{GL}_n(\mathbb{A})}\varphi(g)\varphi'(g)|\det(g)|^{-i\sigma}\,dg$$

and at $s=1-i\sigma$ we can write the residue as

$$\operatorname*{Res}_{s=1-i\sigma} I(s;\varphi,\varphi',\Phi)=c\hat{\Phi}(0)\int_{Z_n(\mathbb{A})\,\mathrm{GL}_n(k)\backslash \mathrm{GL}_n(\mathbb{A})}\widetilde{\varphi}(g)\widetilde{\varphi}'(g)|\det(g)|^{i\sigma}\,dg.$$

Therefore these residues define $\mathrm{GL}_n(\mathbb{A})$ invariant pairings between π and $\pi'\otimes|\det|^{-i\sigma}$ or equivalently between $\widetilde{\pi}$ and $\widetilde{\pi}'\otimes|\det|^{i\sigma}$. Hence a residues can be nonzero only if $\pi\simeq\widetilde{\pi}'\otimes|\det|^{i\sigma}$ and in this case we can find φ, φ', and Φ such that indeed the residue does not vanish.

We have yet to check that our integrals are Eulerian. To this end we take the integral, replace the Eisenstein series by its definition, and unfold we find

$$I(s;\varphi,\varphi',\Phi)=\int_{P_n(k)\backslash \mathrm{GL}_n(\mathbb{A})}\varphi(g)\varphi'(g)\Phi(e_n g)|\det(g)|^s\,dg.$$

We next replace φ by its Fourier expansion and unfold as before to find

$$\begin{aligned}I(s;\varphi,\varphi',\Phi)&=\Psi(s;W_\varphi,W'_{\varphi'},\Phi)\\&=\int_{N_n(\mathbb{A})\backslash \mathrm{GL}_n(\mathbb{A})}W_\varphi(g)W'_{\varphi'}(g)\Phi(e_n g)|\det(g)|^s\,dg.\end{aligned}$$

This expression converges for $\mathrm{Re}(s)\gg 0$.

To continue, we assume that φ, φ' and Φ are decomposable tensors under the isomorphisms $\pi\simeq\otimes'\pi_v$, $\pi'\simeq\otimes'\pi'_v$, and $\mathcal{S}(\mathbb{A}^n)\simeq\otimes'\mathcal{S}(k_v^n)$ so that we have $W_\varphi(g)=\prod_v W_{\varphi_v}(g_v)$, $W'_{\varphi'}(g)=\prod_v W'_{\varphi'_v}(g_v)$ and $\Phi(g)=\prod_v\Phi_v(g_v)$. Then, since the domain of integration also naturally factors we can decompose this last integral into an Euler product and now write

$$\Psi(s;W_\varphi,W'_{\varphi'},\Phi)=\prod_v\Psi_v(s;W_{\varphi_v},W'_{\varphi'_v},\Phi_v),$$

where

$$\Psi_v(s;W_{\varphi_v},W'_{\varphi'_v},\Phi_v)=\int_{N_n(k_v)\backslash \mathrm{GL}_n(k_v)}W_{\varphi_v}(g_v)W'_{\varphi'_v}(g_v)\Phi_v(e_n g_v)|\det(g_v)|^s\,dg_v,$$

still with convergence for $\mathrm{Re}(s)\gg 0$ by the local gauge estimates. Once again we see that the Euler factorization is a direct consequence of the uniqueness of the Whittaker models.

Theorem 2.2. *Let $\varphi \in V_\pi$ and $\varphi' \in V_{\pi'}$ cusp forms on $\mathrm{GL}_n(\mathbb{A})$ and let $\Phi \in \mathcal{S}(\mathbb{A}^n)$. Then the family of integrals $I(s; \varphi, \varphi', \Phi)$ define meromorphic functions of s, bounded in vertical strips away from the poles. The only possible poles are simple and occur iff $\pi \simeq \tilde{\pi}' \otimes |\det|^{i\sigma}$ with σ real and are then at $s = -i\sigma$ and $s = 1 - i\sigma$ with residues as above. They satisfy the functional equation*

$$I(s; \varphi, \varphi', \Phi) = I(1 - s; \widetilde{W}_\varphi, \widetilde{W}'_{\varphi'}, \hat{\Phi}).$$

Moreover, for φ, φ', and Φ factorizable we have that the integrals are Eulerian and we have

$$I(s; \varphi, \varphi', \Phi) = \prod_v \Psi_v(s; W_{\varphi_v}, W'_{\varphi'_v}, \Phi_v)$$

with convergence absolute and uniform for $\mathrm{Re}(s) \gg 0$.

We remark in passing that the right-hand side of the functional equation also unfolds as

$$\begin{aligned} I(1 - s; \tilde{\varphi}, \tilde{\varphi}', \hat{\Phi}) &= \int_{\mathrm{N}_n(\mathbb{A})\backslash \mathrm{GL}_n(\mathbb{A})} \widetilde{W}_\varphi(g)\widetilde{W}'_{\varphi'}(g)\hat{\Phi}(e_n g)|\det(g)|^{1-s}\, dg \\ &= \prod_v \Psi_v(1 - s; \widetilde{W}_\varphi, \widetilde{W}'_{\varphi'}, \hat{\Phi}) \end{aligned}$$

with convergence for $\mathrm{Re}(s) \ll 0$.

3 Local L-functions

If (π, V_π) is a cuspidal representation of $\mathrm{GL}_n(\mathbb{A})$ and $(\pi', V_{\pi'})$ is a cuspidal representation of $\mathrm{GL}_m(\mathbb{A})$ we have associated to the pair (π, π') a family of Eulerian integrals $\{I(s; \varphi, \varphi')\}$ (or $\{I(s; \varphi, \varphi', \Phi)\}$ if $m = n$) and through the Euler factorization we have for each place v of k a family of local integrals $\{\Psi_v(s; W_v, W'_v)\}$ (or $\{\Psi_v(s; W_v, W'_v, \Phi_v)\}$) attached to the pair of local components (π_v, π'_v). In this section we would like to attach a local L-function (or local Euler factor) $L(s, \pi_v \times \pi'_v)$ to such a pair of local representations through the family of local integrals and analyze its basic properties, including the local functional equation.

3.1 *Nonarchimedean local factors*

For this section let k denote a nonarchimedean local field. We will let $\mathfrak{o}$ denote the ring of integers of k and $\mathfrak{p}$ the unique prime ideal of $\mathfrak{o}$. Fix a generator ϖ of $\mathfrak{p}$. We let q be the residue degree of k, so $q = |\mathfrak{o}/\mathfrak{p}| = |\varpi|^{-1}$. We fix a nontrivial continuous additive character ψ of k. (π, V_π) and $(\pi', V_{\pi'})$ will now be the smooth vectors in irreducible admissible unitary generic representations of $\mathrm{GL}_n(k)$ and $\mathrm{GL}_m(k)$, respectively, as is true for local components of cuspidal representations. We will let ω and ω' denote their central characters. The basic reference here is the paper of Jacquet, Piatetski-Shapiro, and Shalika [23].

For each pair of Whittaker functions $W \in \mathcal{W}(\pi, \psi)$ and $W' \in \mathcal{W}(\pi', \psi^{-1})$ and in the case $n = m$ each Schwartz–Bruhat function $\Phi \in \mathcal{S}(k^n)$ we have defined local integrals $\Psi(s; W, W')$, $\widetilde{\Psi}(s; W, W')$ in the case $m < n$ and $\Psi(s; W, W', \Phi)$ in the case $n = m$, both convergent for $\mathrm{Re}(s) \gg 0$. To make the notation more convenient for what follows, in the case $m < n$ for any $0 \leq j \leq n - m - 1$ let us set

$$\Psi_j(s : W, W')$$
$$= \int_{\mathrm{N}_m(k)\backslash \mathrm{GL}_m(k)} \int_{M_{j,m}(k)} W\begin{pmatrix} h & & \\ x & I_j & \\ & & I_{n-m-j} \end{pmatrix} dx\; W'(h)|\det(h)|^{s-(n-m)/2}\, dh,$$

so that $\Psi(s; W, W') = \Psi_0(s; W, W')$ and $\widetilde{\Psi}(s; W, W') = \Psi_{n-m-1}(s; W, W')$, which is still absolutely convergent for $\mathrm{Re}(s) \gg 0$.

We need to understand what type of functions of s these local integrals are. To this end, we need to understand the local Whittaker functions. So let $W \in \mathcal{W}(\pi, \psi)$. Since W is smooth there is a compact open subgroup K so that $W(gk) = W(g)$ for all $k \in K$. W transforms on the left under $\mathrm{N}_n(k)$ via ψ. So the Iwasawa decomposition on $\mathrm{GL}_n(k)$ gives that it suffices to understand a general Whittaker function on the torus. Let α_i, $i = 1, \ldots, n-1$, denote the standard simple roots of GL_n, so that if $a = \begin{pmatrix} a_1 & & \\ & \ddots & \\ & & a_n \end{pmatrix} \in \mathrm{A}_n(k)$, then $\alpha_i(a) = a_i/a_{i+1}$. The fundamental result on the asymptotics of Whittaker functions [22] is that there is a finite set of finite functions $X(\pi) = \{\chi_i\}$ on $\mathrm{A}_n(k)$, depending only on π, so that for every $W \in \mathcal{W}(\pi, \psi)$ there are Schwartz–Bruhat functions $\phi_i \in \mathcal{S}(k^{n-1})$ such that for all $a \in \mathrm{A}_n(k)$ with $a_n = 1$ we have

$$W(a) = \sum_{X(\pi)} \chi_i(a)\phi_i(\alpha_1(a), \ldots, \alpha_{n-1}(a)).$$

By a finite function on $\mathrm{A}_n(k)$ we mean a continuous function whose translates span a finite-dimensional vector space [21, 22, Section 2.2]. For the field $k^\times$ itself the finite functions are spanned by products of characters and powers of the valuation map. The finite set of finite functions $X(\pi)$ which occur in the asymptotics near 0 of the Whittaker functions come from analyzing the Jacquet module $\mathcal{W}(\pi, \psi)/\langle \pi(n)W - W | n \in \mathrm{N}_n \rangle$ which is naturally an $\mathrm{A}_n(k)$-module. Note that due to the Schwartz–Bruhat functions, the Whittaker functions vanish whenever any simple root $\alpha_i(a)$ becomes large.

Several nice consequences follow from inserting these formulas for W and W' into the local integrals $\Psi_j(s; W, W')$ or $\Psi(s; W, W', \Phi)$ [22, 23].

1. Each integral converges for $\mathrm{Re}(s) \gg 0$. For π and π' unitary, as we have assumed, they converge absolutely for $\mathrm{Re}(s) \geq 1$. For π and π' tempered, we have absolute convergence for $\mathrm{Re}(s) > 0$.

2. Each integral defines a rational function in q^{-s} and hence meromorphically extends to all of $\mathbb{C}$.

3. Each such rational function can be written with a common denominator which depends only on the finite functions $X(\pi)$ and $X(\pi')$ and hence only on π and π'.

Let $\mathcal{I}_j(\pi, \pi')$ denote the complex linear span of the local integrals $\Psi_j(s; W, W')$ if $m < n$ and $\mathcal{I}(\pi, \pi')$ the complex linear span of the $\Psi(s; W, W', \Phi)$ if $m = n$. In the case $m < n$ one can show by a rather elementary although somewhat involved manipulation of the integrals that all of the ideals $\mathcal{I}_j(\pi, \pi')$ are the same [23], so we will write this ideal as $\mathcal{I}(\pi, \pi')$. These are then subspaces of $\mathbb{C}(q^{-s})$ which have "bounded denominators" in the sense of (3). In fact, these subspaces have more structure – each $\mathcal{I}(\pi, \pi')$ is a fractional $\mathbb{C}[q^s, q^{-s}]$-ideal of $\mathbb{C}(q^{-s})$. Since $\mathbb{C}[q^s, q^{-s}]$ is a principal ideal domain each fractional ideal $\mathcal{I}(\pi, \pi')$ has a single generator and since each of these fractional ideals contain 1 we can always normalize our generator to be of the form $P_{\pi,\pi'}(q^{-s})^{-1}$ where the polynomial $P(X)$ satisfies $P(0) = 1$. This gives us the definition of our local L-function:

$$L(s, \pi \times \pi') = P_{\pi,\pi'}(q^{-s})^{-1}$$

is the normalized generator of the fractional ideal $\mathcal{I}(\pi, \pi')$ formed by the family of local integrals. If $\pi' = \mathbf{1}$ is the trivial representation of $GL_1(k)$, then we write $L(s, \pi) = L(s, \pi \times \mathbf{1})$.

One can show easily that the ideal $\mathcal{I}(\pi, \pi')$ is independent of the character ψ used in defining the Whittaker models, so that $L(s, \pi \times \pi')$ is independent of the choice of ψ. So it is not included in the notation. Also, note that for $\pi' = \chi$ an automorphic representation (character) of $GL_1(\mathbb{A})$ we have the identity $L(s, \pi \times \chi) = L(s, \pi \otimes \chi)$ where $\pi \otimes \chi$ is the representation of $GL_n(\mathbb{A})$ on V_π given by $\pi \otimes \chi(g)\xi = \chi(\det(g))\pi(g)\xi$.

We summarize the above in the following theorem.

Theorem 3.1. *Let π and π' be as above. The family of local integrals form a $\mathbb{C}[q^s, q^{-s}]$-fractional ideal $\mathcal{I}(\pi, \pi')$ in $\mathbb{C}(q^{-s})$ with generator the local L-function $L(s, \pi \times \pi')$.*

Another useful way of thinking of the local L-function is the following. $L(s, \pi \times \pi')$ is the minimal (in terms of degree) function of the form $P(q^{-s})^{-1}$, with $P(X)$ a polynomial satisfying $P(0) = 1$, such that the ratios $\frac{\Psi(s; W, W')}{L(s, \pi \times \pi')}$ (resp., $\frac{\Psi(s; W, W', \Phi)}{L(s, \pi \times \pi')}$) are entire for all $W \in \mathcal{W}(\pi, \psi)$ and $W' \in \mathcal{W}(\pi', \psi^{-1})$, and if necessary $\Phi \in \mathcal{S}(k^n)$. That is, $L(s, \pi \times \pi')$ is the standard Euler factor determined by the poles of the functions in $\mathcal{I}(\pi, \pi')$.

One should note that since the L-factor is a generator of the ideal $\mathcal{I}(\pi, \pi')$, then in particular it lies in $\mathcal{I}(\pi, \pi')$. Since this ideal is spanned by our local integrals, then we can always write

$$L(s, \pi \times \pi') = \sum_i \Psi(s; W_i, W_i') \quad \text{or} \quad L(s, \pi \times \pi') = \sum_i \Psi(s; W_i, W_i', \Phi_i)$$

with a finite collection of $W_i \in \mathcal{W}(\pi, \psi)$, $W_i' \in \mathcal{W}(\pi', \psi^{-1})$, and if necessary $\Phi_i \in \mathcal{S}(k^n)$. This will be necessary for the global theory.

Either by analogy with Tate's thesis or from the corresponding global statement, we would expect our local integrals to satisfy a local functional equation. From the functional equations for our global integrals, we would expect these to relate the integrals $\Psi(s; W, W')$ and $\widetilde{\Psi}(1-s; \rho(w_{n,m})\widetilde{W}, \widetilde{W}')$ when $m < n$ and $\Psi(s; W, W', \Phi)$ and $\Psi(1-s; \widetilde{W}, \widetilde{W}', \hat{\Phi})$ when $m = n$. This will indeed be the case. These functional equations will come from interpreting the local integrals as families (in s) of quasiinvariant bilinear forms on $\mathcal{W}(\pi, \psi) \times \mathcal{W}(\pi', \psi^{-1})$ or trilinear forms on $\mathcal{W}(\pi, \psi) \times \mathcal{W}(\pi', \psi^{-1}) \times \mathcal{S}(k^n)$ depending on the case. One is able to analyze such functional using Bruhat's theory and one shows that, except for a finite number if exceptional values of q^{-s} such bilinear functionals are unique [23]. Hence we obtain the following local functional equation.

Theorem 3.2. *There is a rational function $\gamma(s, \pi \times \pi', \psi) \in \mathbb{C}(q^{-s})$ such that we have*

$$\widetilde{\Psi}(1-s; \rho(w_{n,m})\widetilde{W}, \widetilde{W}') = \omega'(-1)^{n-1}\gamma(s, \pi \times \pi', \psi)\Psi(s; W, W') \quad \textit{if } m < n$$

or

$$\Psi(1-s; \widetilde{W}, \widetilde{W}', \hat{\Phi}) = \omega'(-1)^{n-1}\gamma(s, \pi \times \pi', \psi)\Psi(s; W, W', \Phi) \quad \textit{if } m = n$$

for all $W \in \mathcal{W}(\pi, \psi)$, $W' \in \mathcal{W}(\pi', \psi^{-1})$, and if necessary all $\Phi \in \mathcal{S}(k^n)$.

An equally important local factor, which occurs in the current formulations of the local Langlands correspondence [2, 16, 19], is the local ε-factor, defined as the ratio

$$\varepsilon(s, \pi \times \pi', \psi) = \frac{\gamma(s, \pi \times \pi', \psi)L(s, \pi \times \pi')}{L(1-s, \widetilde{\pi} \times \widetilde{\pi}')}.$$

With the local ε-factor the local functional equation can be written in the form

$$\frac{\widetilde{\Psi}(1-s; \rho(w_{n,m})\widetilde{W}, \widetilde{W}')}{L(1-s, \widetilde{\pi} \times \widetilde{\pi}')} = \omega'(-1)^{n-1}\varepsilon(s, \pi \times \pi', \psi)\frac{\Psi(s; W, W')}{L(s, \pi \times \pi')} \quad \text{if } m < n$$

or

$$\frac{\Psi(1-s; \widetilde{W}, \widetilde{W}', \hat{\Phi})}{L(1-s, \widetilde{\pi} \times \widetilde{\pi}')} = \omega'(-1)^{n-1}\varepsilon(s, \pi \times \pi', \psi)\frac{\Psi(s; W, W', \Phi)}{L(s, \pi \times \pi')} \quad \text{if } m = n.$$

Analyzing the local functional equation in this form allows one to prove that $\varepsilon(s, \pi \times \pi', \psi)$ is a monomial function of the form cq^{-fs}. If we consider a single π, take ψ unramified, and write $\varepsilon(s, \pi, \psi) = \varepsilon(0, \pi, \psi)q^{-f(\pi)s}$, then in [24] it is shown that $f(\pi)$ is a nonnegative integer, $f(\pi) = 0$ iff π is unramified, and in general the integer $f(\pi)$ is the *conductor of* π.

Let us now turn to the calculation of the local L-functions. The first case to consider is that where both π and π' are unramified. Since they are assumed generic, they are both full induced representations from unramified characters of the Borel subgroup [49]. So let us write $\pi \simeq \mathrm{Ind}_{\mathrm{B}_n}^{\mathrm{GL}_n}(\mu_1 \otimes \cdots \otimes \mu_n)$ and

$\pi' \simeq \mathrm{Ind}_{\mathrm{B}_m}^{\mathrm{GL}_m}(\mu_1' \otimes \cdots \otimes \mu_m')$ with the μ_i and μ_j' unramified characters of $k^\times$. The Satake parameterization of unramified representations associates to each of these representation the semisimple conjugacy classes $[A_\pi] \in \mathrm{GL}_n(\mathbb{C})$ and $[A_{\pi'}] \in \mathrm{GL}_m(\mathbb{C})$ given by

$$A_\pi = \begin{pmatrix} \mu_1(\varpi) & & \\ & \ddots & \\ & & \mu_n(\varpi) \end{pmatrix}, \qquad A_{\pi'} = \begin{pmatrix} \mu_1'(\varpi) & & \\ & \ddots & \\ & & \mu_m'(\varpi) \end{pmatrix}.$$

(Recall that ϖ is a uniformizing parameter for k, that is, a generator of $\mathfrak{p}$.) In the Whittaker models there will be unique normalized $\mathrm{K} = \mathrm{GL}(\mathfrak{o})$-fixed Whittaker functions, $W_\circ \in \mathcal{W}(\pi, \psi)$ and $W_\circ' \in \mathcal{W}(\pi', \psi^{-1})$, normalized by $W_\circ(e) = W_\circ'(e) = 1$. There is an explicit formula for $W_\circ$ in terms of the Satake parameter A_π due to Shintani [43]. Utilizing this formula, one obtains the following explicit computation of the local L-factor in this case.

Theorem 3.3. *If π, π', and ψ are all unramified, then*

$$L(s, \pi \times \pi') = \det(I - q^{-s} A_\pi \otimes A_{\pi'})^{-1} = \begin{cases} \Psi(s; W_\circ, W_\circ') & m < n \\ \Psi(s; W_\circ, W_\circ', \Phi_\circ) & m = n \end{cases}$$

and $\varepsilon(s, \pi \times \pi', \psi) \equiv 1$.

The other basic case is when both π and π' are supercuspidal. For this calculation, one must analyze the local integrals in terms of the Kirillov models of the representations [11, 8].

Theorem 3.4. *If π and π' are both (unitary) supercuspidal, then $L(s, \pi \times \pi') \equiv 1$ if $m < n$ and if $m = n$ we have*

$$L(s, \pi \times \pi') = \prod (1 - \alpha q^{-s})^{-1}$$

with the product over all $\alpha = q^{s_0}$ with $\tilde{\pi} \simeq \pi' \otimes |\det|^{s_0}$.

In the other cases, we must rely on the Bernstein–Zelevinsky classification of generic representations of $\mathrm{GL}_n(k)$ [49]. All generic representations can be realized as prescribed constituents of representations parabolically induced from supercuspidals. One can proceed by analyzing the Whittaker functions of induced representations in terms of Whittaker functions of the inducing data as in [23] or by analyzing the poles of the local integrals in terms of quasi invariant pairings of derivatives of π and π' as in [8] to compute $L(s, \pi \times \pi')$ in terms of L-functions of pairs of supercuspidal representations. We refer you to those papers or [28] for the explicit formulas.

To conclude this section, let us mention two results on the γ-factors. One is used in the computations of L-factors in the general case. This is the *multiplicativity of γ-factors* [23]. The second is the *stability of γ-factors* [26]. Both of these results are necessary in applications of the Converse Theorem to liftings.

Multiplicativity of γ-factors. *If $\pi = \mathrm{Ind}(\pi_1 \otimes \pi_2)$, with π_i an irreducible admissible representation of $\mathrm{GL}_{r_i}(k)$, then*

$$\gamma(s, \pi \times \pi', \psi) = \gamma(s, \pi_1 \times \pi', \psi)\gamma(s, \pi_2 \times \pi', \psi)$$

and similarly for π'. Moreover $L(s, \pi \times \pi')^{-1}$ divides $[L(s, \pi_1 \times \pi')L(s, \pi_2 \times \pi')]^{-1}$.

Stability of γ-factors. *If π_1 and π_2 are two irreducible admissible generic representations of $\mathrm{GL}_n(k)$, having the same central character, then for every sufficiently highly ramified character η of $\mathrm{GL}_1(k)$ we have*

$$\gamma(s, \pi_1 \times \eta, \psi) = \gamma(s, \pi_2 \times \eta, \psi)$$

and

$$L(s, \pi_1 \times \eta) = L(s, \pi_2 \times \eta) \equiv 1.$$

More generally, if in addition π' is an irreducible generic representation of $\mathrm{GL}_m(k)$, then for all sufficiently highly ramified characters η of $\mathrm{GL}_1(k)$ we have

$$\gamma(s, (\pi_1 \otimes \eta) \times \pi', \psi) = \gamma(s, (\pi_2 \otimes \eta) \times \pi', \psi)$$

and

$$L(s, (\pi_1 \otimes \eta) \times \pi') = L(s, (\pi_2 \otimes \eta) \times \pi') \equiv 1.$$

3.2 *The archimedean local factors*

The treatment of the archimedean local factors parallels that of the nonarchimedean in many ways, but there are some significant differences. The major work on these factors is that of Jacquet and Shalika in [27], which we follow for the most part without further reference, and in the archimedean parts of [25].

One significant difference in the development of the archimedean theory is that the local Langlands correspondence was already in place when the theory was developed [30]. The correspondence is very explicit in terms of the usual Langlands classification. Thus to π is associated an n-dimensional semisimple representation $\rho = \rho_\pi$ of the Weil group W_k of k and to π' an m-dimensional semisimple representation $\rho' = \rho'_\pi$ of W_k. Then $\rho \otimes \rho'$ is an nm-dimensional representation of W_k and to this representation of the Weil group is attached Artin–Weil L- and ε-factors [47], denoted $L(s, \rho \otimes \rho')$ and $\varepsilon(s, \rho \otimes \rho', \psi)$. In essence, Jacquet and Shalika *define*

$$L(s, \pi \times \pi') = L(s, \rho_\pi \otimes \rho'_\pi) \quad \text{and} \quad \varepsilon(s, \pi \times \pi', \psi) = \varepsilon(s, \rho_\pi \otimes \rho'_\pi, \psi)$$

and then set

$$\gamma(s, \pi \times \pi', \psi) = \frac{\varepsilon(s, \pi \times \pi', \psi)L(1 - s, \widetilde{\pi} \times \widetilde{\pi}')}{L(s, \pi \times \pi')}.$$

They then proceed to show that these functions have the expected relation to the local integrals. To this end, they define $\mathcal{M}(\pi \times \pi')$ to be the space of all meromorphic functions $\phi(s)$ with the property that if $P(s)$ is a polynomial function such that $P(s)L(s, \pi \times \pi')$ is holomorphic in a vertical strip $S[a, b] = \{s\ a \leq \mathrm{Re}(s) \leq b\}$, then $P(s)\phi(s)$ is bounded in $S[a, b]$. Note in particular that if $\phi \in \mathcal{M}(\pi \times \pi')$, then the quotient $\phi(s)L(s, \pi \times \pi')^{-1}$ is entire. One then analyzes the local integrals $\Psi_j(s; W, W')$ and $\Psi(s; W, W', \Phi)$, defined as in the nonarchimedean case for $W \in \mathcal{W}(\pi, \psi)$, $W' \in \mathcal{W}(\pi', \psi^{-1})$, and $\Phi \in \mathcal{S}(k^n)$, using methods that are direct analogues of those used in [23] for the nonarchimedean case.

Theorem 3.5. *The integrals $\Psi_j(s; W, W')$ or $\Psi(s; W, W', \Phi)$ extend to meromorphic functions of s which lie in $\mathcal{M}(\pi \times \pi')$. In particular, the ratios*

$$e_j(s; W, W') = \frac{\Psi_j(s; W, W')}{L(s, \pi \times \pi')} \quad \text{or} \quad e(s; W, W', \Phi) = \frac{\Psi(s; W, W', \Phi)}{L(s, \pi \times \pi')}$$

are entire and in fact are bounded in vertical strips.

This statement has more content than just the continuation and "bounded denominator" statements in the nonarchimedean case. Since it prescribes the "denominator" to be the L factor $L(s, \pi \times \pi')^{-1}$ it is bound up with the actual computation of the poles of the local integrals. In fact, a significant part of the paper of Jacquet and Shalika [27] is taken up with the simultaneous proof of this and the local functional equations.

Theorem 3.6. *We have the local functional equations*

$$\Psi_{n-m-j-1}(1 - s; \rho(w_{n,m})\widetilde{W}, \widetilde{W}') = \omega'(-1)^{n-1}\gamma(s, \pi \times \pi', \psi)\Psi_j(s; W, W')$$

or

$$\Psi(1 - s; \widetilde{W}, \widetilde{W}', \hat{\Phi}) = \omega'(-1)^{n-1}\gamma(s, \pi \times \pi', \psi)\Psi(s; W, W', \Phi).$$

The one fact that we are missing is the statement of "minimality" of the L-factor. That is, we know that $L(s, \pi \times \pi')$ is a standard archimedean Euler factor (i.e., a product of Γ-functions of the standard type) and has the property that the poles of all the local integrals are contained in the poles of the L-factor, even with multiplicity. But we have not established that the L-factor cannot have extraneous poles. In particular, we do not know that we can achieve the local L-function as a finite linear combination of local integrals. Towards this end, Jacquet and Shalika enlarge the allowable space of local integrals. Let Λ and Λ' be the Whittaker functionals on V_π and $V_{\pi'}$ associated with the Whittaker models $\mathcal{W}(\pi, \psi)$ and $\mathcal{W}(\pi', \psi^{-1})$. Then $\hat{\Lambda} = \Lambda \otimes \Lambda'$ defines a continuous linear functional on the algebraic tensor product $V_\pi \otimes V_{\pi'}$ which extends continuously to the topological tensor product $V_{\pi\otimes\pi'} = V_\pi \hat{\otimes} V_{\pi'}$, viewed as representations of $\mathrm{GL}_n(k) \times \mathrm{GL}_m(k)$. Now let

$$\mathcal{W}(\pi\hat{\otimes}\pi', \psi) = \{W(g, h) = \hat{\Lambda}(\pi(g) \otimes \pi'(h)\xi) | \xi \in V_{\pi\otimes\pi'}\}.$$

Then $\mathcal{W}(\pi\hat{\otimes}\pi', \psi)$ contains the algebraic tensor product $\mathcal{W}(\pi, \psi)\otimes\mathcal{W}(\pi', \psi^{-1})$ and is again equal to the topological tensor product. Now we can extend all our local integrals to the space $\mathcal{W}(\pi\hat{\otimes}\pi', \psi)$ by setting

$$\Psi_j(s; W) = \int\int W\left(\begin{pmatrix} h & & \\ x & I_j & \\ & & I_{n-m-j} \end{pmatrix}, h\right) dx \, |\det(h)|^{s-(n-m)/2} \, dh$$

and

$$\Psi(s; W, \Phi) = \int W(g, g)\Phi(e_n g)|\det(g)|^s \, dh$$

for $W \in \mathcal{W}(\pi\hat{\otimes}\pi', \psi)$. Since the local integrals are continuous with respect to the topology on the topological tensor product, all of the above facts remain true, in particular the convergence statements, the local functional equations, and the fact that these integrals extend to functions in $\mathcal{M}(\pi \times \pi')$. At this point, let $\mathcal{I}_j(\pi, \pi') = \{\Psi_j(s; W) | W \in \mathcal{W}(\pi\hat{\otimes}\pi')\}$ and let $\mathcal{I}(\pi, \pi')$ be the span of the local integrals $\{\Psi(s; W, \Phi) | W \in \mathcal{W}(\pi\hat{\otimes}\pi', \psi), \ \Phi \in \mathcal{S}(k^n)\}$. Once again, in the case $m < n$ we have that the space $\mathcal{I}_j(\pi, \pi')$ is independent of j and we denote it also by $\mathcal{I}(\pi, \pi')$. These are still independent of ψ. So we know from above that $\mathcal{I}(\pi, \pi') \subset \mathcal{M}(\pi \times \pi')$. The remainder of [27] is then devoted to showing $\mathcal{I}(\pi, \pi') = \mathcal{M}(\pi \times \pi')$.

In the cases of $m = n - 1$ or $m = n$, Stade [44, 45] and Jacquet and Shalika (see [9]) have shown that one can indeed get the local L-function as a finite linear combination of integrals involving only K-finite functions in $\mathcal{W}(\pi, \psi)$ and $\mathcal{W}(\pi', \psi^{-1})$, that is, without going to the completion of $\mathcal{W}(\pi, \psi)\otimes\mathcal{W}(\pi', \psi^{-1})$.

4 Global L-functions

Once again, we let k be a global field, $\mathbb{A}$ its ring of adeles, and fix a nontrivial continuous additive character $\psi = \otimes\psi_v$ of $\mathbb{A}$ trivial on k.

Let (π, V_π) be an cuspidal representation of $\mathrm{GL}_n(\mathbb{A})$ and $(\pi', V_{\pi'})$ a cuspidal representation of $\mathrm{GL}_m(\mathbb{A})$. Since they are irreducible we have restricted tensor product decompositions $\pi \simeq \otimes'\pi_v$ and $\pi' \simeq \otimes'\pi'_v$ with (π_v, V_{π_v}) and $(\pi'_v, V_{\pi'_v})$ irreducible admissible smooth generic unitary representations of $\mathrm{GL}_n(k_v)$ and $\mathrm{GL}_m(k_v)$ [10, 13]. Let $\omega = \otimes'\omega_v$ and $\omega' = \otimes'\omega'_v$ be their central characters. These are both continuous characters of $k^\times\backslash\mathbb{A}^\times$. Let S be the finite set of places of k, containing the archimedean places S_∞, such that for all $v \notin S$ we have that π_v, π'_v, and ψ_v are unramified.

For each place v of k we have defined the local factors $L(s, \pi_v \times \pi'_v)$ and $\varepsilon(s, \pi_v \times \pi'_v, \psi_v)$. Then we can at least formally define

$$L(s, \pi \times \pi') = \prod_v L(s, \pi_v \times \pi'_v) \quad \text{and} \quad \varepsilon(s, \pi \times \pi') = \prod_v \varepsilon(s, \pi_v \times \pi'_v, \psi_v).$$

The product defining the L-function is absolutely convergent for $\mathrm{Re}(s) \gg 0$. For the ε-factor, $\varepsilon(s, \pi_v \times \pi'_v, \psi_v) \equiv 1$ for $v \notin S$ so that the product is in fact a finite product and there is no problem with convergence. The fact that $\varepsilon(s, \pi \times \pi')$ is independent of ψ can either be checked by analyzing how the local ε-factors vary as you vary ψ, as is done in [5, Lemma 2.1], or it will follow from the global functional equation presented below.

4.1 *Basic analytic properties*

Our first goal is to show that these L-functions have nice analytic properties.

Theorem 4.1. *The global L-functions $L(s, \pi \times \pi')$ are nice in the sense that*

1. *$L(s, \pi \times \pi')$ has a meromorphic continuation to all of $\mathbb{C}$,*
2. *the extended function is bounded in vertical strips (away from its poles),*
3. *they satisfy the functional equation*

$$L(s, \pi \times \pi') = \varepsilon(s, \pi \times \pi') L(1 - s, \tilde{\pi} \times \tilde{\pi}').$$

To do so, we relate the L-functions to the global integrals.

Let us begin with continuation. In the case $m < n$ for every $\varphi \in V_\pi$ and $\varphi' \in V_{\pi'}$ we know the integral $I(s; \varphi, \varphi')$ converges absolutely for all s. From the unfolding in Section 2 and the local calculation of Section 3 we know that for $\mathrm{Re}(s) \gg 0$ and for appropriate choices of φ and φ' we have

$$\begin{aligned} I(s; \varphi, \varphi') &= \prod_v \Psi_v(s; W_{\varphi_v}, W_{\varphi'_v}) = \left(\prod_{v \in S} \Psi_v(s; W_{\varphi_v}, W_{\varphi'_v}) \right) L^S(s, \pi \times \pi') \\ &= \left(\prod_{v \in S} \frac{\Psi_v(s; W_{\varphi_v}, W_{\varphi'_v})}{L(s, \pi_v \times \pi'_v)} \right) L(s, \pi \times \pi') \\ &= \left(\prod_{v \in S} e_v(s; W_{\varphi_v}, W_{\varphi'_v}) \right) L(s, \pi \times \pi') \end{aligned}$$

We know that each ratio $e_v(s; W_v, W'_v)$ is entire and hence $L(s, \pi \times \pi')$ has a meromorphic continuation. If $m = n$, then for appropriate $\varphi \in V_\pi$, $\varphi' \in V_{\pi'}$, and $\Phi \in \mathcal{S}(\mathbb{A}^n)$ we again have

$$I(s; \varphi, \varphi', \Phi) = \left(\prod_{v \in S} e_v(s; W_{\varphi_v}, W'_{\varphi'_v}, \Phi_v) \right) L(s, \pi \times \pi').$$

and so $L(s, \pi \times \pi')$ has a meromorphic continuation.

Let us next turn to the functional equation. This will follow from the functional equation for the global integrals and the local functional equations. We will consider only the case where $m < n$ since the other case is entirely analogous. The functional equation for the global integrals is simply

$$I(s;\varphi,\varphi') = \tilde{I}(1-s;\widetilde{\varphi},\widetilde{\varphi}').$$

Once again we have for appropriate φ and φ'

$$I(s;\varphi,\varphi') = \left(\prod_{v\in S}\frac{\Psi(s;W_v,W_v')}{L(s,\pi_v\times\pi_v')}\right)L(s,\pi\times\pi')$$

while on the other side

$$\tilde{I}(1-s;\widetilde{\varphi},\widetilde{\varphi}') = \left(\prod_{v\in S}\frac{\widetilde{\Psi}(1-s;\rho(w_{n,m})\widetilde{W}_v,\widetilde{W}_v')}{L(1-s,\widetilde{\pi}_v\times\widetilde{\pi}_v')}\right)L(1-s,\widetilde{\pi}\times\widetilde{\pi}').$$

However, by the local functional equations, for each $v \in S$ we have

$$\frac{\widetilde{\Psi}(1-s;\rho(w_{n,m})\widetilde{W}_v,\widetilde{W}_v')}{L(1-s,\widetilde{\pi}_v\times\widetilde{\pi}_v')} = \omega_v'(-1)^{n-1}\varepsilon(s,\pi_v\times\pi_v',\psi_v)\frac{\Psi(s;W_v,W_v')}{L(s,\pi_v\times\pi_v')}.$$

Combining these, we have

$$L(s,\pi\times\pi') = \left(\prod_{v\in S}\omega_v'(-1)^{n-1}\varepsilon(s,\pi_v\times\pi_v',\psi_v)\right)L(1-s,\widetilde{\pi}\times\widetilde{\pi}').$$

Now, for $v \notin S$ we know that π_v' is unramified, so $\omega_v'(-1) = 1$, and also that $\varepsilon(s,\pi_v\times\pi_v',\psi_v) \equiv 1$. Therefore

$$\prod_{v\in S}\omega_v'(-1)^{n-1}\varepsilon(s,\pi_v\times\pi_v',\psi_v) = \varepsilon(s,\pi\times\pi')$$

and we indeed have

$$L(s,\pi\times\pi') = \varepsilon(s,\pi\times\pi')L(1-s,\widetilde{\pi}\times\widetilde{\pi}').$$

Note that this implies that $\varepsilon(s,\pi\times\pi')$ is independent of ψ as well.

Let us now turn to the boundedness in vertical strips. For the global integrals $I(s;\varphi,\varphi')$ or $I(s;\varphi,\varphi,\Phi)$ this simply follows from the absolute convergence. For the L-function itself, the paradigm is the following. For every finite place $v \in S$ we know that there is a choice of $W_{v,i}$, $W_{v,i}'$, and $\Phi_{v,i}$ if necessary such that

$$L(s,\pi_v\times\pi_v') = \sum\Psi(s;W_{v,i},W_{v'i}')$$

or

$$L(s,\pi_v\times\pi_v') = \sum\Psi(s;W_{v,i},W_{v'i}',\Phi_{v,i}).$$

If $m = n - 1$ or $m = n$, then by the work of Stade and Jacquet and Shalika we know that we have similar statements for $v \in S_\infty$. Hence if $m = n - 1$ or $m = n$ there are global choices φ_i, φ'_i, and if necessary Φ_i such that

$$L(s, \pi \times \pi') = \sum I(s; \varphi_i, \varphi'_i) \quad \text{or} \quad L(s, \pi \times \pi') = \sum I(s; \varphi_i, \varphi'_i, \Phi_i).$$

Then the boundedness in vertical strips for the L-functions follows from that of the global integrals.

However, if $m < n - 1$, then all we know at those $v \in S_\infty$ is that there is a function $W_v \in \mathcal{W}(\pi_v \hat{\otimes} \pi'_v, \psi_v) = \mathcal{W}(\pi_v, \psi_v) \hat{\otimes} \mathcal{W}(\pi'_v, \psi_v^{-1})$ or a finite collection of such functions $W_{v,i}$ and of $\Phi_{v,i}$ such that

$$L(s, \pi_v \times \pi'_v) = I(s; W_v) \quad \text{or} \quad L(s, \pi_v \times \pi'_v) = \sum I(s; W_{v,i}, \Phi_{v,i}).$$

To make the above paradigm work for $m < n - 1$ we would need to rework the theory of global Eulerian integrals for cusp forms in $V_\pi \hat{\otimes} V_{\pi'}$. This is naturally the space of smooth vectors in an irreducible unitary cuspidal representation of $\mathrm{GL}_n(\mathbb{A}) \times \mathrm{GL}_m(\mathbb{A})$. So we would need extend the global theory of integrals parallel to Jacquet and Shalika's extension of the local integrals in the archimedean theory. There seems to be no obstruction to carrying this out, but we have not done this.

We should point out that if one approaches these L-function by the method of constant terms and Fourier coefficients of Eisenstein series, then Gelbart and Shahidi have shown a wide class of automorphic L-functions, including ours, to be bounded in vertical strips [12].

4.2 *Poles of L-functions*

Let us determine where the global L-functions can have poles. The poles of the L-functions will be related to the poles of the global integrals. Recall from Section 2 that in the case of $m < n$ we have that the global integrals $I(s; \varphi, \varphi')$ are entire and that when $m = n$, then $I(s; \varphi, \varphi', \Phi)$ can have at most simple poles and they occur at $s = -i\sigma$ and $s = 1 - i\sigma$ for σ real when $\pi \simeq \tilde{\pi}' \otimes |\det|^{i\sigma}$. As we have noted above, the global integrals and global L-functions are related, for appropriate φ, φ', and Φ, by

$$I(s; \varphi, \varphi') = \left(\prod_{v \in S} e_v(s; W_{\varphi_v}, W'_{\varphi'_v}) \right) L(s, \pi \times \pi')$$

or

$$I(s; \varphi, \varphi', \Phi) = \left(\prod_{v \in S} e_v(s; W_{\varphi_v}, W'_{\varphi'_v}, \Phi_v) \right) L(s, \pi \times \pi').$$

On the other hand, for any $s_0 \in \mathbb{C}$ and any v there is a choice of local W_v, W'_v, and Φ_v such that the local ratios $e_v(s_0; W_v, W'_v) \neq 0$ or $e_v(s_0; W_v, W'_v, \Phi_v) \neq 0$. So as we vary φ, φ' and Φ at the places $v \in S$ we see that division by these factors can introduce no extraneous poles in $L(s, \pi \times \pi')$, that is, in keeping with the local characterization of the L-factor in terms of poles of local integrals, globally the poles of $L(s, \pi \times \pi')$ are precisely the poles of the family of global integrals $\{I(s; \varphi, \varphi')\}$ or $\{I(s; \varphi, \varphi', \Phi)\}$. Hence from Theorems 2.1 and 2.2 we have.

Theorem 4.2. *If $m < n$, then $L(s, \pi \times \pi')$ is entire. If $m = n$, then $L(s, \pi \times \pi')$ has at most simple poles and they occur iff $\pi \simeq \widetilde{\pi}' \otimes |\det|^{i\sigma}$ with σ real and are then at $s = -i\sigma$ and $s = 1 - i\sigma$.*

There are two useful observationss that follow from this.

1. $L(s, \pi \times \widetilde{\pi})$ has simple poles at $s = 0$ and $s = 1$.
2. For π and π' cuspidal representations of $\mathrm{GL}_n(\mathbb{A})$, $L(s, \pi \times \widetilde{\pi}')$ has a pole at $s = 1$ iff $\pi \simeq \pi'$.

As a consequence of this, we obtain the analytic proof of the Strong Multiplicity One Theorem [31, 25].

Strong multiplicity one. *Let (π, V_π) and $(\pi', V_{\pi'})$ be two cuspidal representations of $GL_n(\mathbb{A})$. Suppose there is a finite set of places S such that for all $v \notin S$ we have $\pi_v \simeq \pi'_v$. Then $\pi = \pi'$.*

5 Converse theorems

Let us return first to Hecke. Recall that to a modular form

$$f(\tau) = \sum_{n=1}^{\infty} a_n e^{2\pi i n \tau}$$

for say $\mathrm{SL}_2(\mathbb{Z})$ Hecke attached an L function $L(s, f)$ and they were related via the Mellin transform

$$\Lambda(s, f) = (2\pi)^{-s}\Gamma(s)L(s, f) = \int_0^\infty f(iy)y^s\, d^\times y$$

and derived the functional equation for $L(s, f)$ from the modular transformation law for $f(\tau)$ under the modular transformation law for the transformation $\tau \mapsto -1/\tau$. In his fundamental paper [17] he inverted this process by taking a Dirichlet series

$$D(s) = \sum_{n=1}^{\infty} \frac{a_n}{n^s}$$

and assuming that it converged in a half plane, had an entire continuation to a function of finite order, and satisfied the same functional equation as the L-function of a modular form of weight k, then he could actually reconstruct a modular form from $D(s)$ by Mellin inversion

$$f(iy) = \sum_n a_n e^{-2\pi n y} = \frac{1}{2\pi i} \int_{2-i\infty}^{2+i\infty} (2\pi)^{-s} \Gamma(s) D(s) y^s \, ds$$

and obtain the modular transformation law for $f(\tau)$ under $\tau \mapsto -1/\tau$ from the functional equation for $D(s)$ under $s \mapsto k-s$. This is Hecke's Converse Theorem.

In this Section we will present some analogues of Hecke's theorem in the context of L-functions for GL_n. Surprisingly, the technique is exactly the same as Hecke's, i.e., inverting the integral representation. This was first done in the context of automorphic representation for GL_2 by Jacquet and Langlands [21] and then extended and significantly strengthened for GL_3 by Jacquet, Piatetski-Shapiro, and Shalika [22]. This section is taken mainly from our survey [7]. Further details can be found in [5, 6].

Let k be a global field, $\mathbb{A}$ its adele ring, and ψ a fixed nontrivial continuous additive character of $\mathbb{A}$ which is trivial on k. We will take $n \geq 3$ to be an integer.

To state these Converse Theorems, we begin with an irreducible admissible representation Π of $\mathrm{GL}_n(\mathbb{A})$. In keeping with the conventions of these notes, we will assume that Π is unitary and generic, but this is not necessary. It has a decomposition $\Pi = \otimes' \Pi_v$, where Π_v is an irreducible admissible generic representation of $\mathrm{GL}_n(k_v)$. By the local theory of Section 3, to each Π_v is associated a local L-function $L(s, \Pi_v)$ and a local ε-factor $\varepsilon(s, \Pi_v, \psi_v)$. Hence formally we can form

$$L(s, \Pi) = \prod_v L(s, \Pi_v) \quad \text{and} \quad \varepsilon(s, \Pi, \psi) = \prod_v \varepsilon(s, \Pi_v, \psi_v).$$

We will always assume the following two things about Π:

1. $L(s, \Pi)$ converges in some half plane $\mathrm{Re}(s) \gg 0$,

2. the central character ω_Π of Π is automorphic, that is, invariant under $k^\times$.

Under these assumptions, $\varepsilon(s, \Pi, \psi) = \varepsilon(s, \Pi)$ is independent of our choice of ψ [5].

Our Converse Theorems will involve twists by cuspidal automorphic representations of $\mathrm{GL}_m(\mathbb{A})$ for certain m. For convenience, let us set $\mathcal{A}(m)$ to be the set of automorphic representations of $\mathrm{GL}_m(\mathbb{A})$, $\mathcal{A}_0(m)$ the set of cuspidal representations of $\mathrm{GL}_m(\mathbb{A})$, and $\mathcal{T}(m) = \coprod_{d=1}^m \mathcal{A}_0(d)$. If we fix a finite set of S of finite places, then we let $\mathcal{T}(S; m)$ denote the subset of $\mathcal{T}(m)$ consisting of representations that are *unramified* at all places $v \in S$.

Let $\pi' = \otimes' \pi'_v$ be a cuspidal representation of $\mathrm{GL}_m(\mathbb{A})$ with $m < n$. Then again we can formally define

$$L(s, \Pi \times \pi') = \prod_v L(s, \Pi_v \times \pi'_v) \quad \text{and} \quad \varepsilon(s, \Pi \times \pi') = \prod_v \varepsilon(s, \Pi_v \times \pi'_v, \psi_v)$$

since again the local factors make sense whether Π is automorphic or not. A consequence of (1) and (2) above and the cuspidality of π' is that both $L(s, \Pi \times \pi')$

and $L(s, \widetilde{\Pi} \times \widetilde{\pi}')$ converge absolutely for $\mathrm{Re}(s) \gg 0$, where $\widetilde{\Pi}$ and $\widetilde{\pi}'$ are the contragredient representations, and that $\varepsilon(s, \Pi \times \pi')$ is independent of the choice of ψ.

We say that $L(s, \Pi \times \pi')$ is *nice* if it satisfies the same analytic properties it would if Π were cuspidal, i.e.,

1. $L(s, \Pi \times \pi')$ and $L(s, \widetilde{\Pi} \times \widetilde{\pi}')$ have analytic continuations to entire functions of s,
2. these entire continuations are bounded in vertical strips of finite width,
3. they satisfy the standard functional equation
$$L(s, \Pi \times \pi') = \varepsilon(s, \Pi \times \pi') L(1 - s, \widetilde{\Pi} \times \widetilde{\pi}').$$

The basic Converse Theorem for GL_n is the following [5, 4].

Theorem 5.1. *Let Π be an irreducible admissible representation of $\mathrm{GL}_n(\mathbb{A})$ as above. Let S be a finite set of finite places. Suppose that $L(s, \Pi \times \pi')$ is nice for all $\pi' \in \mathcal{T}(S; n-1)$. Then Π is quasiautomorphic in the sense that there is an automorphic representation Π' such that $\Pi_v \simeq \Pi'_v$ for all $v \notin S$. If S is empty, then, in fact, Π is a cuspidal automorphic representation of $\mathrm{GL}_n(\mathbb{A})$.*

This result is of course valid for $n = 2$ as well.

For applications [3], it is desirable to twist by as little as possible. There are essentially two ways to restrict the twisting. One is to restrict the rank of the groups that the twisting representations live on. The other is to restrict ramification.

When we restrict the rank of our twists, we can obtain the following result [6].

Theorem 5.2. *Let Π be an irreducible admissible representation of $\mathrm{GL}_n(\mathbb{A})$ as above. Let S be a finite set of finite places. Suppose that $L(s, \Pi \times \pi')$ is nice for all $\pi' \in \mathcal{T}(S; n-2)$. Then Π is quasiautomorphic in the sense that there is an automorphic representation Π' such that $\Pi_v \simeq \Pi'_v$ for all $v \notin S$. If S is empty, then, in fact, Π is a cuspidal automorphic representation of $\mathrm{GL}_n(\mathbb{A})$.*

This result is stronger than Theorem 5.1, but its proof is more delicate.

In order to apply these theorems, one uses a good choice of the set S in conjunction with twisting by a highly ramified character η. The set S usually consists of the finite places where Π_v is ramified. η is used in conjunction with the local stability of γ-factors mentioned above. Then the following observation is a key ingredient in applying either of the above theorems [7].

Observation . *Let Π be as in Theorem 5.1 or 5.2. Suppose that η is a fixed (highly ramified) character of $k^\times \backslash \mathbb{A}^\times$. Suppose that $L(s, \Pi \times \pi')$ is nice for all $\pi' \in \mathcal{T} \otimes \eta$, where $\mathcal{T}$ is either of the twisting sets of Theorem 5.1 or 5.2. Then Π is quasiautomorphic as in those theorems.*

The only thing to observe, say by looking at the local or global integrals, is that if $\pi' \in \mathcal{T}$, then $L(s, \Pi \times (\pi' \otimes \eta)) = L(s, (\Pi \otimes \eta) \times \pi')$ so that applying the

Converse Theorem for Π with twisting set $\mathcal{T} \otimes \eta$ is equivalent to applying the Converse Theorem for $\Pi \otimes \eta$ with the twisting set $\mathcal{T}$. So, by either Theorem 5.1 or 5.2, whichever is appropriate, $\Pi \otimes \eta$ is quasiautomorphic and hence Π is as well.

The second way to restrict our twists is to restrict the ramification at all but a finite number of places [5]. Now fix a nonempty finite set of places S which in the case of a number field contains the set S_∞ of all archimedean places. Let $\mathcal{T}_S(m)$ denote the subset consisting of all representations π' in $\mathcal{T}(m)$ which are *unramified for all* $v \notin S$. Note that we are placing a grave restriction on the ramification of these representations.

Theorem 5.3. *Let Π be an irreducible admissible representation of* $\mathrm{GL}_n(\mathbb{A})$ *as above. Let S be a nonempty finite set of places, containing S_∞, such that the class number of the ring $\mathfrak{o}_S$ of S-integers is one. Suppose that $L(s, \Pi \times \pi')$ is nice for all $\pi' \in \mathcal{T}_S(n-1)$. Then Π is quasiautomorphic in the sense that there is an automorphic representation Π' such that $\Pi_v \simeq \Pi'_v$ for all $v \in S$ and all $v \notin S$ such that both Π_v and Π'_v are unramified.*

There are several things to note here. First, there is a class number restriction. However, if $k = \mathbb{Q}$, then we may take $S = S_\infty$ and we have a Converse Theorem with "level 1" twists. As a practical consideration, if we let S_Π be the set of finite places v where Π_v is ramified, then for applications we usually take S and S_Π to be disjoint. Once again, we are losing all information at those places $v \notin S$ where we have restricted the ramification unless Π_v was already unramified there.

The proof of Theorem 5.1 with S empty essentially follows the lead of Hecke, Weil, and Jacquet–Langlands. It is based on the integral representations of L-functions, Fourier expansions, Mellin inversion, and finally a use of the weak form of Langlands spectral theory. For Theorems 5.1, 5.2, and 5.3, where we have restricted our twists, we must impose certain local conditions to compensate for our limited twists. For Theorems 5.1 and 5.2 there are a finite number of local conditions and for Theorem 5.3 an infinite number of local conditions. We must then work around these by using results on generation of congruence subgroups and either weak or strong approximation.

References

[1] J.W. Cogdell, Notes on L-functions for GL_n. ICTP Lectures Notes, to appear.

[2] J.W. Cogdell, Langlands conjectures for GL_n, in J. Bernstein and S. Gelbart, eds., *An Introduction to the Langlands Program*, Birkhäuser Boston, Boston, 2003, 229–249 (this volume).

[3] J.W. Cogdell, Dual groups and Langlands functoriality, in J. Bernstein and S. Gelbart, eds., *Introduction to the Langlands Program*, Birkhäuser Boston, Boston, 2003, 251–269 (this volume).

[4] J.W. COGDELL, H. KIM, I.I. PIATETSKI-SHAPIRO, AND F. SHAHIDI, On lifting from classical groups to GL_N, *Publ. Math. IHES*, **93**(2001), 5–30.

[5] J.W. COGDELL AND I.I. PIATETSKI-SHAPIRO, Converse theorems for GL_n, *Publ. Math. IHES* **79**(1994), 157–214.

[6] J.W. COGDELL AND I.I. PIATETSKI-SHAPIRO, Converse theorems for GL_n, II *J. reine angew. Math.*, **507**(1999), 165–188.

[7] J.W. COGDELL AND I.I. PIATETSKI-SHAPIRO, Converse theorems for GL_n and their applications to liftings. *Cohomology of Arithmetic Groups, Automorphic Forms, and L-functions, Mumbai 1998*, Tata Institute of Fundamental Research, Narosa, 2001, 1–34.

[8] J.W. COGDELL AND I.I. PIATETSKI-SHAPIRO, Derivatives and L-functions for GL_n. *The Heritage of B. Moishezon*, IMCP, to appear.

[9] J.W. COGDELL AND I.I. PIATETSKI-SHAPIRO, Remarks on Rankin–Selberg convolutions. *Contributions to Automorphic Forms, Geometry and Number Theory (Shalikafest 2002)*, H. Hida, D. Ramakrishnan, and F. Shahidi, eds., Johns Hopkins University Press, Baltimore, to appear.

[10] D. FLATH, Decomposition of representations into tensor products, *Proc. Sympos. Pure Math.*, **33**, part 1, (1979), 179–183.

[11] S. GELBART AND H. JACQUET, A relation between automorphic representations of GL(2) and GL(3), *Ann. Sci. École Norm. Sup.* (4) **11**(1978), 471–542.

[12] S. GELBART AND F. SHAHIDI, Boundedness of automorphic L-functions in vertical strips, *J. Amer. Math. Soc.*, **14**(2001), 79–107.

[13] I.M. GELFAND, M.I. GRAEV, AND I.I. PIATETSKI-SHAPIRO, *Representation Theory and Automorphic Functions*, Academic Press, San Diego, 1990.

[14] I.M. GELFAND AND D.A. KAZHDAN, Representations of $GL(n, K)$ where K is a local field, in *Lie Groups and Their Representations*, edited by I.M. Gelfand. John Wiley & Sons, New York–Toronto, 1971, 95–118.

[15] R. GODEMENT AND H. JACQUET, *Zeta Functions of Simple Algebras*, Springer Lecture Notes in Mathematics, No.260, Springer-Verlag, Berlin, 1972.

[16] M. HARRIS AND R. TAYLOR, On the Geometry and Cohomology of Some Simple Shimura Varieties. *Annals of Math Studies* **151**, Princeton University Press, 2001.

[17] E. HECKE, Über die Bestimmung Dirichletscher Reihen durch ihre Funktionalgleichung, *Math. Ann.*, **112**(1936), 664–699.

[18] E. HECKE, *Mathematische Werke*, Vandenhoeck & Ruprecht, Göttingen, 1959.

[19] G. HENNIART, Une preuve simple des conjectures de Langlands pour GL(n) sur un corps p-adique, *Invent. Math.*, **139**(2000), 439–455.

[20] H. JACQUET, *Automorphic Forms on $GL(2)$, II*, Springer Lecture Notes in Mathematics No.278 , Springer-Verlag, Berlin, 1972.

[21] H. JACQUET AND R.P. LANGLANDS, *Automorphic Forms on $GL(2)$*, Springer Lecture Notes in Mathematics No.114, Springer Verlag, Berlin, 1970.

[22] H. JACQUET, I.I. PIATETSKI-SHAPIRO, AND J. SHALIKA, Automorphic forms on $GL(3)$, I & II, *Ann. Math.* **109**(1979), 169–258.

[23] H. JACQUET, I.I. PIATETSKI-SHAPIRO, AND J. SHALIKA, Rankin–Selberg convolutions, *Amer. J. Math.*, **105**(1983), 367–464.

[24] H. JACQUET, I.I. PIATETSKI-SHAPIRO, AND J. SHALIKA, Conducteur des représentations du groupe linéaire. *Math. Ann.*, **256**(1981), 199–214.

[25] H. JACQUET AND J. SHALIKA, On Euler products and the classification of automorphic representations, *Amer. J. Math.* I: **103**(1981), 499–588; II: **103**(1981), 777–815.

[26] H. JACQUET AND J. SHALIKA, A lemma on highly ramified ϵ-factors, *Math. Ann.*, **271**(1985), 319–332.

[27] H. JACQUET AND J. SHALIKA, Rankin–Selberg convolutions: Archimedean theory, in *Festschrift in Honor of I.I. Piatetski-Shapiro*, Part I, Weizmann Science Press, Jerusalem, 1990, 125–207.

[28] S. KUDLA, The local Langlands correspondence: the non-Archimedean case, *Proc. Sympos. Pure Math.*, **55**, part 2, (1994), 365–391.

[29] R.P. LANGLANDS, *Euler Products*, Yale Univ. Press, New Haven, 1971.

[30] R.P. LANGLANDS On the classification of irreducible representations of real algebraic groups, in *Representation Theory and Harmonic Analysis on Semisimple Lie Groups*, AMS Mathematical Surveys and Monographs, No.31, 1989, 101–170.

[31] I.I. PIATETSKI-SHAPIRO, Euler Subgroups, in *Lie Groups and Their Representations*, edited by I.M. Gelfand. John Wiley & Sons, New York–Toronto, 1971, 597–620.

[32] I.I. PIATETSKI-SHAPIRO, Multiplicity one theorems, *Proc. Sympos. Pure Math.*, **33**, Part 1 (1979), 209–212.

[33] R.A. RANKIN, Contributions to the theory of Ramanujan's function $\tau(n)$ and similar arithmetical functions, I and II, *Proc. Cambridge Phil. Soc.*, **35**(1939), 351–372.

[34] A. SELBERG, Bemerkungen über eine Dirichletsche Reihe, die mit der Theorie der Modulformen nahe verbunden ist, *Arch. Math. Naturvid.* **43**(1940), 47–50.

[35] F. SHAHIDI, Functional equation satisfied by certain L-functions. *Compositio Math.*, **37**(1978), 171–207.

[36] F. SHAHIDI, On non-vanishing of L-functions, *Bull. Amer. Math. Soc., N.S.*, **2**(1980), 462–464.

[37] F. SHAHIDI, On certain L-functions. *Amer. J. Math.*, **103**(1981), 297–355.

[38] F. SHAHIDI, Fourier transforms of intertwining operators and Plancherel measures for $\mathrm{GL}(n)$. *Amer. J. Math.*, **106**(1984), 67–111.

[39] F. SHAHIDI Local coefficients as Artin factors for real groups. *Duke Math. J.*, **52**(1985), 973–1007.

[40] F. SHAHIDI, On the Ramanujan conjecture and finiteness of poles for certain L-functions. *Ann. of Math.*, **127**(1988), 547–584.

[41] F. SHAHIDI, A proof of Langlands' conjecture on Plancherel measures; complementary series for p-adic groups. *Ann. of Math.*, **132**(1990), 273–330.

[42] J. SHALIKA, The multiplicity one theorem for $\mathrm{GL}(n)$, *Ann. Math.* **100**(1974), 171–193.

[43] T. SHINTANI, On an explicit formula for class-1 "Whittaker functions" on GL_n over $\mathfrak{P}$-adic fields. *Proc. Japan Acad.* **52**(1976), 180–182.

[44] E. STADE Mellin transforms of $GL(n,\mathbb{R})$ Whittaker functions, *Amer. J. Math.* **123**(2001), 121–161.

[45] E. STADE, Archimedean L-factors on $GL(n)\times GL(n)$ and generalized Barnes integrals, *Israel J. Math.* **127**(2002), 201–219.

[46] J. TATE, Fourier Analysis in Number Fields and Hecke's Zeta-Functions (Thesis, Princeton, 1950), in *Algebraic Number Theory*, edited by J.W.S. Cassels and A. Frolich, Academic Press, London, 1967, 305–347.

[47] J. TATE, Number theoretic background, *Proc. Symp. Pure Math.*, **33**, part 2, 3–26.

[48] A. WEIL Über die Bestimmung Dirichletscher Reihen durch Funktionalgleichungen, *Math. Ann.*, **168**(1967), 149–156.

[49] A. ZELEVINSKY, Induced representations of reductive $\mathfrak{p}$-adic groups, II. Irreducible representations of $GL(n)$, *Ann. scient. Éc. Norm. Sup.*, 4e série, **13**(1980), 165–210.

10

Langlands Conjectures for GL_n

J.W. Cogdell

One of the principle goals of modern number theory is to understand the Galois group $\mathcal{G}_k = Gal(\bar{k}/k)$ of a local or global field k, such as $\mathbb{Q}$ for example. One way to try to understand the group $\mathcal{G}_k$ is by understanding its finite dimensional representation theory. In the case of a number field, to every finite dimensional representation $\rho : \mathcal{G}_k \to \mathrm{GL}_n(\mathbb{C})$ Artin attached a complex analytic invariant, its L-function $L(s, \rho)$. One approach to understanding ρ is through this invariant. For one dimensional ρ this idea was fundamental for the analytic approach to abelian class field theory and the understanding of $\mathcal{G}_k^{ab}$. To obtain a more complete understanding of $\mathcal{G}_k$ we would hope for a more complete understanding of the $L(s, \rho)$ for higher dimensional representations.

There is another class of objects which possess similar analytic invariants. These are the automorphic representations π of $\mathrm{GL}_n(\mathbb{A})$, where $\mathbb{A}$ is the adele ring of k. The analytic properties of the L-functions $L(s, \pi)$ attached to automorphic representations are well understood [13].

The Langlands conjectures predict the existence of a correspondence between the n-dimensional representations of $\mathcal{G}_k$ and the automorphic representations of $\mathrm{GL}_n(\mathbb{A})$ which preserves these analytic invariants. There is a concomitant correspondence between n-dimensional representations of $\mathcal{G}_k$ for a local field k and the admissible representations of $\mathrm{GL}_n(k)$, the local Langlands conjecture. There are two ways to view such correspondences. If one views the passage of information from the automorphic side to the Galois side, as we have done above, this is a local or global nonabelian class field theory. If one views the passage if information from the Galois side to the automorphic side this is an arithmetic parameterization of admissible or automorphic representations.

Over the past ten years there has been significant progress made in the understanding of these Langlands conjectures. It began in the early nineties with the proof of the local Langlands conjecture for local fields k of characteristic p by Laumon, Rapoport, and Stuhler [44]. In the late nineties it was followed by a proof of the local Langlands conjecture for nonarchimedean fields of characteristic zero by Harris and Taylor [27], followed quickly by a simplified proof due to Henniart [30]. Around the same time, following the program of Drinfeld from his proof of the Langlands conjecture for GL_2 over a global function field [23], L. Lafforgue established the global Langlands conjecture for GL_n in the function field case [35].

In this survey we would like to present an overview of these results, emphasizing their common features. There are already several excellent surveys on the individual works, namely those of Carayol for the local conjectures [9, 10] and Laumon for the global conjectures [42, 43] and we refer the reader to these for more in depth coverage. The first two sections of this paper discuss Galois representations, automorphic representations, and their L-functions. We next discuss the local Langlands conjectures in both the representation theoretic version, proved by Langlands in the archimedean case around 1973 [38], and the L-function version, which was the version established by Laumon, Rapoport, Stuhler, Harris, Taylor, and Henniart in the nonarchimedean case. Finally we discuss the version of the global Langlands conjecture established by Drinfeld and Lafforgue in characteristic p.

Although there has been little general progress on the global Langlands conjecture for number fields, there have been spectacular special cases established recently. Most notable among these is the proof by Wiles of the modularity of certain 2-dimensional ℓ-adic representations of $\mathcal{G}_{\mathbb{Q}}$ associated to elliptic curves over $\mathbb{Q}$, which he established on his way to the proof of Fermat's last theorem [59], and related results. Unfortunately, we will not discuss these results here.

1 Galois representations and their L-functions

If G is a topological group and F is a topological field, then let $\mathrm{Rep}_n(G; F)$ denote the set of equivalence classes of continuous representations $\rho : G \to \mathrm{GL}_n(F)$. Let $\mathrm{Rep}_n^0(G; F)$ be the subset of irreducible representations. For the most part we will be interested in complex representations and so we will use $\mathrm{Rep}_n(G)$ for $\mathrm{Rep}_n(G; \mathbb{C})$ and similarly for Rep_n^0. At times we will be interested in $F = \overline{\mathbb{Q}}_\ell$ and when we do, we will use the coefficient field in the notation.

If k is either a local or global field we will let $\bar{k}$ be a separable algebraic closure of k. Let $\mathcal{G}_k = Gal(\bar{k}/k)$ be the (absolute) Galois group and W_k the (absolute) Weil group [51].

Let k be a nonarchimedean local field. Let p be the characteristic and q the order of its residue field κ. Let $I \subset \mathcal{G}_k$ be the inertia subgroup. If we let Φ denote a choice of geometric Frobenius element of $\mathcal{G}_k$, then W_k can be taken as the subgroup of $\mathcal{G}_k$ algebraically generated by Φ and I but topologized such that I has the induced topology from $\mathcal{G}_k$, I is open, and multiplication by Φ is a homeomorphism. This can also be given the structure of a scheme over $\mathbb{Q}$ [51]. Then we have a continuous homomorphism $W_k \to \mathcal{G}_k$ with dense image. Thus we have a natural inclusion $\mathrm{Rep}_n(\mathcal{G}_k) \to \mathrm{Rep}_n(W_k)$. The image, that is, the those representations that factor through continuous representations of $\mathcal{G}_k$, are the representations of W_k of Galois type. We also have a natural character $\omega^s \in \mathrm{Rep}_1(W_k)$ defined by $\omega^s(I) = 1$ and $\omega^s(\Phi) = q^{-s}$. This is also denoted by $\omega^s(w) = \|w\|^{-s}$ and gives a homomorphism $\nu : W_k \to \mathbb{Z}$ defined by $\|w\| = q^{-\nu(w)}$. Then every irreducible representation ρ of W_k is of the form $\rho = \rho^\circ \otimes \omega^s$ where ρ° is of Galois type [49, 19].

The representations that arise most naturally in arithmetic algebraic geometry, for example those associated with the ℓ-adic cohomology of algebraic varieties, are not complex representations but rather representations in $\mathrm{Rep}_n(\mathcal{G}_k; \overline{\mathbb{Q}}_\ell)$, with

$\ell \neq p$, or $\text{Rep}_n(W_k; \overline{\mathbb{Q}}_\ell)$. The representation theory for ℓ-adic representations is richer than for complex representations due to the difference in topologies in the two fields. Recognizing this, Deligne introduced what is now known as the Weil–Deligne group W_k' of the local field so that its representation theory is essentially algebraic, so in essence it doesn't distinguish between $\mathbb{C}$ and $\overline{\mathbb{Q}}_\ell$, and whose category of representations is the same that of the continuous ℓ-adic representations of $\mathcal{G}_k$ or W_k [19]. Following Tate [51], let us define W_k' to be the group scheme over $\mathbb{Q}$ which is the semidirect product of the Weil group W_k with the additive group $\mathbb{G}_a$, i.e., $W_k' = W_k \ltimes \mathbb{G}_a$, where W_k acts on $\mathbb{G}_a$ by $wxw^{-1} = \|w\|x$. If F is any field of characteristic 0, such as $\overline{\mathbb{Q}}_\ell$ or $\mathbb{C}$, the F-points of W_k' is just $W_k \times F$ with composition $(w_1, x_1)(w_w, x_2) = (w_1w_2, x_1 + \|w_1\|x_2)$. But what is really important is the representation theory of W_k'. An n-dimensional representation of W_k' over F is a pair $\rho' = (\rho, N)$ consisting of (i) an n-dimensional F-vector space V with a group homomorphism $\rho : W_k \to \text{GL}(V)$ whose kernel contains an open subgroup of I, that is, which is continuous for the *discrete* topology on $\text{GL}(V)$, and (ii) a nilpotent endomorphism N of V such that $\rho(w)N\rho(w)^{-1} = \|w\|N$ [19, 49, 51].

If $\rho' = (\rho, N)$ is a representation of W_k', there is a unique unipotent automorphism u of V which commutes with both N and $\rho(W_k)$ and such that $e^{aN}\rho(w)u^{-\nu(w)}$ is semisimple for all $a \in F$ and all $w \in W_k - I$ [19, 51]. The Φ semisimplification of ρ' is then $\rho_{ss}' = (\rho u^{-\nu}, N)$. ρ' is called Φ-semisimple (or Frobenius semisimple) if $\rho' = \rho_{ss}'$, for in this case u is the identity and all the Frobeniuses act semisimply. This is equivalent to the representation ρ being semisimple in the ordinary sense.

We will let $\text{Rep}_n(W_k'; F)$ denote the equivalence classes of n-dimensional Φ-semisimple F-representations of the Weil–Deligne group W_k'. When $F = \mathbb{C}$ we will simply write $\text{Rep}_n(W_k')$ for $\text{Rep}_n(W_k'; \mathbb{C})$.

The importance of the Weil–Deligne group is in that it lets us capture, in an algebraic way, the continuous ℓ-adic representations of $\mathcal{G}_k$ or W_k [19, 49, 51]: for every semisimple ℓ-adic representation $\rho_\ell \in \text{Rep}_n(W_k; \overline{\mathbb{Q}}_\ell)$ there is an open subgroup of the inertia group I on which ρ_ℓ is trivial and hence ρ_ℓ gives rise to an (ordinary) Φ-semisimple $\overline{\mathbb{Q}}_\ell$-representation ρ' of W_k'.

Note that by condition (ii) in the definition of a representation of W_k' the topology on F plays no role, so that if we have a fixed isomorphism $\iota : \overline{\mathbb{Q}}_\ell \to \mathbb{C}$ we may identify $\text{Rep}_n(W_k'; \overline{\mathbb{Q}}_\ell) \simeq \text{Rep}_n(W_k'; \mathbb{C}) = \text{Rep}_n(W_k')$. Furthermore, note that in an irreducible representation of W_k' we must have that $N = 0$, since the kernel of N would be an invariant subspace, and so $\text{Rep}_n^0(W_k') = \text{Rep}_n^0(W_k)$.

If $\rho' = (\rho, N) \in \text{Rep}_n(W_k'; F)$ is an representation of W_k' on the vector space V, let $V_N^I = (Ker\ N)^{\rho(I)}$ be the invariants of the inertia subgroup I on the kernel of N. The we can define the local L-factor by setting

$$Z(t, V) = \det(1 - t\rho(\Phi)|V_N^I)^{-1} \in F(t)$$

to be the inverse of the characteristic polynomial of Φ acting on V_N^I and if we have an embedding $F \hookrightarrow \mathbb{C}$, so if $F = \mathbb{C}$ or we use the isomorphism $\iota : \overline{\mathbb{Q}}_\ell \to \mathbb{C}$, then we view F as a subfield of $\mathbb{C}$ and set

$$L(s,\rho') = Z(q^{-s},\rho').$$

The definition of the local constants $\varepsilon(s,\rho',\psi)$, with ψ an additive character of k, is more delicate and we refer the reader to Deligne [19], Rohrlich [49] or Tate [51] for their precise definition. Of course, in the case $N=0$ these are the usual local Artin–Weil L-functions and ε-factors.

If the local field k is archimedean, so $k=\mathbb{R}$ or $\mathbb{C}$, then we are interested only in complex representations of $\mathcal{G}_k$ or W_k. When $k=\mathbb{C}$ the Weil group is simply $\mathbb{C}^\times$ while if $k=\mathbb{R}$, then $W_\mathbb{R} \simeq \mathbb{C}^\times \coprod j\mathbb{C}^\times$, where $j^2=-1$ and $jcj^{-1}=\bar{c}$ for $c \in \mathbb{C}^\times$. In either case we have

$$1 \longrightarrow \mathbb{C}^\times \longrightarrow W_k \longrightarrow \mathcal{G}_k \longrightarrow 1.$$

There is no archimedean Weil–Deligne group, so for consistency we will set $W_k' = W_k$ in these cases. The L-and ε-factors are then defined in terms of the classical Γ-function and a local functional equation [51].

When k is a global field we will at least be interested in the representations of the global Galois group $\mathcal{G}_k$, the global Weil group W_k, or possibly the conjectural Langlands group $\mathcal{L}_k$ [46].

When $F=\overline{\mathbb{Q}}_\ell$ we will let $\mathrm{Rep}_n(G;\overline{\mathbb{Q}}_\ell)$ denote the set of global ℓ-adic representations in the following sense. They should be continuous, algebraic (in the sense that they take values in $\mathrm{GL}(E_\lambda)$ for a finite dimensional extension $E_\lambda/\mathbb{Q}_\ell$), and almost everywhere unramified (in the sense that there is a finite set of places $S(\rho)$ of k such that for all $v \notin S(\rho)$ the representation ρ is unramified at v).

For any global representation ρ of $\mathcal{G}_k$ or W_k we have a local representation ρ_v for each completion v of k obtained by composing ρ with the natural maps $\mathcal{G}_{k_v} \to \mathcal{G}_k$ or $W_{k_v} \to W_k$. The conjectural Langlands group $\mathcal{L}_k$ should have similar local-global compatibility with the local Weil–Deligne groups.

To any n-dimensional complex or ℓ-adic representation of either the Galois group or the Weil group we have attached a global complex analytic invariant, the global L-function $L(\rho,s)$ defined by the Euler product

$$L(s,\rho) = \prod_v L(s,\rho_v), \qquad \varepsilon(s,\rho) = \prod_v \varepsilon(s,\rho_v,\psi_v),$$

where $\psi = \prod_v \psi_v$ is an additive character of k.

These global analytic invariants are conjectured to be nice in the sense that

1. $L(s,\rho)$ should have a meromorphic continuation with at most a finite number of poles, entire if ρ is irreducible but not trivial;

2. these continuations should be bounded in vertical strips;

3. they satisfy the functional equation $L(s,\rho) = \varepsilon(s,\rho)L(1-s,\tilde{\rho})$.

For $G = \mathcal{G}_k$ or W_k and $F = \mathbb{C}$ these are the classical Artin–Weil L-functions and by Brauer's Theorem are known to converge in a right half plane, have meromorphic continuation to $\mathbb{C}$, and satisfy a functional equation.

If k is a global function field, with constant field of order q, and ρ is an ℓ-adic representation of $\mathcal{G}_k$ as above, then Grothendieck has shown that the L-function is in fact a rational function of q^{-s} and satisfies a functional equation and Deligne later showed that the ε-factor of the functional equation had a local factorization and was given as above [51].

2 Automorphic representations and their L-functions

On the automorphic side, if k is a local field, we let $\mathcal{A}_n(k)$ denote the set of equivalence classes of irreducible admissible complex representations of $GL_n(k)$. When k is nonarchimedean local, we let $\mathcal{A}_n^0(k)$ denote the subset of equivalence classes of supercuspidal representations of $GL_n(k)$. By the theory of Godement–Jacquet [24], or the theory of Jacquet–Piatetski-Shapiro–Shalika outlined in [13], there a complex analytic invariant attached to every $\pi \in \mathcal{A}_n(k)$, namely its L-function $L(s,\pi)$ and a local ε-factor $\varepsilon(s,\pi,\psi)$ depending on a choice of additive character. If in addition we have an irreducible admissible representation π' of $GL_m(k)$, then we have the local Rankin–Selberg convolution L-functions $L(s,\pi\times\pi')$ and ε-factor $\varepsilon(s,\pi\times\pi',\psi)$.

If k is a global field we let $\mathbb{A}$ denote its ring of adeles. Let $\mathcal{A}_n(k)$ denote the set of irreducible automorphic representations of $GL_n(\mathbb{A})$ and $\mathcal{A}_n^0(k)$ the subset of cuspidal automorphic representations. If $\pi = \otimes'\pi_v$ is an automorphic representation of $GL_n(\mathbb{A})$ and $\pi' = \otimes'\pi'_v$ an automorphic representation of $GL_m(\mathbb{A})$, then we have its associated L-function and ε-factor defined by Euler products

$$L(s,\pi\times\pi') = \prod_v L(s,\pi_v\times\pi'), \qquad \varepsilon(s,\pi\times\pi') = \prod_v \varepsilon(s,\pi_v\times\pi'_v,\psi_v).$$

As we have seen [13], these invariants are known to be nice, that is if π and π' are unitary cuspidal representations, then

1. $L(s,\pi\times\pi')$ has an analytic continuation to all of $\mathbb{C}$ with at most simple poles at $s = 0, 1$ iff $\pi' = \tilde{\pi}$;
2. these continuations are bounded in vertical strips;
3. they satisfy the functional equation

$$L(s,\pi\times\pi') = \varepsilon(s,\pi\times\pi')L(1-s,\tilde{\pi}\times\tilde{\pi}').$$

When considering representations that occur in ℓ-adic cohomologies it is most natural to use $\overline{\mathbb{Q}}_\ell$-valued automorphic forms and representations, which we denote by $\mathcal{A}_n(k;\overline{\mathbb{Q}}_\ell)$. For example, we will need to consider the space of $\overline{\mathbb{Q}}_\ell$-valued cuspidal representations whose central character is of finite order, which we will denote by $\mathcal{A}_n^0(k;\overline{\mathbb{Q}}_\ell)_f$. These are the representations of $\mathrm{GL}_n(\mathbb{A})$, or the associated Hecke algebra $\mathcal{H}$ of locally constant $\overline{\mathbb{Q}}_\ell$-valued functions of compact support on $\mathrm{GL}_n(\mathbb{A})$, in the space of certain $\overline{\mathbb{Q}}_\ell$-valued cusp forms on $\mathrm{GL}_n(\mathbb{A})$. For the convenience of the reader, we will review the definition from [42] for the case of function fields over finite fields. The $\overline{\mathbb{Q}}_\ell$-valued cusp form on $\mathrm{GL}_n(\mathbb{A})$ of interest is a function $\varphi : \mathrm{GL}_n(\mathbb{A}) \to \overline{\mathbb{Q}}_\ell$ such that

(i) $\varphi(\gamma g) = \varphi(g)$ for all $\gamma \in \mathrm{GL}_n(k)$.

(ii) There is a compact open subgroup $K_\varphi \subset K = \mathrm{GL}_n(\mathcal{O}_k)$ such that $\varphi(gk) = \varphi(g)$ for all $k \in K_\varphi$.

(iii) There is an $a \in \mathbb{A}^\times$ with $\deg(a) \neq 0$ such that $\varphi(ag) = \varphi(g)$.

(iv) φ is cuspidal in the usual sense that the integral $\int_U \varphi(ug)\,du \equiv 0$ for each unipotent radical U of a maximal parabolic subgroup of $\mathrm{GL}_n(\mathbb{A})$.

Note that the condition (iii) implies that the central character of φ is of finite order. The theory can essentially be identified with the complex theory through the isomorphism $\iota : \overline{\mathbb{Q}}_\ell \to \mathbb{C}$ and the natural L-functions can be identified with the usual complex analytic ones or they can be left as ℓ-adic valued rational functions as in the appendix of [35] .

3 The local Langlands conjecture

In its most basic form, the local Langlands conjecture is a nonabelian generalization of (abelian) local class field theory. The conjecture as first formulated by Langlands was in terms of the Weil group. An early formulation, possibly the first, can be found in [36]. Langlands never restricted himself to GL_n but always formulated in terms of reductive algebraic groups in general. Deligne first pointed the necessity of passing to what is now known as the Weil–Deligne group to be able to include the special representations of $GL_2(k)$ for a local field k [18]. The current formulation of the conjecture which is closest to Langlands original is to be found in Borel's article in Corvallis [4].

Local Langlands Conjecture I. *Let k be a local field. Then there are a series of natural bijections*

$$\mathrm{Rep}_n(W_k') \leftrightarrow \mathcal{A}_n(k), \qquad \rho = \rho_\pi \leftrightarrow \pi = \pi_\rho$$

satisfying a set of representation theoretic desiderata, including the following:

(i) *For $n = 1$ it should be given by the local class field theory isomorphism.*

(ii) *The central character of π_ρ corresponds to the determinant* $\det(\rho)$ *under the $n = 1$ correspondence.*

(iii) *Compatibility with twisting, i.e., if χ is a character of $k^\times$, then $\rho_{\pi\otimes\chi} = \rho\otimes\chi$.*

(iv) *π_ρ is square integrable iff $\rho(W_k')$ does not lie in a proper Levi subgroup of* $\mathrm{GL}_n(\mathbb{C})$.

(v) *π_ρ is tempered iff $\rho(W_k)$ is bounded.*

(vi) *If H is a reductive connected k-group and $H(k) \to \mathrm{GL}_n(k)$ is a k morphism with commutative kernel and co-kernel, then there is a required compatibility between these bijections for $\mathrm{GL}_n(k)$ and similar maps for $H(k)$.*

For more details on the the compatibility condition (vi), see the article of Borel in Corvallis [4] or the accompanying article [14]. This is related to Langlands' general functoriality conjecture. Langlands himself never separated this version of his conjecture from his general principle of functoriality [40]. Note that there is no mention of L-functions in this formulation.

For nonarchimedean local fields, it was in the book by Jacquet and Langlands [31] and then in the work of Deligne [18] that a version of the local Langlands conjecture was phrased not in terms of representation theoretic properties, but rather in terms of the complex analytic invariants, or L-functions, of the two sets in question. Deligne gave the complete formulation for GL_2. It was in this paper that he utilized for the first time the Weil–Deligne group, which he had introduced in [19] in the context of ℓ-adic representations, in order to have a correct formulation on the case of GL_2 over a nonarchimedean local field.

Local Langlands Conjecture II. *Let k be a nonarchimedean local field. For each $n \geq 1$ there exists a bijective map $\mathcal{A}_n(k) \to \mathrm{Rep}_n(W_k')$ denoted $\pi \mapsto \rho_\pi$ with the following properties.*

(i) *For $n = 1$ the bijection is given by local class field theory, normalized so that the uniformizer of k corresponds to the geometric Frobenius.*

(ii) *For any $\pi \in \mathcal{A}_n(k)$ and $\pi' \in \mathcal{A}_{n'}(k)$, we have*

$$L(s, \rho_\pi \otimes \rho_{\pi'}) = L(s, \pi \times \pi'), \qquad \varepsilon(s, \rho_\pi \otimes \rho_{\pi'}, \psi) = \varepsilon(s, \pi \times \pi', \psi).$$

(iii) *For any $\pi \in \mathcal{A}_n(k)$ the determinant of ρ_π corresponds to the central character of π under local class field theory.*

(iv) *For any π in $\mathcal{A}_n(k)$ we have $\rho_{\tilde{\pi}} = \widetilde{\rho_\pi}$.*

(v) *For any $\pi \in \mathcal{A}_n(k)$ and any character χ of $k^\times$ of finite order, we have $\rho_{\pi\otimes\chi} = \rho_\pi \otimes \chi$.*

There are two ways to think about what these conjectures offer. If one views the primary passage of information to be from $\mathcal{A}_n(k)$ to $\mathrm{Rep}_n(W_k')$, then this can be thought of as Langlands formulation of a *nonabelian local class field theory*. If one views the primary passage of information from $\mathrm{Rep}_n(W_k')$ to $\mathcal{A}_n(k)$, then this gives an arithmetic parameterization of irreducible admissible representations of $\mathrm{GL}_n(k)$. This is the *arithmetic Langlands classification* of $\mathcal{A}_n(k)$.

3.1 *k local archimedean, i.e., $k = \mathbb{R}$ or $\mathbb{C}$*

In this case $\mathcal{G}_k$ is well understood; it is either $\mathbb{Z}/2\mathbb{Z}$ or trivial. So the passage of information in this case is in the opposite direction. This was done in great generality by Langlands about 1973 [38], and not only for GL_n but for general real reductive groups. For archimedean local fields there is no Weil–Deligne group. The representation theoretic version is what is now known as the Langlands classification or the Langlands parameters for representations of real groups. In fact, Langlands did this in conjunction with the *arithmetic* parameterization in terms of $\mathrm{Rep}_n(W_k)$ for GL_n (or admissible homomorphisms $W_k \to {}^L G$ for general G). The deep and interesting part is the classification of representations in term of the information obtained from these maps, particularly their relation with the construction of the discrete series.

Theorem 3.1. *Let k be $\mathbb{R}$ or $\mathbb{C}$. Then there are a are a series of natural bijections*

$$\mathrm{Rep}_n(W_k) \leftrightarrow \mathcal{A}_n(k)$$

satisfying the properties (i)–(vi) *of version* I *of the local Langlands conjecture.*

For the precise relation with the usual Langlands classification for real algebraic groups, see [38]. The statement proved is of course that originally given by Langlands and this may well have motivated the precise conditions in the conjecture. Note again that the conditions are representation theoretic and the L-functions and ε-factors play no role.

3.2 *k local nonarchimedean*

Recently, this second version of the local Langlands conjecture has been established for nonarchimedean local fields, first by Laumon, Rapoport, and Stuhler in the positive characteristic case in 1993 [44] and then in the characteristic 0 case by by Harris and Taylor [27] in 1999 and by Henniart [30] in 2000. In both cases, the correspondence is established at the level of a correspondence between irreducible Galois representations and supercuspidal representations. Much of the original representation theoretic desiderata of the original conjecture has been replaced by an equality of twisted L-functions, i.e., of the associated families of complex analytic invariants.

Let $\mathcal{A}_n^0(k)_f$ denote the set of isomorphism classes of irreducible admissible representations of $\mathrm{GL}_n(k)$ having central character of finite order. Then the theorem of Laumon, Rapoport, and Stuhler is the following [44].

Theorem 3.2. *Let k be a local field of characteristic $p > 0$. For each $n \geq 1$ there exists a bijective map $\mathcal{A}_n^0(k)_f \to \mathrm{Rep}_n^0(\mathcal{G}_k)$ denoted $\pi \mapsto \rho_\pi$ satisfying the conditions* (i)–(v) *of version* II *of the local Langlands conjecture.*

When the local field k is of characteristic 0 the local Langlands conjecture established by Harris and Taylor [27] and Henniart [28] has precisely the same statement.

Theorem 3.3. *Let k be a local field of characteristic* 0. *For each* $n \geq 1$ *there exists a bijective map* $\mathcal{A}_n^0(k)_f \to \mathrm{Rep}_n^0(\mathcal{G}_k)$ *denoted* $\pi \mapsto \rho_\pi$ *satisfying conditions* (i)–(v) *of version* II *of the local Langlands conjecture.*

The proofs involve the use of $\overline{\mathbb{Q}}_\ell$-representations on both the Galois and automorphic side, and is translated into the statements above in terms of complex analytic L-functions through the isomorphism ι of $\overline{\mathbb{Q}}_\ell$ with $\mathbb{C}$.

Reductions and constructions. In any of the nonarchimedean local cases, the proof passes through a chain of identical reductions which reduces one to proving the existence of a single map having the desired properties. The three proofs then differ in the constructions used to prove the existence of at least one correspondence.

There are essentially three steps in the reduction. Assume that we have any correspondence $\mathcal{A}_n^0(k)_f \to \mathrm{Rep}_n^0(\mathcal{G}_k)$, still denoted $\pi \mapsto \rho_\pi$, which satisfies (i)–(v) of the theorem.

1. *Injectivity: Poles of L-functions.* For ρ and ρ' in $\mathrm{Rep}_n^0(\mathcal{G}_k)$ we have that $L(s, \rho \otimes \rho')$ has a pole at $s = 0$ iff $\rho' \cong \tilde{\rho}$. Similarly, if π and π' are both in $\mathcal{A}_n^0(k)$, then $L(s, \pi \times \pi')$ has a pole at $s = 0$ iff $\pi' \simeq \tilde{\pi}$. Thus we see that any such correspondence satisfying (ii) is automatically injective.

2. *Bijectivity: Numerical local Langlands.* For $\rho \in \mathrm{Rep}_n^0(\mathcal{G}_k)$ let $a(\rho)$ denote the exponent of the Artin conductor of ρ [51]. This is determined by the ε-factor $\varepsilon(s, \rho, \psi)$. Let $\mu : k^\times \to \mathbb{C}^\times$ be identified with a character of the Galois group via local class field theory. If we let $\mathrm{Rep}_n^0(\mathcal{G}_k)_{m,\mu}$ denote the set of $\rho \in \mathrm{Rep}_n^0(\mathcal{G}_k)$ with $a(\rho) = m$ and $\det(\rho) = \mu$, then this set is finite. Similarly, we let $\mathcal{A}_n^0(k)_{m,\mu}$ denote the set of $\pi \in \mathcal{A}_n^0(k)$ with $f(\pi) = m$ and central character $\omega_\pi = \mu$, where now $f(\pi)$ is the exponent of the conductor of Jacquet, Piatetski-Shapiro, and Shalika [13], and this set is also finite. The statement of the numerical local Langlands conjecture, which had been established by Henniart in 1988 [30], is that for fixed $m \in \mathbb{Z}_+$ and multiplicative character μ of finite order we have $|\mathrm{Rep}_n^0(\mathcal{G}_k)_{m,\mu}| = |\mathcal{A}_n^0(k)_{m,\mu}|$. Since (ii) and (iii) guarantee that our correspondence preserves the character μ and the conductors, then once we know that the correspondence is injective, the numerical local Langlands conjecture gives that the correspondence is surjective and hence bijective.

3. *Uniqueness: The local converse theorem.* The uniqueness of a correspondence satisfying (i)–(v) is a consequence of the *local converse theorem for* $GL(n)$. This result was first stated by Jacquet, Piatetski-Shapiro, and Shalika [32] but the first published proof was by Henniart [29] with precisely this application in mind. The statement is the following. Suppose that π and π' are both elements of $\mathcal{A}_n^0(k)$ and that the twisted ε-factors agree, that is,

$$\varepsilon(s, \pi \times \tau, \psi) = \varepsilon(s, \pi' \times \tau, \psi),$$

for all $\tau \in \mathcal{A}_m^0(k)$ with $1 \leq m \leq n-1$. Then $\pi \cong \pi'$. (Note that the corresponding twisted L-functions are all identically 1 [13].) From this, by induction on n, one sees that any such (now bijective) correspondence satisfying (i)–(v) must be unique.

These three steps then reduce the local Langlands conjecture to the question of existence of some correspondence satisfying (i)–(v). It is this existence problem that was solved by Laumon, Rapoport, and Stuhler in positive characteristic and by Harris and Taylor and then Henniart in the characteristic zero case.

4. *Existence: Global geometric constructions.* In all cases, the local existence is based on establishing certain instances of a *global* correspondence of Galois representations and automorphic representations. Note that we now work with ℓ-adic representations on both the Galois and automorphic side.

For k of characteristic $p > 0$, Laumon, Rapoport, and Stuhler begin with a local representation $\pi \in \mathcal{A}_n^0(k)$. They realize k as a local component of a global field K of characteristic p, so $k = K_v$ for some place v of K, and then embed π as the local component at v of a cuspidal representation Π of a global division algebra $D(\mathbb{A})$ of rank n such that $D^\times(K_v) = \mathrm{GL}_n(k)$ and $\Pi_v = \pi$. They globally realize an action of $\mathcal{G}_K \times D^\times(\mathbb{A})$ on the ℓ-adic cohomology of the moduli space of $\mathcal{D}$-elliptic modules ($\mathcal{D}$ an order in D) such that in the decomposition of this cohomology a representation $R \otimes \Pi$ of $\mathcal{G}_K \times D^\times(\mathbb{A})$ occurs. By construction $\Pi_v = \pi$ and they take $R_v = \rho_\pi$. By the nature of their construction they are able to verify that (i)–(v) are satisfied. Thus they establish the needed local existence statement via a global geometric construction and a limited global correspondence.

The proof of Harris and Taylor of the local Langlands conjecture for nonarchimedean fields of characteristic 0 is similar in spirit to that of Laumon, Rapoport, and Stuhler in characteristic p. They replace the moduli space of $\mathcal{D}$-elliptic modules with certain "simple Shimura varieties" associated to unitary groups U_n of Kottwitz. They realize k as a local component of a number field K, so $k = K_v$ for some place v of K, and then embed π as the local component at v of a cuspidal representation Π of a certain (twisted) unitary group of rank n. They then realize a global correspondence between these cuspidal representations and global Galois representations in the ℓ-adic cohomology of the associated Shimura varieties. By studying the resulting correspondence locally at a place of *bad* reduction they find a local representation on which they have an irreducible action of $\mathrm{GL}_n(k) \times D_n^\times(k) \times W_k$, where D is the division algebra over k of rank n and Hasse invariant $1/n$, by $\pi \otimes JL(\pi) \otimes \rho_\pi$, where $JL(\pi)$ is the image of π under the local Jacquet–Langlands correspondence and ρ_π is thus defined. Again, from their construction they can verify that this correspondence satisfies conditions (i)–(v). Note that not only do they get a geometric realization of the local Langlands correspondence, they get a simultaneous realization of the local Jacquet–Langlands correspondence.

Henniart, in his proof of the local Langlands conjecture for nonarchimedean fields of characteristic 0, again uses a global construction but in a far less serious way. In particular, he does not give a geometric realization of the correspondence. For Henniart, both the statement and the proof most naturally give a bijection from the Galois side to the automorphic side $\rho \mapsto \pi_\rho$. Henniart begins with an irreducible representation ρ of $\mathcal{G}_k$ with finite order determinant. This then factors through a representation of $Gal(F/k)$ for a finite dimensional extension F of k. Using Brauer induction, he writes ρ as a sum of monomial representations. The characters can be lifted to the automorphic side by local class field theory, and so he must show that the corresponding sum of automorphically induced representations exists and is supercuspidal. The resulting supercuspidal representation is then π_ρ. This he does again by embedding the local situation into a global one and then appealing to certain weak cases of global automorphic induction that had been earlier established by Harris [25]. Harris's result relies on the theory of base change and the association of ℓ-adic representations to automorphic representations of $GL_n(\mathbb{A})$ by Clozel [12], which in turn relies on the work of Kottwitz on the good reduction of certain unitary Shimura varieties. So at the bottom there is in fact a global geometric construction, but it is of a simpler type than used by Harris and Taylor. Henniart's proof makes more use of L-functions and less use of geometry. His proof is shorter and more analytic, but does not give a geometric realization of the correspondence.

A more complete synopsis of these results can be found in the Séminaire Bourbaki reports of Carayol [9, 10].

3.3 *Complements*

In order to complete the local Langlands correspondence one needs to consider all suitable representations of the Weil–Deligne group on the Galois theoretic side and all irreducible admissible representations of $GL_n(k)$ on the automorphic side. In order to do this, the first step is to remove the condition of finite-order on the central character. This is obtained by simply replacing the Galois group by the Weil group on the Galois side of the correspondence. On the automorphic side one still has supercuspidal representations. Then to pass to all admissible representations of $GL_n(k)$ one uses the representations of the Weil–Deligne group. Representation theoretically, the passage from representations of the Weil group to representations of the Weil–Deligne group on the Galois side mirrors the passage from supercuspidal representations to irreducible representations on the automorphic side, as was shown by Bernstein and Zelevinsky (see [1, 60], particularly Section 10 of [60], or [47]). Thus from the results of Laumon, Rapoport, Stuhler, Harris, Taylor, and Henniart the full local Langlands correspondence follows.

In spite of these results, the work on the local Langlands conjecture continues. The proofs above give the existence of the correspondence and in some cases provides an explicit geometric model. It provides a matching of certain invariants, like the conductor, and the local L-functions. However, for applications, it would be desirable to have an explicit version of the local Langlands correspondence, particularly for the supercuspidal representations of GL_n in terms of the Bushnell–Kutzko compact induction data [6]. The search for an explicit local Langlands correspondence is currently being pursued by Bushnell, Henniart, Kutzko, and others.

4 The global Langlands conjecture

As in the local case, in its most basic form, the global Langlands conjecture should be a nonabelian generalization of (abelian) global class field theory. When Deligne pointed out the necessity of introducing the Weil–Deligne group in the local nonarchimedean regime, it was realized that there seemed to be no natural global version of the Weil–Deligne group. This lead to a search for a global group to replace the Weil–Deligne group. This was one of the purposes of Langlands' article [39]. It is now believed that this group, which Ramakrishnan calls the conjectural Langlands group $\mathcal{L}_k$, should be related to the equally conjectural motivic Galois group of k, $\mathcal{M}_k$ [46].

4.1 *k a global field of characteristic* $p > 0$

In spite of these difficulties, Drinfeld formulated and proved a version of the global Langlands conjecture for global function fields [23] which related the irreducible 2-dimensional representations of the Galois group itself with the irreducible cuspidal representations of $\mathrm{GL}_2(\mathbb{A})$. This is the global analogue of the local theorem of Laumon, Rapoport, and Stuhler for which the Weil–Deligne group was not needed. We should emphasize that the results of Drinfeld were obtained in the 1970's and so predate those of Laumon, Rapoport, and Stuhler by several years. Recently the work of Drinfeld has been extended by L. Lafforgue to give a proof of the global Langlands conjecture for GL_n over a function field [35].

The formulation of the global Langlands conjecture established by Drinfeld and Lafforgue is essentially the same as in the local nonarchimedean case above with a few modifications that we would now like to explain. Take k to be the function field of a smooth, projective, geometrically connected curve X over a finite field $\mathbb{F}$ of characteristic p. Fix a prime ℓ which is different from p and an isomorphism $\iota : \overline{\mathbb{Q}}_\ell \to \mathbb{C}$.

On the Galois side, they consider isomorphism classes of irreducible continuous ℓ-adic representations $\rho : \mathcal{G}_k \to \mathrm{GL}_n(\overline{\mathbb{Q}}_\ell)$ which are unramified outside a finite number of places, as described in Section 1, and whose determinant is of finite order. We will denote these by $\mathrm{Rep}_n^0(\mathcal{G}_k; \overline{\mathbb{Q}}_\ell)_f$. On the automorphic side they consider the space of $\overline{\mathbb{Q}}_\ell$-valued cuspidal representations whose central character is of finite order $\mathcal{A}_n^0(k; \overline{\mathbb{Q}}_\ell)_f$, as described in Section 2. A reasonable formulation of a global Langlands conjecture in analogy with what we have in the local situation is the following.

Global Langlands Conjecture in characteristic p. *For each $n \geq 1$ there exists a bijective map $\mathcal{A}_n^0(k; \overline{\mathbb{Q}}_\ell)_f \to \mathrm{Rep}_n^0(\mathcal{G}_k; \overline{\mathbb{Q}}_\ell)_f$ denoted $\pi \mapsto \rho_\pi$ with the following properties.*

(i) *For $n = 1$ the bijection is given by global class field theory.*

(ii) *For any $\pi \in \mathcal{A}_n^0(k; \overline{\mathbb{Q}}_\ell)_f$ and $\pi' \in \mathcal{A}_{n'}^0(k; \overline{\mathbb{Q}}_\ell)_f$, we have*

$$L(s, \rho_\pi \otimes \rho_{\pi'}) = L(s, \pi \times \pi') \qquad \varepsilon(s, \rho_\pi \otimes \rho_{\pi'}) = \varepsilon(s, \pi \times \pi').$$

(iii) *For any $\pi \in \mathcal{A}_n^0(k; \overline{\mathbb{Q}}_\ell)_f$ the determinant of ρ_π corresponds to the central character of π under global class field theory.*

(iv) *For any π in $\mathcal{A}_n^0(k:\overline{\mathbb{Q}}_\ell)_f$, we have $\rho_{\tilde{\pi}} = \widetilde{\rho_\pi}$.*

(v) *for any $\pi \in \mathcal{A}_n^0(k;\overline{\mathbb{Q}}_\ell)_f$ and any character χ of $k^\times$ of finite order, we have $\rho_{\pi\otimes\chi} = \rho_\pi \otimes \chi$.*

(vi) *The global bijections should be compatible with the local bijections of the local Langlands conjecture.*

If we take a cuspidal representation $\pi \in \mathcal{A}_n^0(k;\overline{\mathbb{Q}}_\ell)_f$, let $S(\pi)$ denote the finite set of places $x \in |X|$ such that π_x is unramified for all $x \notin S(\pi)$. For $x \notin S(\pi)$ the local representation π_x is completely determined by a semisimple conjugacy class A_{π_x} in $\mathrm{GL}_n(\overline{\mathbb{Q}}_\ell)$, which we identify with $\mathrm{GL}_n(\mathbb{C})$ via ι, called the Satake class or Satake parameter of π_x [4]. As this parameter determines and is determined by a character of the associated unramified Hecke algebra $\mathcal{H}_x$ at x [50] the eigenvalues of A_{π_x}, denoted $z_1(\pi_x), \ldots, z_n(\pi_x)$, are also called the *Hecke eigenvalues* of π_x. These Hecke eigenvalues completely determine π_x. Then by the strong multiplicity one theorem for GL_n [13] the collection of Hecke eigenvalues $\{z_1(\pi_x), \ldots, z_n(\pi_x)\}$ for almost all $x \notin S(\pi)$ completely determine π.

If we take a Galois representation $\rho \in \mathrm{Rep}_n^0(\mathcal{G}_k;\overline{\mathbb{Q}}_\ell)_f$, then we also have a finite set of places $S(\rho)$ such that ρ is unramified at all $x \notin S(\rho)$. For $x \notin S(\rho)$ the image $\rho(\Phi_x)$ of a geometric Frobenius Φ_x at x is a well-defined semisimple conjugacy class in $\mathrm{GL}_n(\overline{\mathbb{Q}}_\ell) \simeq \mathrm{GL}_n(\mathbb{C})$. The eigenvalues of $\rho(\Phi_x)$, denoted $z_1(\rho_x), \ldots, z_n(\rho_x)$, are called the *Frobenius eigenvalues* of ρ_x. These Frobenius eigenvalues completely determine ρ_x and by the Chebotarev density theorem the collection of Frobenius eigenvalues $\{z_1(\rho_x), \ldots, z_n(\rho_x)\}$ for almost all $x \notin S(\rho)$ completely determine ρ itself.

The result established by Drinfeld for $n = 2$ [23] and Lafforgue for $n \geq 3$ [35], which as we will outline below is equivalent to the statement above, is the following.

Theorem 4.1. *Let k be a global function field of characteristic p as above. For each $n \geq 1$ there exists a unique bijective map $\mathcal{A}_n^0(k;\overline{\mathbb{Q}}_\ell)_f \to \mathrm{Rep}_n^0(\mathcal{G}_k;\overline{\mathbb{Q}}_\ell)_f$ denoted $\pi \mapsto \rho_\pi$ such that for every cuspidal $\pi \in \mathcal{A}_n^0(k;\overline{\mathbb{Q}}_\ell)_f$ we have the equality of Hecke and Frobenius eigenvalues*

$$\{z_1(\pi_x), \ldots, z_n(\pi_x)\} = \{z_1(\rho_{\pi,x}), \ldots, z_n(\rho_{\pi,x})\}$$

for all $x \notin S$, a finite set of places containing $S(\pi) \cup S(\rho_\pi)$, or equivalently we have the equality of the partial complex analytic L-functions

$$L^S(s,\pi) = L^S(s,\rho_\pi).$$

To see that this does indeed give the statements of the conjecture as presented above, note first that for all $x \notin S$, the equality of the equality of the associated Hecke and Frobenius eigenvalues at these places is consistent with the local Langlands conjecture at the unramified places. Hence (vi) is satisfied at these places. Next, for $\pi \in \mathcal{A}_n^0(k;\overline{\mathbb{Q}}_\ell)_f$ we have an Euler product factorization for the L-function $L(s,\pi) = \prod_{x\in|X|} L(s,\pi_x)$ and for all $x \notin S(\pi)$ the local L-function is given by $L(s,\pi_x) = \det(1-\iota(A_{\pi_x})q^{-s})^{-1}$ [13]. Hence the Hecke eigenvalues for $\pi_x, x \notin S(\pi)$, are determined by the local L-factor at these places and conversely. Similarly, for $\rho \in \mathrm{Rep}_n^0(\mathcal{G}_k; \overline{\mathbb{Q}}_\ell)_f$ we again have a factorization of the global L-function $L(s,\rho) = \prod_{x\in|X|} L(s,\rho_x)$ where now $L(s,\rho_x) = \det(1-\iota(\rho(\Phi_x))q_x^{-s})^{-1}$. Hence now the local eigenvalues of Frobenius for $\rho(\Phi_x)$, $x \notin S(\rho)$, are determined by the local L-factor at these places and conversely. Hence we do have the equality of partial L-functions $L^S(s,\pi) = L^S(s,\rho_\pi)$ for a finite set $S \supset S(\pi)\cup S(\rho_\pi)$ as stated. Using the global functional equation for both $L(s,\pi)$ and $L(s,\rho_\pi)$ and the local factorization of the global ε-factors, standard L-function techniques give that in fact $S(\pi) = S(\rho_\pi)$, that $L(s,\pi_x) = L(s,\rho_{\pi,x})$ at these places, and that in general the restriction of ρ_π to the local Galois group of k at x corresponds to π_x under the local Langlands conjecture. Thus (vi) is satisfied in general and from this (ii)–(v) follow from the Euler product factorizations and the analogous statements from the local conjecture.

Reductions and constructions. As in the local situation, the proof passes through a chain of reductions that reduces one to proving the existence of a single map having the desired properties. The existence is then established by a global construction using the cohomology of a certain moduli scheme on which GL_n acts.

1. *Uniqueness and bijectivity*. Given the existence of one global bijection as above, the uniqueness of the bijection has long been known to follow from an application of the strong multiplicity one theorem on the automorphic side [13] and the Chebotarev density theorem on the Galois side. These strong uniqueness principles also imply that any such maps $\pi \mapsto \rho_\pi$ and $\rho \mapsto \pi_\rho$ satisfying the conditions of the theorem must be reciprocal bijections.

2. *The inductive procedure of Piatetski-Shapiro and Deligne*. This inductive principle was outlined by Deligne in an IHES seminar in 1980 and then later recorded in [41]. It reduces the proof of the theorem to the following seemingly weaker existence statement.

Theorem 4.2. *For each $n \geq 1$ there exists a map $\mathcal{A}_n^0(k; \overline{\mathbb{Q}}_\ell)_f \to \mathrm{Rep}_n^0(\mathcal{G}_k; \overline{\mathbb{Q}}_\ell)_f$ denoted $\pi \mapsto \rho = \rho_\pi$ such that we have the equality of Hecke and Frobenius eigenvalues*

$$\{z_1(\pi_x), \ldots, z_n(\pi_x)\} = \{z_1(\rho_{\pi,x}), \ldots, z_n(\rho_{\pi,x})\}$$

for almost all $x \notin S(\pi) \cup S(\rho_\pi)$.

Indeed, suppose one has established the existence of the map $\pi \mapsto \rho_\pi$ for $\pi \in A_r^0(k; \overline{\mathbb{Q}}_\ell)_f$ for $r = 1, \ldots, n-1$. Then utilizing the global functional equation of Grothendieck, the factorization of the global Galois ε-factor [41], and the converse theorem for GL_n [13, 15, 16] one obtains for free the inverse map $\rho \mapsto \pi_\rho$ for $\rho \in \mathrm{Rep}_r^0(\mathcal{G}_k; \overline{\mathbb{Q}}_\ell)_f$ for $r = 1, \ldots, n$.

3. *Existence* ([23, 35]). It had been known since Weil that there is a natural moduli problem associated to GL_n over a function field, namely the set of isomorphism

classes of rank n vector bundles on the curve X are parameterized by the double cosets $\mathrm{GL}_n(k)\backslash \mathrm{GL}_n(\mathbb{A})/\mathrm{GL}_n(\mathcal{O})$. To obtain the maps in question, one needs a bit more structure, and so Drinfeld and then Lafforgue considered the (compactified) Deligne-Mumford stack V of rank n shtukas (with level structure), which is actually a stack over $X \times X$. There is a natural action of the global Hecke algebra $\mathcal{H}$ on this stack by correspondences and the corresponding ℓ-adic cohomology $H_c^*(\overline{k} \otimes_k V; \overline{\mathbb{Q}}_\ell)$ then affords a simultaneous representation of $\mathcal{H}$ and $\mathcal{G}_k \times \mathcal{G}_k$. One then uses the geometric Grothendieck–Lefshetz trace formula to compute the trace of this representation. One then compares this with the output of the Arthur-Selberg trace formula to prove that indeed the derived representation $\pi \otimes \rho_\pi \otimes \widetilde{\rho_\pi}$ of $\mathcal{H} \times \mathcal{G}_k \times \mathcal{G}_k$ occurs in this cohomology. The construction is inductive and essentially uses everything.

Complements. An immediate consequence of this result is the Ramanujan–Petersson conjecture for GL_n. This had been earlier established by Drinfeld for $n = 2$ [22] and partially by Lafforgue for $n \geq 3$ [34]. The complete solution follows from the global Langlands conjecture.

Theorem 4.3. *For every* $\pi \in A_n^0(k)_f$ *and every place* $x \notin S(\pi)$, *we have* $|z_i(\pi_x)| = 1$.

In addition, Lafforgue [35] is able to deduce the following conjecture of Deligne [20].

Theorem 4.4. *Every irreducible local system* ρ *over a curve whose determinant is of finite order is pure of weight* 0; *moreover, the symmetric polynomials in the eigenvalues of Frobenius generate a finite extension of* $\mathbb{Q}$.

In addition, Lafforgue is able to conclude that over a curve the notion of an irreducible local ℓ-adic system does not depend on the choice of ℓ and to verify the assertion of *descent* in the "geometric Langlands correspondence."

A more complete synopsis of these results can be found in the reports of Laumon [42, 43].

4.2 *k a global field of characteristic* 0

There is very little known of a general nature in the number field case. However, there are some rather spectacular examples of such global correspondences.

General conjectures. Recall that for $n = 1$ from global class field theory we have a canonical bijection between the continuous characters of $\mathcal{G}_k$ and characters of finite order of $k^\times\backslash\mathbb{A}^\times$. To obtain all characters of $k^\times\backslash\mathbb{A}^\times$ we must again replace the Galois group by the global Weil group W_k.

For $n \geq 2$, by analogy with the local Langlands conjecture, we need a global analogue of the Weil–Deligne group. But unfortunately no such analogue is available. Instead the conjectures are envisioned in terms of a conjectural *Langlands group* $\mathcal{L}_k$ [46]. At best, one hopes that $\mathcal{L}_k$ fits into an exact sequence

$$1 \longrightarrow \mathcal{L}_k^0 \longrightarrow \mathcal{L}_k \longrightarrow \mathcal{G}_k \longrightarrow 1$$

with $\mathcal{L}_k^0$ complex pro-reductive. This should fit into a commutative diagram

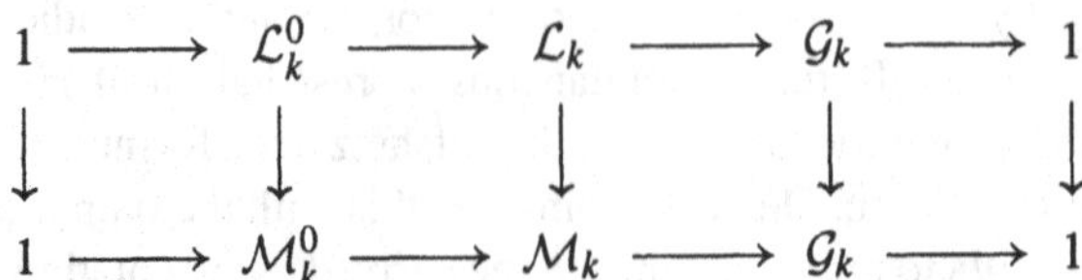

where $\mathcal{M}_k$ is the equally conjectural *motivic Galois group* [46].

In these terms, in general one expects/conjectures the following types of global correspondences [11, 46]:

(i) The irreducible n-dimensional representations of $\mathcal{G}_k$ should be in bijective correspondence with the cuspidal representations of $\mathrm{GL}_n(\mathbb{A})$ of *Galois type*. (This is a restriction on π_∞.)

(ii) The irreducible n-dimensional representations of $\mathcal{M}_k$ should be in bijective correspondence with the *algebraic* cuspidal representations of $\mathrm{GL}_n(\mathbb{A})$. These are the analogues of algebraic Hecke characters.

(iii) The irreducible n-dimensional representations of $\mathcal{L}_k$ should be in bijective correspondence with all cuspidal representations of $\mathrm{GL}_n(\mathbb{A})$.

Of course, all of these correspondences should satisfy properties similar to those on the local conjectures, particularly the preservation of L- and ε-factors (with twists), compatibility with the local correspondences, etc.

In reality, very little is known of a truly general nature. One problem for the current methods seems to be that there is no natural moduli problem for GL_n over a number field.

Known results. There are many partial results of a general nature if one starts on the automorphic side and tries to construct the associated Galois representation.

When $n = 2$ and $k = \mathbb{Q}$ we have the fundamental result of Deligne [17], based on foundational work of Eichler and Shimura, which associates to every cuspidal representation π of $\mathrm{GL}_2(\mathbb{A}_\mathbb{Q})$ which corresponds to a holomorphic new form of weight ≥ 2 a compatible system of ℓ-adic representations $\rho = \rho_\pi$ such that $L(s, \pi) = L(s, \rho)$. The Ramanujan–Petersson conjecture for such forms followed. This was extended to weight one forms over $\mathbb{Q}$ in the classical context by Deligne and Serre [21]. These results were extended to totally real fields k, still with $n = 2$, by a number of people, including Rogawski-Tunnell [48], Ohta [45], Carayol [8], Wiles [58], Taylor [52, 54], and Blasius–Rogawski [3]. For imaginary quadratic fields there is the work of Harris–Soudry–Taylor [26] and Taylor [53]. For more complete surveys, the reader can consult the surveys of Blasius [2] and Taylor [55].

For general GL_n and k a totally real number field Clozel has been able to attach a compatible system of ℓ-adic representations to cuspidal, algebraic, regular, self-dual representation of $GL_n(\mathbb{A}_k)$ having local components of a certain type at one or two finite places [12].

More spectacular are the results which go in the opposite direction, that is, starting with a specific Galois representation and showing that it is modular. The results we have in mind are those of Langlands [37] and Tunnell [57], with partial results by Taylor et al. [7, 56], on the modularity of degree 2 complex Galois representations (the strong Artin conjecture) and the results of Wiles [59] and then Breuil, Conrad, Diamond, and Taylor [5] on the modularity of (the two dimensional Galois representation on the ℓ-adic Tate module of) elliptic curves over $\mathbb{Q}$.

References

[1] J. Bernstein and A. Zelevinsky, Induced representations of $\mathfrak{p}$-adic groups, I. *Ann. scient. Éc. Norm. Sup.*, 4e série, **10**(1977), 441-472.

[2] D. Blasius, Automorphic forms and Galois representations: some examples. *Automorphic forms, Shimura varieties, and L-functions, Vol. II (Ann Arbor, MI, 1988). Perspect. Math.* **11**, Academic Press, Boston, MA, 1990, 1–13.

[3] D.Blasius, and J. Rogawski, Motives for Hilbert modular forms. *Invent. math.* **114**(1993), 55–87.

[4] A. Borel, Automorphic L-functions, *Proc. Symp. Pure math.* **33**, Part 2, (1979), 27–61.

[5] C. Breuil, B. Conrad, F. Diamond, and R. Taylor, On the modularity of elliptic curves over $\mathbb{Q}$. *J. Amer. Math. Soc.* **14**(2001), 843–939.

[6] C. Bushnell and P. Kutzko, *The Admissible Dual of* $GL(N)$ *via Compact Open Subgroups*. Annals of Mathematics Studies **129**. Princeton University Press, Princeton, NJ, 1993.

[7] K.Buzzard, M. Dickinson, N. Shepherd-Barron, and R. Taylor, On icosahedral Artin representations. *Duke Math. J.* **109**(2001), 283–318.

[8] H. Carayol, Sur les représentations l-adiques associées aux formes modulaires de Hilbert. *Ann. Sci. École Norm. Sup.* (4) **19**(1986), 409–468.

[9] H. Carayol, Variétés de Drinfeld compactes (d'après Laumon, Rapoport, et Stuhler). Séminaire Bourbaki No. 756. *Astérisque* **206**(1992), 369–409.

[10] H. Carayol, Preuve de la conjecture de Langlands locale pour GL_n: Travaux de Harris–Taylor et Henniart. Séminaire Bourbaki No. 857. *Astérisque* **266**(2000), 191–243.

[11] L. CLOZEL, Motifs et formes automorphes: applications du principe de fonctorialité. *Automorphic forms, Shimura varieties, and L-functions, Vol. I (Ann Arbor, MI, 1988)*, Perspect. Math. **10**, Academic Press, Boston, MA, 1990, 77–159.

[12] L. CLOZEL, Représentations galoisiennes associées aux représentations automorphes autoduales de GL(n). *Publ. Math. IHES* **73**(1991), 97–145.

[13] J.W. COGDELL, Analytic theory of L-functions for GL_n, in J. Bernstein and S. Gelbart, eds., *An Introduction to the Langlands Program*, Birkhäuser Boston, Boston, 2003, 197–228 (this volume).

[14] J.W. COGDELL, Dual groups and Langlands functoriality, in J. Bernstein and S. Gelbart, eds., *An Introduction to the Langlands Program*, Birkhäuser Boston, Boston, 2003, 251–269 (this volume).

[15] J.W. COGDELL AND I.I. PIATETSKI-SHAPIRO, Converse Theorems for GL_n. *Publ. Math. IHES* **79**(1994), 157–214.

[16] J.W. COGDELL AND I.I. PIATETSKI-SHAPIRO, Converse Theorems for GL_n, II. *J. reine angew. Math.* **507**(1999), 165–188.

[17] P. DELIGNE, Formes modulaires et représentations ℓ-adiques. *Séminaire Bourbaki no. 355*, Février 1969.

[18] P. DELIGNE, Formes modulaires et représentations de GL(2), in *Modular Forms of One Variable, II*, Lecture Notes in Mathematics No. 349, Springer–Verlag, 1973, 55–105 .

[19] P. DELIGNE, Les constantes des équations fonctionnelles des fonctions L. *Modular Forms of One Variaable, II*, Lecture Notes in Mathematics No. 349, Springer–Verlag, 1973, 501–597 .

[20] P. DELIGNE, La conjecture de Weil, II. *Publ. Math. IHES* **52**(1980), 137–252.

[21] P. DELIGNE AND J-P. SERRE, Formes modulaires de poids 1. *Ann. Sci. Ec. Norm. Sup* **7**(1974), 507–530.

[22] V. DRINFELD, Proof of the Petersson conjecture for GL(2) over a global field of characteristic p. *Funct. Anal. and its Appl.* **22**(1988), 28–43.

[23] V. DRINFELD, Cohomology of compactified modules of F-sheaves of rank 2. *Journal of Soviet Mathematics* **46**(1989), 1789–1821.

[24] R. GODEMENT AND H. JACQUET, *Zeta functions of simple algebras*. Lecture Notes in Mathematics, Vol. 260. Springer-Verlag, Berlin-New York, 1972

[25] M. HARRIS, The local Langlands conjecture for GL(n) over a p-adic field, $n < p$, *Invent. math.* **134**(1998), 177–210.

[26] M. Harris, D. Soudry, and R. Taylor, l-adic representations associated to modular forms over imaginary quadratic fields. I. *Invent. math.* **112**(1993), 377–411.

[27] M. Harris and R. Taylor, The geometry and cohomology of certain simple Shimura varieties, *Annals of Math Studies* **151**, Princeton University Press, 2001.

[28] G. Henniart, La conjecture de Langlands numérique pour GL(n), *Ann. Scient. Éc. Norm. Sup.* (4) **21**(1988), 497–544.

[29] G. Henniart, Caractérization de la correspondance de Langlands locale par les facteurs ε de pairs, *Invent. math.* **113**(1993), 339–350.

[30] G. Henniart, Une preuve simple des correspondance de Langlands pour GL(n) sur un corps p-adic, *Invent. math.* **139**(2000), 439–455.

[31] H. Jacquet and R.P. Langlands, *Automorphic Forms on* GL(2). Springer Lecture Notes in Mathematics No. 114, Springer Verlag, Berlin, 1970.

[32] H. Jacquet, I.I. Piatetski-Shapiro, and J. Shalika, Facteurs L et ε du groupe linéaire. *C. R. Acad. Sci. Paris Sér. A-B* **289**(1979), no. 2, A59–A61.

[33] S. Kudla, The local Langlands correspondence: the non-Archimedean case. *Motives (Seattle, WA, 1991). Proc. Sympos. Pure Math.* **55** Part 2, (1994), 365–391.

[34] L. Lafforgue, Chtoucas de Drinfeld et Conjecture de Ramanujan–Petersson. *Astérisque* **243**, Sociéte Mathématique de France, 1997.

[35] L. Lafforgue, Chtoucas de Drinfeld et correspondance de Langlands. *Invent. math.* **147**(2002), 1–241.

[36] R.P. Langlands, Problems in the theory of automorphic forms, in *Lectures in Modern Analysis and Applications*. Lecture Notes in Mathematics No. 349, Springer, New York, 1970, 18–86.

[37] R.P. Langlands, Base change for GL(2). *Annals of Mathematics Studies* **96**. Princeton University Press, Princeton, 1980.

[38] R.P. Langlands, On the classification of irreducible representations of real algebraic groups, in *Representation Theory and Harmonic Analysis on Semisimple Lie Groups*, Mathematical Surveys and Monographs, No. 13, AMS, Providence, 1989, 101–170.

[39] R.P. Langlands, Automorphic representations, Shimura varieties, and motives. Ein Märchen. *Proc. Symp. Pure math.* **33**, part 2, (1979), 205–246.

[40] R.P. Langlands, Endoscopy and beyond. in *Contributions to Automorphic Forms, Geometry, and Number Theory* (*Shalikafest* 2002), H. Hida, D. Ramakrishnan, and F. Shahidi, eds., Johns Hopkins University Press, Baltimore, to appear.

[41] G. Laumon, Transformation de Fourier, constantes d'équations fonctionelles, et conjecture de Weil. *Publ. Math, IHES* **65**(1987), 131–210.

[42] G. Laumon, The Langlands correspondence for function fields (following Laurent Lafforgue), in *Current Developments in Mathematics 1999* (preliminary edition). International Press, Sommerville, 1999, 69–87.

[43] G. Laumon, La correspondance de Langlands sur les corps de fonctions (d'après Laurent Lafforgue), *Séminaire Bourbaki*, No. 873, Mars 2000.

[44] G. Laumon, M. Rapoport, and U. Stuhler, $\mathcal{D}$-elliptic sheaves and the Langlands correspondence, *Invent. math.* **113**(1993), 217–338.

[45] M. Ohta, Hilbert modular forms of weight one and Galois representations. *Automorphic forms of several variables (Katata, 1983). Progr. Math.* **46**, Birkhäuser Boston, Boston, MA, 1984, 333–352.

[46] D. Ramakrishnan, Pure motives and automorphic forms, in Motives (Seattle, WA, 1991). *Proc. Sympos. Pure Math.* **55**, Part 2, (1994) 411–446.

[47] F. Rodier, Représentations de GL(n, k) où k est un corps p-adique. Séminaire Bourbaki No. 587, *Astérisque*, **92–92**(1982), 201–218.

[48] J. Rogawski and J. Tunnell, On Artin L-functions associated to Hilbert modular forms of weight one. *Invent. math.* **74**(1983), 1–42.

[49] D. Rohrlich, Elliptic curves and the Weil–Deligne group. *CRM Proc. & Lecture Notes* **4**(1994), 125–157.

[50] I. Satake, Theory of spherical functions on reductive algebraic groups over a $\mathfrak{p}$-adic field. *Publ. Math. IHES* **18**(1963), 5–70.

[51] J.T. Tate, Number theoretic background, *Proc. Symp. Pure math.* **33**, Part 2, (1979), 3–26.

[52] R. Taylor, On Galois representations associated to Hilbert modular forms. *Invent. math.* **98**(1989), 265–280.

[53] R. Taylor, ℓ-adic representations associated to modular forms over imaginary quadratic fields II. *Invent. Math.* **116**(1994), 619–643.

[54] R. Taylor, On Galois representations associated to Hilbert modular forms. II. *Elliptic curves, modular forms, & Fermat's last theorem (Hong Kong, 1993)*, Internat. Press, Cambridge, MA, 1995, 185–191.

[55] R. TAYLOR, Representations of Galois groups associated to modular forms. *Proceedings of the International Congress of Mathematicians, Vol. 1, 2 (Zürich, 1994)*, Birkhäuser, Basel, 1995, 435–442.

[56] R. TAYLOR, On icosahedral Artin representations. II. *Amer. J. Math.*, to appear.

[57] J. TUNNELL, Artin's conjecture for representations of octahedral type. *Bull. Amer. Math. Soc. (N.S.)* **5**(1981), 173–175.

[58] A. WILES, On ordinary λ-adic representations associated to modular forms. *Invent. math.* **94**(1988), no. 3, 529–573.

[59] A. WILES, Modular elliptic curves and Fermat's last theorem. *Ann. of Math.* **141**(1995), 443–551.

[60] A. ZELEVINSKY, Induced representations of $\mathfrak{p}$-adic groups II. *Ann. Sci. École Norm. Sup.*, 4e série, **13**(1980), 165–210.

11
Dual Groups and Langlands Functoriality

J.W. Cogdell

Langlands never separated the Langlands conjectures for GL_n from his general principle of functoriality [30]. In particular, he formulated a correspondence between certain Galois representations and admissible or automorphic representations for any connected reductive algebraic group G. For GL_n there was a correspondence between certain n-dimensional Galois representations, that is, representations into $GL_n(\mathbb{C})$, and admissible representations of $GL_n(k)$ or automorphic representations of $GL_n(\mathbb{A})$ [4]. For general G we understand what to replace the automorphic side with: admissible representations of $G(k)$ or automorphic representations of $G(\mathbb{A})$. But what replaces the target $GL_n(\mathbb{C})$ on the Galois side? Based on the Satake parameterization of unramified representations [33] and his classification of representations of algebraic tori [24] Langlands introduced his idea of a dual group, now known as the Langlands dual group or L-group , ${}^LG$ to play the role of $GL_n(\mathbb{C})$. The role of the n-dimensional Galois representations is taken by certain admissible homomorphisms of the Galois group into this L-group. For the purposes of functoriality, it is most convenient to view these local and global correspondences for G as giving an arithmetic parameterization of the admissible or automorphic representations of G in terms of these admissible Galois homomorphisms to ${}^LG$.

Langlands principle of functoriality states that any L-homomorphism ${}^LH \to {}^LG$ should determine a *transfer* or *lifting* of admissible or automorphic representations of H to admissible or automorphic representations of G. Once one has a parameterization, then this is conceptually done by composing the parameterizing homomorphism for the representation of H with the L-homomorphism to obtain a parameterizing homomorphism for a representation of G. If one takes $H = \{1\}$, then ${}^LH$ is simply the Galois group or a closely related group and one in essence recovers the local or global Langlands correspondence for G from this principle of functoriality.

There have been many fundamental examples of functoriality established by trace formula methods: cyclic base change, cyclic automorphic induction, lifting between inner forms. Recently however there has been much progress in global functorialities to GL_n obtained using the converse theorem for GL_n. These include the tensor product lifting from $GL_2 \times GL_2$ to GL_4 by Ramakrishnan [31] and from

$GL_2 \times GL_3$ to GL_6 by Kim and Shahidi [21], the symmetric cube and symmetric fourth power lifts from GL_2 to GL_4 and GL_5 by Kim and Shahidi [20, 21, 22], and the lifting from split classical groups to GL_N with Kim, Piatetski-Shapiro, and Shahidi [5, 6].

In this paper we first describe the construction of the L-group and the formulation of the local and global Langlands conjectures for a general reductive group G [2]. We next outline Langlands' principle of functoriality and its relation to the local and global Langlands correspondences. We then turn to examples. We briefly consider some of the examples of functoriality mentioned above that were established using the trace formula. We then give a more detailed description of the new functorialities to GL_n and how one uses the converse theorem as a means for establishing these liftings.

I would like to thank the referee for helping to clarify certain issues related to this chapter.

1 The dual group

Begin with G a connected reductive algebraic group defined over k, k a local or global field. Let $\bar{k}$ be a separable algebraic closure of k and $\mathcal{G}_k = \mathrm{Gal}(\bar{k}/k)$ the Galois group. Over $\bar{k}$, G becomes split and is classified by its root data [2, 36]. Take in $G_{/\bar{k}}$ a Borel subgroup B and maximal torus T, both defined and split over $\bar{k}$. Let $X = \mathfrak{X}^*(T)$ denote the set of $\bar{k}$-rational characters of T, $\Phi = \Phi(G, T) \subset X$, the root system associated to G and T, and $\Delta \subset \Phi$, the set of simple roots corresponding to B. Dual to the triple (X, Φ, Δ) we have the triple $(X^\vee, \Phi^\vee, \Delta^\vee)$ consisting of the lattice $X^\vee = \mathfrak{X}_*(T)$ of co-characters, or $\bar{k}$-rational one-parameter subgroups, the co-root system $\Phi^\vee$, and the simple co-roots $\Delta^\vee$. The quadruple $\Psi(G) = (X, \Phi, X^\vee, \Phi^\vee)$ is the root data for G over $\bar{k}$ and the quadruple $\Psi_0(G) = (X, \Delta, X^\vee, \Delta^\vee)$ is the based root data for G over $\bar{k}$ [2, 36]. The basic structure for connected reductive $\bar{k}$-groups is the following [36].

Theorem 1.1. *The root data $\Psi(G)$ determines G up to $\bar{k}$-isomorphism.*

For the relative structure theory, there is a split exact sequence

$$1 \longrightarrow \mathrm{Int}(G) \longrightarrow \mathrm{Aut}(G) \longrightarrow \mathrm{Aut}(\Psi_0(G)) \longrightarrow 1.$$

A splitting is given by making a choice of root vector x_α for each $\alpha \in \Delta$, which then defines a splitting $(G, B, T, \{x_\alpha\}_{\alpha\in\Delta})$ of G and gives a canonical isomorphism

$$\mathrm{Aut}(\Psi_0(G)) \to \mathrm{Aut}(G, B, T, \{x_\alpha\}) \subset \mathrm{Aut}(G).$$

If G is defined over k, there is an action of $\mathcal{G}_k$ on $G_{/\bar{k}}$ giving the k-structure. Hence we have homomorphisms

$$\mathcal{G}_k \to \mathrm{Aut}(G_{/\bar{k}}) \to \mathrm{Aut}(\Psi_0(G)).$$

So $G_{/k}$ determines the two pieces of data consisting of the root data $\Psi(G)$, determining the group over $\bar{k}$, and the homomorphism $\mathcal{G}_k \to \mathrm{Aut}(\Psi_0(G))$.

To define ${}^LG$ one simply dualizes this structure theory. Let $\Psi_0(G)^\vee = (X^\vee, \Delta^\vee, X, \Delta)$ be the dual based root data. This defines a connected reductive algebraic group ${}^LG^0$ over $\mathbb{C}$. We can transfer the Galois structure since

$$\mathrm{Aut}(\Psi_0({}^LG^0)) = \mathrm{Aut}(\Psi_0(G)^\vee) = \mathrm{Aut}(\Psi_0(G))$$

and a splitting of the exact sequence above for ${}^LG^0$ gives a map $\mu : \mathcal{G}_k \to \mathrm{Aut}(\Psi_0({}^LG^0)) \to \mathrm{Aut}({}^LG^0)$ which fixes the corresponding splitting $({}^LG^0, {}^LB^0, {}^LT^0, \{x_{\alpha^\vee}\}_{\alpha^\vee \in \Delta^\vee})$ of ${}^LG^0$ and hence a $\mathcal{G}_k$ action on the complex reductive group ${}^LG^0$ which encodes some of the original k-structure of G.

Definition 1.1. The (Langlands) dual group, or L-group, of G is

$${}^L(G_{/k}) = {}^LG = {}^LG^0 \rtimes \mathcal{G}_k.$$

Remarks.

1. Sometimes it is convenient use the Weil form of the dual group. Since there is a natural map $W_k \to \mathcal{G}_k$ one may form instead ${}^LG = {}^LG^0 \rtimes W_k$, but there is no essential difference. One could also use a Weil–Deligne form for certain purposes.

2. If G' is a k-group which is isomorphic to G over $\bar{k}$, then G and G' are inner forms of each other iff ${}^LG$ is isomorphic to ${}^LG'$ over $\mathcal{G}_k$ [2]. So the dual group does not quite distinguish between k-forms; it distinguishes only up to inner forms. It does completely determine a quasisplit form.

In practice, this duality preserves the types A_n and D_n and interchanges the types B_n and C_n. In addition it interchanges the adjoint and simply connected forms of the relevant groups.

G	${}^LG^0$
GL_n	$GL_n(\mathbb{C})$
SO_{2n+1}	$Sp_{2n}(\mathbb{C})$
Sp_{2n}	$SO_{2n+1}(\mathbb{C})$
SO_{2n}	$SO_{2n}(\mathbb{C})$
adjoint type	simply connected
simply connected	adjoint type

The local and global constructions are compatible. So if G is defined over a global field k, v is a place of k, and we let G_v denote G as a group over k_v, then there are natural maps ${}^LG_v \to {}^LG$.

2 Langlands conjectures for G

2.1 *Local Langlands conjecture*

Let k be a local field and let W'_k be the associated Weil–Deligne group [4]. If k is archimedean, we simply take $W'_k = W_k$ to be the Weil group. Following Borel [2] a homomorphism $\phi : W'_k \to {}^LG$ is called *admissible* if we have the following:

(i) ϕ is a homomorphism over $\mathcal{G}_k$, i.e., the following diagram commutes:

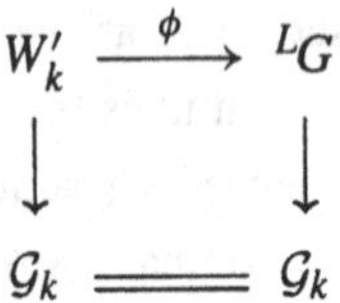

(ii) ϕ is continuous, $\phi(\mathbb{G}_a)$ is unipotent in ${}^LG^0$, and ϕ maps semisimple elements to semisimple elements.

(iii) If $\phi(W'_k)$ is contained in a Levi subgroup of a proper parabolic subgroup P of ${}^LG$, then P is relevant.

For all undefined concepts, such as *relevant*, we refer the reader to Borel [2]. If $G = \mathrm{GL}_n$ the admissible homomorphisms are precisely the Frobenius semisimple complex representations of W'_k [4].

Following Borel [2] and Langlands [26], we let $\Phi(G)$ denote the set of all admissible homomorphisms $\phi : W'_k \to {}^LG$ modulo inner automorphisms by elements of ${}^LG^0$ (not to be confused with the earlier [4] use of Φ as a geometric Frobenius). Note that if G and G' are inner forms of one another, so that ${}^LG = {}^LG'$, it need not be true that $\Phi(G) = \Phi(G')$ since the condition (iii) above sees the k structures. If G is the quasisplit form, then one does have $\Phi(G') \subset \Phi(G)$.

To state the local Langlands conjecture for G, there are two supplemental constructions that are needed, for which we refer the reader to Borel [2]. First, for every $\phi \in \Phi(G)$ there is a way to construct a character ω_ϕ of the center $C(G)$ of G. Next, if we let $C({}^LG^0)$ denote the center of ${}^LG^0$, then to every $\alpha \in H^1(W'_k; C({}^LG^0))$ there is associated a character χ_α of $G(k)$. If we write $\phi \in \Phi(G)$ as $\phi = (\phi_1, \phi_2)$ with $\phi_1(w) \in {}^LG^0$ and $\phi_2(w) \in \mathcal{G}_k$, then ϕ_1 is a cocycle on W'_k with values in ${}^LG^0$ and the map $\phi \mapsto \phi_1$ gives an embedding of $\Phi(G) \hookrightarrow H^1(W'_k; {}^LG^0)$. Then $H^1(W'_k; C({}^LG^0))$ acts naturally on $H^1(W'_k; {}^LG^0)$ and this action preserves $\Phi(G)$.

With these constructions, we can state the local Langlands conjecture for G [2]. As before, let $\mathcal{A}(G) = \mathcal{A}(G(k))$ denote the set of equivalence classes of irreducible admissible complex representations of $G(k)$.

Local Langlands Conjecture. *Let k be a local field. Then there is a surjective map $\mathcal{A}(G) \to \Phi(G)$ with finite fibres which partitions $\mathcal{A}(G)$ into disjoint finite sets $\mathcal{A}_\phi = \mathcal{A}_\phi(G)$ satisfying the following:*

(i) *If $\pi \in \mathcal{A}_\phi$, then the central character ω_π of π is equal to ω_ϕ.*

(ii) *Compatibility with twisting, i.e., if $\alpha \in H^1(W'_k; C({}^LG^0))$ and χ_α is the associated character of $G(k)$, then $\mathcal{A}_{\alpha\cdot\phi} = \{\pi\chi_\alpha | \pi \in \mathcal{A}_\phi\}$.*

(iii) *One element $\pi \in \mathcal{A}_\phi$ is square integrable modulo $C(G)$ iff all $\pi \in \mathcal{A}_\phi$ are square integrable modulo $C(G)$ iff $\phi(W'_k)$ does not lie in a proper Levi subgroup of ${}^LG$.*

(iv) *One element $\pi \in \mathcal{A}_\phi$ is tempered iff all $\pi \in \mathcal{A}_\phi$ are tempered iff $\phi(W_k)$ is bounded.*

(v) *If H is a reductive connected k-group and $\eta : H(k) \to G(k)$ is a k morphism with commutative kernel and co-kernel, then there is a required compatibility between decompositions for $G(k)$ and $H(k)$. Namely, η induces a natural map ${}^L\eta : {}^LG \to {}^LH$, and if we set $\phi' = {}^L\eta \circ \phi$ for $\phi \in \Phi(G)$, then any $\pi \in \mathcal{A}_\phi(G)$, when viewed as a $H(k)$ module, decomposes into a direct sum of finitely many irreducible admissible representations belonging to $\mathcal{A}_{\phi'}(H)$.*

The sets $\mathcal{A}_\phi(G)$ for $\phi \in \Phi(G)$ are called *L-packets*. The version I of the local Langlands conjecture in [4] was the specialization of this to the group GL_n. In that case, the L-packets are all singletons and the map from $\mathcal{A}(G)$ to $\Phi(G)$ was a bijection. This conjecture gives an *arithmetic parameterization* of the irreducible admissible representations of $G(k)$.

Other than the results for GL_n, the following is known towards this conjecture.

1. If the local field k is archimedean, i.e., $k = \mathbb{R}$ or $\mathbb{C}$, then this was completely established by Langlands [26].

2. If k is nonarchimedean and G is quasisplit over k and split over a finite Galois extension, then one knows how to parameterize the unramified representations of $G(k)$ via the unramified admissible homomorphisms [2]. This is a rephrasing in this language of the Satake classification [33].

3. If k is nonarchimedean, then Kazhdan and Lusztig have shown how to parameterize those representations of $G(k)$ having an Iwahori fixed vector in terms of admissible homomorphisms of the Weil–Deligne group [19].

4. Recently, in the case of k nonarchimedean of characteristic zero and G the split SO_{2n+1}, Jiang and Soudry have given the parameterization of generic representations of $\mathrm{SO}_{2n+1}(k)$ in terms of admissible homomorphisms of the Weil–Deligne group [16, 17]. They obtain this parameterization as an outgrowth of recent work on global functoriality from split SO_{2n+1} to GL_{2n}, to be discussed later, by pulling back the parameterization for $\mathrm{GL}_{2n}(k)$.

If one thinks of this version of the local Langlands conjecture as providing an arithmetic parameterization of the irreducible admissible representations of $G(k)$, then one can use this parameterization to define local L-functions associated to arbitrary $\pi \in \mathcal{A}(G)$. One needs a second parameter, namely a representation $r : {}^LG \to \mathrm{GL}_n(\mathbb{C})$, by which we mean a continuous homomorphism whose restriction to ${}^LG^0$ is a morphism of complex Lie groups. Then for any admissible homomorphism $\phi \in \Phi(G)$ the composition $r \circ \phi : W'_k \to \mathrm{GL}_n(\mathbb{C})$ is a continuous complex representation of the Weil–Deligne group as considered in [4] and to it we can associate an L-factor $L(s, r \circ \phi)$ and ε-factor $\varepsilon(s, r \circ \phi, \psi)$ for an additive character ψ of k.

Definition 2.1. If $\pi \in \mathcal{A}_\phi$ is in the L-packet defined by the admissible homomorphism ϕ, then we set

$$L(s, \pi, r) = L(s, r \circ \phi) \qquad \text{and} \qquad \varepsilon(s, \pi, r, \psi) = \varepsilon(s, r \circ \phi, \psi).$$

According to this definition, one cannot distinguish between the representations π lying in a given L-packet $\mathcal{A}_\phi$ in terms of their L-functions and ε-factors, hence the terminology. At present these L-functions are well defined only for those π for which the parameterization is known, for example if π is unramified.

If one takes this as the definition of the local L-functions attached to an admissible representation, then version II of the local Langlands conjecture presented in [4] would be phrased in terms of matching L- and ε-factors defined in an analytic nature, as in [3] for GL_n, with those defined here. I have not seen a formulation in these terms for general reductive groups, however in the work of Jiang and Soudry cited above this is what they achieve. To each generic representation π of $\mathrm{SO}_{2n+1}(k)$ they attach an admissible homomorphism ϕ_π such that for the standard embedding $r : \mathrm{Sp}_{2n}(\mathbb{C}) \hookrightarrow \mathrm{GL}_{2n}(\mathbb{C})$ they have an equality

$$L(s, \pi \times \pi') = L(s, \pi \times \pi', r \otimes id) = L(s, (r \circ \phi_\pi) \otimes \rho_{\pi'})$$

with the similar equality of ε-factors where π' is an irreducible admissible representation of $\mathrm{GL}_m(k)$, $\rho_{\pi'}$ is the associated representation of W'_k from the local Langlands conjecture for GL_m, and $L(s, \pi \times \pi')$ is the analytic L-function defined by Shahidi [34].

2.2 *Global Langlands conjecture*

Now take k to be a global field and $\mathbb{A}$ its ring of adeles. For G a reductive algebraic group over k, let $\mathcal{A}(G) = \mathcal{A}(G(\mathbb{A}))$ denote the set of irreducible automorphic representations of $G(\mathbb{A})$. As with GL_n, to formulate a global Langlands conjecture we would replace the Weil–Deligne group W'_k by the conjectural Langlands group $\mathcal{L}_k$ and consider the set of admissible homomorphisms $\phi : \mathcal{L}_k \to {}^LG$. These homomorphisms should then parameterize irreducible automorphic representations of $G(\mathbb{A})$ in some way. The exact form this would take is quite speculative at the moment.

Not knowing what this should look like, one still expects to have global–local compatibility. If one begins an irreducible automorphic representation $\pi = \otimes' \pi_v$ of $G(\mathbb{A})$, then, assuming the local Langlands conjecture for each local group $G(k_v)$, one can attach to π the collection $\{\phi_v\}$ of local parameters $\phi_v = \phi_{\pi_v} : W'_{k_v} \to {}^LG_v$ given by the local components π_v. If we compose these with the natural compatibility maps for the dual groups $\iota_v : {}^LG_v \to {}^LG$ one gets a collection $\{\iota_v \circ \phi_v\}$ of local parameters $\iota_v \circ \phi_v : W'_{k_v} \to {}^LG$.

Such a system of maps must come out of a global parameter $\phi : \mathcal{L}_k \to {}^LG$ for the local and global theories to be consistent. This system of local parameters can often be used as a substitute for a global parameter ϕ. For example, this collection of local data is sufficient to define the global L-function and ε-factor attached to π. If $r : {}^LG \to \mathrm{GL}_n(\mathbb{C})$, then the composition $r_v = r \circ \iota_v : {}^LG_v \to \mathrm{GL}_n(\mathbb{C})$ gives representations of the local dual groups.

Definition 2.2. If $\pi = \otimes' \pi_v$ is an irreducible automorphic representation of $G(\mathbb{A})$ and $r : {}^LG \to \mathrm{GL}_n(\mathbb{C})$, we set

$$L(s, \pi, r) = \prod_v L(s, \pi_v, r_v) = \prod_v L(s, r \circ \iota_v \circ \phi_v)$$

and

$$\varepsilon(s, \pi, r) = \prod_v \varepsilon(s, \pi_v, r_v, \psi_v) = \prod_v \varepsilon(s, r \circ \iota_v \circ \phi_v, \psi_v)$$

where $\psi = \otimes \psi_v$ is an additive character of $\mathbb{A}$ trivial on k.

To define the full L-function as above requires the solution of the local Langlands conjecture at all places, something only known for GL_n. However, for any irreducible automorphic representation π, there is a finite set of places $S = S(\pi)$ such that, for all $v \notin S$, the representation π_v is unramified and hence the local parameterization problem has been solved. Then the partial L-function

$$L^S(s, \pi, r) = \prod_{v \notin S} L(s, \pi_v, r_v)$$

is always well defined and Langlands has shown that this Euler product is always absolutely convergent in a right half plane [25].

3 Functoriality

As one can tell from his recent writings [29, 30] Langlands has always viewed the "principle of functoriality" as central to his view of automorphic representations. It encompasses what is referred to above as the "local and global Langlands conjectures" as special cases of this principle.

Let k denote either a local or global field and let H and G be two connected reductive groups defined over k. We have defined their associated dual groups ${}^LH$ and ${}^LG$. A homomorphism $u : {}^LH \to {}^LG$ is called an *L-homomorphism* if (i) it is a homomorphism over $\mathcal{G}_k$, that is, we have the commutation of the diagram

$$\begin{array}{ccc} {}^LH & \xrightarrow{u} & {}^LG \\ \downarrow & & \downarrow \\ \mathcal{G}_k & = & \mathcal{G}_k \end{array}$$

(ii) u is continuous, and (iii) the restriction of u to ${}^LH^0$ is a complex analytic homomorphism $u : {}^LH^0 \to {}^LG^0$.

If in addition G is quasisplit, then for any admissible homomorphism $\phi \in \Phi(H)$ the composition $u \circ \phi$ is again an admissible homomorphism in $\Phi(G)$. So the map $\phi \mapsto u \circ \phi$ defines a map $\Phi(u) : \Phi(H) \to \Phi(G)$. If k is a global field and v a place of k, then, since $\mathcal{G}_{k_v}$ can be viewed naturally as a subgroup of $\mathcal{G}_k$, we can view ${}^LG_v$ as a subgroup of ${}^LG$. Then, upon restriction to ${}^LH_v$, u will induce an L-homomorphism of the local dual groups $u_v : {}^LH_v \to {}^LG_v$ and hence a local map $\Phi(u_v) : \Phi(H_v) \to \Phi(G_v)$.

The principle of functoriality can now be roughly formulated as follows [30].

The Principle of Functoriality. *If k is a local (resp., global) field, H and G connected reductive k-groups with G quasisplit, then to each L-homomorphism $u : {}^LH \to {}^LG$ there is associated a* transfer *or* lifting *of admissible (resp., automorphic) representations of H to admissible (resp., automorphic) representations of G.*

If we assume the local and global Langlands conjectures, so that we have an arithmetic parameterization of $\mathcal{A}(H)$ and $\mathcal{A}(G)$, then this process of lifting is easy to describe.

3.1 *Local functoriality*

First, take k to be a local field and $u : {}^LH \to {}^LG$ a local L-homomorphism. If we take $\pi \in \mathcal{A}(H)$ to be an irreducible admissible representation of $H(k)$, then this is parameterized by an admissible homomorphism $\phi = \phi_\pi : W_k' \to {}^LH$. In fact, ϕ parameterizes an entire local L-packet $\mathcal{A}_\phi(H)$. If we compose ϕ with u, we obtain $\phi' = \Phi(u)(\phi) = u \circ \phi \in \Phi(G)$, an admissible homomorphism of W_k' to ${}^LG$. Then ϕ' parameterizes a local L-packet $\mathcal{A}_{\phi'}(G)$ and this L-packet (or sometimes any element Π of it) is the functorial lift (or transfer, or Langlands lift, etc.) of π or of the packet $\mathcal{A}_\phi(H)$.

In general, we then "understand" the local functoriality in the cases where we understand the local parameterization:

1. $k = \mathbb{R}$ or $\mathbb{C}$, H any connected reductive k-group and G any quasisplit connected reductive k-group.

2. k a nonarchimedean local field, $H = \mathrm{GL}_m$ and $G = \mathrm{GL}_n$ (and related examples—see Section 4).

3. Suppose that k is nonarchimedean with a ring of integers $\mathcal{O}$. Suppose both H and G are quasisplit and there is a finite extension K of k such that both H and G split over K and have an $\mathcal{O}$ structure so that both $H(\mathcal{O})$ and $G(\mathcal{O})$ are special maximal compact subgroups. Let π be an unramified representation of $H(k)$ with a nontrivial $H(\mathcal{O})$-fixed vector and unramified parameter $\phi = \phi_\pi \in \Phi(H)$. Then for any L-homomorphism $u : {}^LH \to {}^LG$ the parameter $\phi' = u \circ \phi$ is unramified and defines an L-packet $\mathcal{A}_{\phi'}(G)$ which contains a (unique) representation Π of $G(k)$ which is unramified with respect to $G(\mathcal{O})$ [2]. Π is called the natural unramified lift of π.

3.2 *Global functoriality*

If we now consider k to be a global field, then, in principle, functorial lifting should work as it does in the local situation in terms of global parameterization. But now we are again at a disadvantage since we don't really understand the parameterizing group $\mathcal{L}_k$. In its stead, we fall back on the desired local-global compatibility. So let H be a connected reductive k-group, G a quasisplit connected reductive k-group and $u : {}^LH \to {}^LG$ an L-homomorphism. For each place v of k we have the associated local L-homomorphism $u_v : {}^LH_v \to {}^LG_v$ described above. Now let $\pi \in \mathcal{A}(G)$, $\pi = \otimes' \pi_v$, be an irreducible automorphic representation of $H(\mathbb{A})$. If v is archimedean, then by the work of Langlands we know how to parameterize π_v with a local parameter $\phi_v : W_{k_v}' \to {}^LH_v$. If v is a nonarchimedean place, then for almost all v the local group H_v is quasisplit, split over a finite extension of k_v, and the representation π_v is unramified with respect to a special maximal compact subgroup. So we are in the situation where we have a local parameter $\phi_v : W_{k_v}' \to {}^LH_v$ for π_v. In either of these situations, we can form a local lift Π_v as a representation of $G(k_v)$ associated to the parameter $\phi_v' = u_v \circ \phi_v$, that is, a local lift as defined above.

Definition 3.1. Let H be a connected reductive k-group and let $\pi = \otimes' \pi_v$ be an irreducible automorphic representation of $H(\mathbb{A})$. Let G be a quasisplit connected reductive k-group and let $u : {}^LH \to {}^LG$ be an L-homomorphism. Then an automorphic representation $\Pi = \otimes' \Pi_v$ of $G(\mathbb{A})$ is a (weak) functorial lift of π (with respect to u) if for all archimedean places and almost all finite places where π_v is unramified we have that Π_v is a local functorial lift with respect to u_v as described above. Π is a (strong) functorial lift of π if Π_v is a local functorial lift of π_v for all places of k.

Note that as a consequence of this definition, if π is an automorphic representation of $H(\mathbb{A})$, $u : {}^LH \to {}^LG$ an L-homomorphism, and Π a functorial lift of π to an automorphic representations of $G(\mathbb{A})$, then for every representation $r : {}^LG \to \mathrm{GL}_n(\mathbb{C})$ we have an equality of L-functions and ε-factors

$$L^S(s, \pi, r \circ u) = L^S(s, \Pi, r) \qquad \varepsilon^S(s, \pi, r \circ u, \psi) = \varepsilon^S(s, \Pi, r, \psi)$$

where S is the finite (possibly empty) set of places where we do not know how to locally lift π_v.

In fact, we need to do this on the level of L-packets. This is easy enough to formulate, but given the partial state of our knowledge, there seems to be little gained in doing this at this time. But the ambiguity in the local lifts and hence the global lifts coming from the phenomenon of local and global L-packets should always be kept in mind.

4 Examples

We have noted that Langlands views functoriality as encompassing the local and global Langlands conjectures and their consequences, such as the strong Artin conjecture. One reason for this is the following example.

Consider the case where $H = \{1\}$. Begin with k a local field. Since there is a natural map from the Weil–Deligne group W_k' to $\mathcal{G}_k$ we may consider the Weil–Deligne form of the L-group: ${}^LG = {}^LG^0 \rtimes W_k'$. Then ${}^LH = W_k'$. If we take, for example, $G = \mathrm{GL}_n$, then $u : {}^LH \to {}^LG$ is an admissible homomorphism in $\Phi(G)$ or a complex representation of the Weil–Deligne group and functoriality for these groups encompasses the local Langlands conjectures. If one takes k a global field and leaves ${}^LG$ as the Galois form of the L-group, then again taking $H = \{1\}$ and $G = \mathrm{GL}_n$ we obtain a global Langlands conjecture for GL_n.

The other examples of functoriality I wish to discuss fall into what I view as two types: Galois theoretic and group theoretic. The first include base change, automorphic induction, and lifting between inner forms. The second are all liftings to GL_n and include the tensor product liftings, symmetric powers liftings, and liftings from classical groups. I will not touch on the important class of liftings known as *endoscopic*, even though some of the examples we discuss can be interpreted as examples of (possibly twisted) endoscopy. Endoscopic liftings are those in which the L-homomorphism $u : {}^LH \to {}^LG$ realizes ${}^LH^0$ as the fixed points of an involution in ${}^LG^0$. The significance of these liftings come primarily from their necessity in understanding the trace formula, which we are not in a position to discuss. Instead, we refer the reader to the work of Langlands [28, 30] and of Kottwitz and Shelstad [23] and the references therein.

4.1 *Galois theoretic examples*

In these examples, the L-homomorphisms have their origins in Galois theory.

1. *Base change (or automorphic restriction)*. Suppose that K is a finite extension of k. Then on the level of Weil groups we have $W_K \subset W_k$ so that any representation of W_k gives a representation of W_K by restriction. The analogous lifting on the level of admissible or automorphic representations is the following. Let H be connected, reductive and split over k. Then we may consider H as a group over K as well and if we let $G = R_{K/k}(H)$ be Weil's restriction of scalars from K to k, so $G(k) = H(K)$, and then G is the group over k determined by $H_{/K}$. There is then a natural embedding

$$u : {}^LH = {}^LH^0 \times \mathcal{G}_k \to \Big(\prod_{\mathcal{G}_K\backslash\mathcal{G}_k} {}^LH^0\Big) \rtimes \mathcal{G}_k = {}^LG,$$

where $\mathcal{G}_k$ acts on $\prod {}^LH^0$ via permutations of the index set, which is the diagonal map on ${}^LH^0$ and the identity on $\mathcal{G}_k$. In the case where k is a local field, then the induced map $\Phi(u) : \Phi(H) \to \Phi(G)$ is indeed the restriction map, viewing W'_K as an open subgroup of W'_k. Functoriality coming from this L-homomorphism would begin with a representation π of $H(k)$ or $H(\mathbb{A}_k)$ and produce a representation of $G(k) = H(K)$ or $H(\mathbb{A}_K)$ called the *base change* of π. This program has been carried out when $H = \mathrm{GL}_n$ and the extension K/k is solvable, first for $n = 2$ by Langlands [27] and then for general n by Arthur and Clozel [1]. Their technique was the twisted trace formula. In addition, when $H = \mathrm{GL}_2$ Jacquet, Piatetski-Shapiro, and Shalika have obtained a nonnormal cubic base change by converse theorem methods [14].

2. *Automorphic induction.* We still take K to be a finite separable extension of k of degree d, so that $W_K \subset W_k$. If one starts with a representation of W_K, then one obtains a representation of W_k simply by induction. The analogous lifting on the level of admissible or automorphic representations is now the following. Take $H = R_{K/k}(\mathrm{GL}_n)$ to be $\mathrm{GL}_n(K)$ viewed as a k-group as above and let $G = \mathrm{GL}_{dn}(k)$. Now one has an L-homomorphism

$$u : {}^LH = \left(\prod_{\mathcal{G}_K \backslash \mathcal{G}_k} \mathrm{GL}_n(\mathbb{C}) \right) \rtimes \mathcal{G}_k \to {}^LG = \mathrm{GL}_{dn}(\mathbb{C}) \times \mathcal{G}_k$$

by sending ${}^LH^0 = \mathrm{GL}_n(\mathbb{C}) \times \cdots \times \mathrm{GL}_n(\mathbb{C})$ into ${}^LG^0 = \mathrm{GL}_{dn}(\mathbb{C})$ as block diagonal matrices and extending to an L-homomorphism by letting $\mathcal{G}_k$ act on $\mathrm{GL}_{dn}(\mathbb{C})$ via permutation matrices from $\mathfrak{S}_d$. The local or global functorialities coming from such an L-homomorphism are called *automorphic induction*. The map $\Phi(u)$ on the sets of admissible homomorphisms should be induction. Again, when the extension K/k is solvable, this was analyzed locally and globally by Arthur and Clozel [1] using the twisted trace formula, preceded by Jacquet and Langlands for $n = 2$ [15]. Henniart and Herb, building on earlier work by Kazhdan in the $n = 1$ case [18], gave the first definition and analysis of local automorphic induction for GL_n in terms of local character identities [13]. This work uses a simpler version of the trace formula than either [1] or [18] and allows fields of positive characteristic.

3. *Inner forms.* Let G be connected, reductive, and quasisplit over a local or global k and let H be an inner form of G. Then ${}^LH = {}^LG$, the identity map $u : {}^LH \to {}^LG$ is an L-homomorphism, and we should have a corresponding lifting. Note that if k is a local field we have $\Phi(H) \subset \Phi(G)$, while if k is a global field, we in fact have $H_v = G_v$ for almost all places so that $\Phi(H_v) = \Phi(G_v)$. In the case of $G = \mathrm{GL}_2$ and $H = D^\times$, the multiplicative group of a rank 2 division algebra over k, the lifting from representations of $D^\times$ to representations of GL_2, is the so-called Jacquet–Langlands correspondence, established in [15]. If we take $G = \mathrm{GL}_n$ and $H = \mathrm{GL}_m(D)$, where D is a central simple division algebra of rank d with $dm = n$, then the local functoriality has been analyzed by Rogawski [32] in the case $m = 1$ and by Deligne, Kazhdan, and Vigneras [10] utilizing the trace formula.

4.2 Group theoretic examples

In this set of examples, the groups H involved are all split and the target group G is always a general linear group GL_n, so the Galois theory plays little role. The L-homomorphism is a natural map from group theory. There has been much progress in this family of functorialities recently based on using the converse theorem for GL_n as the primary tool for establishing global functorialities to GL_n.

1. *Tensor products*. Let k be either a local of global field and let $H = \mathrm{GL}_m \times \mathrm{GL}_n$. Then ${}^L H^0 = \mathrm{GL}_m(\mathbb{C}) \times \mathrm{GL}_n(\mathbb{C})$ and ${}^L H = {}^L H^0 \times \mathcal{G}_k$. If we take $G = \mathrm{GL}_{mn}$, then ${}^L G^0 = \mathrm{GL}_{mn}(\mathbb{C})$ and ${}^L G = {}^L G^0 \times \mathcal{G}_k$. The simple tensor product map $\otimes : \mathrm{GL}_m(\mathbb{C}) \times \mathrm{GL}_n(\mathbb{C}) \to \mathrm{GL}_{mn}(\mathbb{C})$, extended by the identity map on $\mathcal{G}_k$, defines an L-homomorphism $u_\otimes : {}^L H \to {}^L G$. The associated functoriality is the *tensor product lifting*. Note that if k is a local field, then the local lifting is now understood in principle since the local parameterization problem (local Langlands conjecture) for GL_n has been solved. So the interesting question is the global functoriality. This has been recently solved in the cases of $\mathrm{GL}_2 \times \mathrm{GL}_2$ to GL_4 by Ramakrishnan [31] and $\mathrm{GL}_2 \times \mathrm{GL}_3$ to GL_6 by Kim and Shahidi [21].

2. *Symmetric powers*. Let k be either a local or global field and let $H = \mathrm{GL}_2$, so ${}^L H^0 = \mathrm{GL}_2(\mathbb{C})$ and ${}^L H = {}^L H^0 \times \mathcal{G}_k$. We take $G = \mathrm{GL}_{n+1}$ for $n \geq 1$, so ${}^L G^0 = \mathrm{GL}_{n+1}(\mathbb{C})$ and ${}^L G = {}^L G^0 \times \mathcal{G}_k$. For each $n \geq 1$ there is the natural symmetric n-th power map $sym^n : \mathrm{GL}_2(\mathbb{C}) \to \mathrm{GL}_{n+1}(\mathbb{C})$. If we extend this symmetric power map by the identity map on the Galois group, we obtain an L-homomorphism $sym^n : {}^L H \to {}^L G$. The associated functoriality is the *symmetric power lifting* from representations of GL_2 to representations of GL_{n+1}. Once again, if k is a local field, the local symmetric powers liftings are understood in principle thanks to the solution of the local Langlands conjecture for GL_n. So once again the interesting functoriality is the global one. The global symmetric square lifting, i.e., GL_2 to GL_3, is an old theorem of Gelbart and Jacquet [11]. Recently, Kim and Shahidi have shown the existence of the global symmetric cube lifting from GL_2 to GL_4 [21, 22] and then Kim followed with the global symmetric fourth power lifting from GL_2 to GL_5 [20, 22]. The achievement of symmetric power functoriality for all n would lead to a proof of the Ramanujan conjecture for GL_2.

3. *Classical groups*. Again, k is either a local or global field. Take H to be a split classical group over k, more specifically, the split form of either SO_{2n+1}, Sp_{2n}, or SO_{2n}. The connected component of the L-group are then $\mathrm{Sp}_{2n}(\mathbb{C})$, $\mathrm{SO}_{2n+1}(\mathbb{C})$, or $\mathrm{SO}_{2n}(\mathbb{C})$ and there are natural embeddings into an appropriate general linear group.

H	${}^L H^0$	$u^0 : {}^L H^0 \hookrightarrow {}^L G^0$	${}^L G^0$	G
SO_{2n+1}	$\mathrm{Sp}_{2n}(\mathbb{C})$	$\mathrm{Sp}_{2n}(\mathbb{C}) \hookrightarrow \mathrm{GL}_{2n}(\mathbb{C})$	$\mathrm{GL}_{2n}(\mathbb{C})$	GL_{2n}
Sp_{2n}	$\mathrm{SO}_{2n+1}(\mathbb{C})$	$\mathrm{SO}_{2n+1}(\mathbb{C}) \hookrightarrow \mathrm{GL}_{2n+1}(\mathbb{C})$	$\mathrm{GL}_{2n+1}(\mathbb{C})$	GL_{2n+1}
SO_{2n}	$\mathrm{SO}_{2n}(\mathbb{C})$	$\mathrm{SO}_{2n}(\mathbb{C}) \hookrightarrow \mathrm{GL}_{2n}(\mathbb{C})$	$\mathrm{GL}_{2n}(\mathbb{C})$	GL_{2n}

These homomorphisms extend to L-homomorphisms by extending them with the identity map on the Galois groups. Associated to each should be a lifting of admissible or automorphic representations from $\mathcal{A}(H)$ to $\mathcal{A}(G)$. In collaboration with Kim, Piatetski-Shapiro, and Shahidi, we established a weak global lift for generic cuspidal representations from SO_{2n+1} to GL_{2n} over a number field k using converse theorem methods [5]. Soon thereafter, Ginzburg, Rallis, and Soudry showed that our weak lift was indeed a strong lift and characterized the image [12]. The results of Jiang and Soudry on the local Langlands conjecture for SO_{2n+1} over a p-adic field cited above [16, 17] were then obtained as a local consequence of this global functoriality. We have been able to extend our functoriality results to the other split classical groups as well [6].

We would like to explain the converse theorem method for obtaining global functorialities to general linear groups. We begin with a group H defined over a number field k. Take $\pi = \otimes\pi_v$, a cuspidal representation of $H(\mathbb{A})$. For each local place v we apply local functoriality to construct a local representation Π_v of $G(k_v) = GL_N(k_v)$ for an appropriate N. If we are in example 1 or 2 above, we can do this for all v since the local Langlands conjecture is known for $GL_n(k_v)$ [4]. For the cases of the classical groups we can perform this at all archimedean places v and at the nonarchimedean places v where π_v is unramified. The method is simply composing the local parameter map ϕ_v for π_v with the L-homomorphism as described above. In the case of classical groups we must finesse the local liftings at the remaining places v to construct a local lift Π_v. But assume for now that we understand the local lifts at all places. Then by construction we have an equality of local L-factors

$$L(s, \pi_v, r_v) = L(s, r_v \circ \phi_v) = L(s, u_v \circ \phi_v) = L(s, \Pi_v, \iota_v)$$

with a similar equality for local ε-factors. Here we may take $r = u^0$ viewed as a complex representation $r : {}^LH \to GL_N(\mathbb{C})$ and $\iota : {}^LG \to GL_N(\mathbb{C})$ is just projection onto the first factor ${}^LG^0$. Hence, if we set $\Pi = \otimes'\Pi_v$ as an irreducible admissible representation of $GL_N(\mathbb{A})$, then we globally have

$$L(s, \pi, r) = L(s, \Pi, \iota) \quad \text{and} \quad \varepsilon(s, \pi, r) = \varepsilon(s, \Pi, \iota).$$

Additionally, if $\pi' = \otimes\pi'_v$ is a cuspidal representation of $GL_m(\mathbb{A})$ with $m \leq N-2$, then we similarly have

$$L(s, \pi_v \times \pi'_v, r_v \otimes \iota_v) = L(s, \Pi_v \times \pi'_v, \iota_v \otimes \iota_v)$$

and hence

$$L(s, \pi \times \pi', r \otimes \iota) = L(s, \Pi \times \pi', \iota \otimes \iota) = L(s, \Pi \times \pi')$$

with similar equalities for local and global ε-factors. As outlined in [3], to apply the converse theorem for GL_N we must control the analytic properties of the twisted L-functions $L(s, \Pi\times\pi') = L(s, \Pi\times\pi', \iota\otimes\iota)$ for a sufficient family of cuspidal twists π'. But from our equality of L- and ε-factors, we have that these are controlled by the analytic properties of the *automorphic* L-functions $L(s, \pi \times \pi', r \otimes \iota)$ for the group $H(\mathbb{A})$ with twisting by $\mathrm{GL}_m(\mathbb{A})$. So once sufficient analytic control of these L-functions is known, one simply applies the converse theorem [3] for GL_N and concludes that Π is automorphic. In most cases to date, this analytic control of the $L(s, \pi \times \pi', r \otimes \iota)$ has been achieved by the Langlands–Shahidi method of analyzing the L-functions through the Fourier coefficients of the Eisenstein series.

Let us now revisit our examples above in light of this sketch.

1. *Tensor products*. In the case of Ramakrishnan [31], the functoriality from $\mathrm{GL}_2 \times \mathrm{GL}_2$ to GL_4, $\pi = \pi_1 \otimes \pi_2$ with each π_i a cuspidal representation of $\mathrm{GL}_2(\mathbb{A})$ and Π is an automorphic representation of $\mathrm{GL}_4(\mathbb{A})$. To apply the converse theorem from [9] Ramakrishnan needs to control the analytic properties of $L(s, \Pi \times \pi')$ for π' cuspidal representations of $\mathrm{GL}_1(\mathbb{A})$ and $\mathrm{GL}_2(\mathbb{A})$, that is, the Rankin triple product L-functions

$$L(s, \pi \times \pi', r \otimes \iota) = L(s, \pi_1 \times \pi_2 \times \pi').$$

This he was able to do using a combination of the integral representation for this L-function due to Garrett and then Rallis and Piatetski-Shapiro and the work of Shahidi on the Langlands–Shahidi method. The case of Kim and Shahidi [21] is similar, now with π_2 a cuspidal representation of $\mathrm{GL}_3(\mathbb{A})$. However, since the lifted representation Π is an automorphic representation of $\mathrm{GL}_6(\mathbb{A})$, to apply the converse theorem of [9] they must control the analytic properties of

$$L(s, \Pi \times \pi') = L(s, \pi_1 \times \pi_2 \times \pi')$$

where now π' must run over appropriate cuspidal representations of $\mathrm{GL}_m(\mathbb{A})$ with $m = 1, 2, 3, 4$. The control of these triple products is an application of the Langlands–Shahidi method of analyzing L-functions and involves coefficients of Eisenstein series on GL_5, Spin_{10}, and the simply connected E_6 and E_7.

2. *Symmetric powers*. The original symmetric square lifting of Gelbart and Jacquet indeed used the converse theorem for GL_3 [11]. One needs only control twists by characters (automorphic forms on GL_1) and the L-function that one must control is the symmetric square L-function for GL_2 since

$$L(s, \Pi) = L(s, \pi, sym^2).$$

This they were able to do via an integral representation due to Shimura. For Kim and Shahidi, the symmetric cube and fourth power liftings were deduced from the functorial $\mathrm{GL}_2 \times \mathrm{GL}_3$ tensor product lift above and the exterior square lift for GL_4 [20, 21, 22].

3. *Classical groups*. Here there is a secondary problem. If we begin with a generic cuspidal representation $\pi = \otimes \pi_v$ of $H(\mathbb{A})$, then there is a finite set of finite places S at which one does not know the local parameterization for π_v in terms of admissible homomorphisms, and hence one does not know what the correct local lift Π_v should be. In this case, one is able to take an *arbitrary* local

lift Π_v at those places, as long as it has trivial central character. To compensate, one applies the form of the converse theorem for GL_N in which one fixes a single highly ramified idele class character η, the ramification depending on the original representation π of $H(\mathbb{A})$ and the constructed representation Π of $\mathrm{GL}_N(\mathbb{A})$ (and actually only on the local components at the places $v \in S$), and then twists by all cuspidal representations π' of $\mathrm{GL}_m(\mathbb{A})$, $m \leq N-1$, of the form $\pi' = \tau \otimes \eta$ where τ is *unramified at all* $v \in S$ [3, 5]. This highly ramified twist plays two roles. First, it helps to control global poles of the twisted L-functions $L(s, \pi \times \pi')$ for $H(\mathbb{A})$ and secondly it allows one to match the local L- and ε-factors at those $v \in S$ through the *stability of the local γ-factors under highly ramified twists* [3, 8, 5]. So for these limited twists, one indeed has

$$L(s, \pi \times \pi') = L(s, \pi \times \pi', r \otimes \iota) = L(s, \Pi \times \pi', \iota \otimes \iota) = L(s, \Pi \times \pi')$$

with similar equalities for ε factors. Since we are able to control the analytic properties of the $L(s, \pi \times \pi')$ via the Langlands–Shahidi method for our family of π' we may apply the converse theorem for GL_N and conclude the existence of an automorphic representation Π' of $\mathrm{GL}_N(\mathbb{A})$ such that $\Pi_v = \Pi'_v$ for all $v \notin S$, that is, a weak lift Π' of π.

Every step in this argument is now valid for the general split classical group of the type we are considering. Originally the local stability of γ-factors was known only for SO_{2n+1} [8, 5]. Now, thanks to recent results of Shahidi expressing his local coefficients as Mellin transforms of Bessel functions [35], the techniques of [8] can be used to establish the stability of the local γ-factors for the other split classical groups as well. This then allows us to extend the functoriality results of [5] to these cases [6].

In the case of SO_{2n+1}, once we have the weak lift then the theory of Ginzburg, Rallis, and Soudry [12] allows one to show that this weak lift is indeed a strong lift in the sense that the local components Π_v at those $v \in S$ are completely determined—there is in fact no possible ambiguity. In conjunction with this they are able to completely characterize the image. Once one knows that these lifts are rigid, then one can begin to define a local lift by setting the lift of π_v to be the Π_v determined globally. This is the content of the papers of Jiang and Soudry [16, 17]. In essence they show that this local lift satisfies the relations on L-functions that one expects from functoriality and then uses this lift to pull back the parameter ϕ_{Π_v} of the local $\mathrm{GL}_N(k_v)$ representation, which we know exists by the local Langlands conjecture, to obtain a parameter ϕ_{π_v} of the correct type, that is, $\phi_{\pi_v} : W'_{k_v} \to {}^L H_v$ and thus deducing the local Langlands conjecture for $H(k_v)$. We refer you to their papers for more detail and precise statements. We expect similar results will be forthcoming for the other split classical groups.

REFERENCES

[1] J. ARTHUR AND L. CLOZEL, *Simple Algebras, Base Change, and the Advanced Theory of the Trace Formula*, Annals of Math. Studies **120**, Princeton Univ. Press, Princeton, 1989.

[2] A. Borel, Automorphic L-functions. *Proc. Symp. Pure Math.* **33**-2(1979), 27–61.

[3] J.W. Cogdell, Analytic theory of L-functions for GL_n, in J. Bernstein and S. Gelbart, eds., *An Introduction to the Langlands Program*, Birkhäuser Boston, Boston, 2003, 197–228 (this volume).

[4] J.W. Cogdell, Langlands conjectures for GL_n, in J. Bernstein and S. Gelbart, eds., *Introduction to the Langlands Program*, Birkhäuser Boston, Boston, 2003, 229–249 (this volume).

[5] J.W. Cogdell, H. Kim, I.I. Piatetski-Shapiro, and F. Shahidi, On lifting from classical groups to GL_n, *Publ. Math. IHES* **93**(2001), 5–30.

[6] J.W. Cogdell, H. Kim, I.I. Piatetski-Shapiro, and F. Shahidi, On lifting from classical groups to GL_n, II, in preparation.

[7] J.W. Cogdell and I.I. Piatetski-Shapiro, Converse theorems for GL_n, I, *Publ. Math. IHES* **79**(1994), 157–214.

[8] J.W. Cogdell and I.I. Piatetski-Shapiro, Stability of gamma factors for $\mathrm{SO}(2n+1)$, *Manuscripta Math.* **95**(1998) 437–461.

[9] J.W. Cogdell and I.I. Piatetski-Shapiro, Converse theorems for GL_n, II, *J. reine angew. Math.* **507**(1999), 165–188.

[10] P. Deligne, D. Kazhdan, and M-F. Vigneras, Représentations des algèbres centrales simples p-adiques. *Représentations des groupes réductifs sur un corps local*, J. Bernstein, P. Deligne, D. Kazhdan, and M-F. Vigneras, eds., Hermann, Paris, 1984, 33–117.

[11] S. Gelbart and H. Jacquet, A relation between automorphic representations of GL(2) and GL(3), *Ann. Sci. École Norm. Sup.* (4) **11**(1978), 471–542.

[12] D. Ginzburg, S. Rallis, and D. Soudry, Generic automorphic forms on $\mathrm{SO}(2n+1)$: functorial lift to $\mathrm{GL}(2n)$, endoscopy, and base change, *Internat. Math. Res. Notices* **2001**-14, 729–764.

[13] G. Henniart and R. Herb, Automorphic induction for $\mathrm{GL}(n)$ (over local non-Archimedean fields), *Duke Math. J.* **78**(1995), 131–192.

[14] H. Jacquet, I. Piatetski-Shapiro, and J. Shalika, Relèvement cubique non normal, *C. R. Acad. Sci. Paris Sér. I Math.* **292**(1981), 567–571.

[15] H. Jacquet and R. Langlands, *Automorphic Forms on* GL(2), Lecture Notes in Mathematics 114, Springer Verlag, Berlin, 1970.

[16] D. Jiang and D. Soudry, The local converse theorem for $SO(2n+1)$ and applications, *Annals of Math.*, **157**(2003), to appear.

[17] D. JIANG AND D. SOUDRY, Generic representations and local Langlands reciprocity law for p-adic SO_{2n+1}, *Contributions to Automorphic Forms, Geometry and Number Theory* (*Shalikafest* 2002), H. Hida, D. Ramakrishnan, and F. Shahidi, eds., Johns Hopkins University Press, Baltimore, to appear.

[18] D. KAZHDAN, On lifting, *Lie Group Representations* II, Lecture Notes in Mathematics 1041, Springer-Verlag, New York, 1984, 209–249.

[19] D. KAZHDAN AND G. LUSZTIG, Proof of the Deligne–Langlands conjecture for Hecke algebras, *Invent. Math.* **87**(1987), 153–215.

[20] H. KIM, Functoriality for the exterior square of GL_4 and the symmetric fourth of GL_2, preprint, 2000.

[21] H. KIM AND F. SHAHIDI, Functorial products for $GL_2 \times GL_3$ and functorial symmetric cube for GL_2, *Ann. of Math.* **155**(2002), 837–893.

[22] H. KIM AND F. SHAHIDI, Cuspidality of symmetric powers with applications, *Duke Math. J.* **112**(2002), 177–197.

[23] R. KOTTWITZ AND D. SHELSTAD, *Foundations of Twisted Endoscopy*, Astérisque **255**, SMF, Paris,1999.

[24] R.P. LANGLANDS, The representation theory of abelian algebraic groups, *Pacific J. Math.* **61**(1998) 231–250 (Olga Taussky-Todd special issue).

[25] R.P. LANGLANDS, *Euler Products*, Yale University Press, New Haven, 1971.

[26] R.P. LANGLANDS, On the classification of irreducible representations of real algebraic groups, *Representation Theory and Harmonic Analysis*, P. Sally and D. Vogan, eds., AMS, Providence, 1989, 101–170.

[27] R.P. LANGLANDS, *Base Change for* GL(2), Annals of Math. Studies **96**, Princeton University Press, Princeton, 1980.

[28] R.P. LANGLANDS, *Les Débuts d'une Formule des Traces Stable*, Publications Mathématiques de l'Université Paris VII 13, Université de Paris VII, U.E.R. de Mathématiques, Paris, 1983.

[29] R.P. LANGLANDS, Where stands functoriality today?, *Representation Theory and Automorphic Forms* (*Edinburgh*, 1996), Proc. Sympos. Pure Math. 61, 1997, 457–471.

[30] R.P. LANGLANDS, Endoscopy and beyond. in *Contributions to Automorphic Forms, Geometry, and Number Theory* (*Shalikafest* 2002), H. Hida, D. Ramakrishnan, and F. Shahidi, eds., Johns Hopkins University Press, Baltimore, to appear.

[31] D. RAMAKRISHNAN, Modularity of the Rankin–Selberg L-series, and multiplicity one for SL(2), *Ann. of Math.* (2) **152**(2000), 45–111.

[32] J. Rogawski, Representations of $GL(n)$ and division algebras over a p-adic field, *Duke Math. J.* **50**(1983), 161–196.

[33] I. Satake, Theory of spherical functions on reductive algebraic groups over $\mathfrak{p}$-adic fields, *Publ. Math. IHES* **18**(1963), 5–69.

[34] F. Shahidi, A proof of Langlands' conjecture on Plancherel measures; complementary series for p-adic groups, *Ann. of Math.* (2) **132**(1990), 273–330.

[35] F. Shahidi, Local coefficients as Mellin transforms of Bessel functions; towards a general stability, *IMRN* **39**(2002), 2075–2119.

[36] T. Springer, *Linear Algebraic Groups*, Progress in Mathematics Vol. 9, Birkhäuser, Boston, 1981.

12
Informal Introduction to Geometric Langlands

D. Gaitsgory*

Since we could not afford to give all the definitions necessary to introduce the subject, we will assume a certain level of familiarity with basic notions of algebraic geometry, number theory and representation theory. To compensate for the missing definitions, at the end we include some bibliographical suggestions.

1 Automorphic forms on GL_n over function fields

1.1

Let $\mathcal{K}$ be a global field of positive characteristic. That is, $\mathcal{K}$ is the field of rational functions over an algebraic curve X over a finite field $\mathbb{F}_q$, and we assume X to be smooth and projective and geometrically irreducible.

We will denote by $|X|$ the set of (closed) points of X, which are the same as valuations of $\mathcal{K}$. For a given point x, we will denote by $\mathcal{K}_x$ the corresponding local field. By $\mathcal{O}_x$ we will denote the ring of integers of $\mathcal{K}_x$, and by k_x the residue field.

Recall that the ring of adeles of K is the restricted product

$$\mathbb{A} := \prod_{x\in|X|}' \mathcal{K}_x.$$

1.2

In this chapter, we will restrict our attention to the algebraic group GL_n; let $GL_n(\mathbb{A})$ be the corresponding locally compact group. The theory of automorphic forms is concerned, roughly speaking, with the space of (complex valued) functions on the quotient $GL_n(\mathcal{K})\backslash GL_n(\mathbb{A})$. We will consider only smooth functions (a function is called smooth if it is invariant with respect to a open compact subgroup), and denote this space by

$$\mathrm{Funct}(GL_n(\mathcal{K})\backslash GL_n(\mathbb{A})).$$

*This chapter is based on the lecture given at the School, and it consists of three sections and an appendix.

By definition, the adele group $GL_n(\mathbb{A})$ acts on $\text{Funct}(GL_n(\mathcal{K})\backslash GL_n(\mathbb{A}))$ by right translations.

1.3

Recall that

$$\mathbb{O} := \prod_{x\in|X|} \mathcal{O}_x \subset \mathbb{A}$$

is called the ring of integral adeles. The corresponding subgroup $GL_n(\mathbb{O}) \subset GL_n(\mathbb{A})$ is a maximal compact subgroup. In what follows, we will fix a Haar measure μ on $GL_n(\mathbb{A})$ such that $\mu(GL_n(\mathbb{O})) = 1$.

By definition, the subspace of $GL_n(\mathbb{O})$-invariants in $\text{Funct}(GL_n(\mathcal{K})\backslash GL_n(\mathbb{A}))$ is the same as the space of functions on the double quotient

$$GL_n(\mathcal{K})\backslash GL_n(\mathbb{A})/GL_n(\mathbb{O}).$$

However, the subspace

$$\text{Funct}(GL_n(\mathcal{K})\backslash GL_n(\mathbb{A})/GL_n(\mathbb{O})) \subset \text{Funct}(GL_n(\mathcal{K})\backslash GL_n(\mathbb{A}))$$

is not stable under the action of $GL_n(\mathbb{A})$. But it carries a natural action of the Hecke algebra $\mathcal{H}(GL_n(\mathbb{A}), GL_n(\mathbb{O}))$.

1.4

First, let us recall the definition of the local Hecke algebra. For a point $x \in |X|$, i.e., a place of $\mathcal{K}$, the Hecke algebra $\mathcal{H}(GL_n(\mathcal{K}_x), GL_n(\mathcal{O}_x))$ is a vector space consisting of compactly supported $GL_n(\mathcal{O}_x)$-biinvariant functions on $GL_n(\mathcal{K}_x)$.

The algebra structure is given by the convolution product $f_1, f_2 \mapsto f_1 \star f_2$, where

$$f_1 \star f_2(g) = \int_{GL_n(\mathcal{K}_x)} f_1(g \cdot g_1) \cdot f_2(g_1^{-1})\, \mu(g_1),$$

where μ is a Haar measure with $\mu(GL_n(\mathcal{O}_x)) = 1$.

By construction, $\mathcal{H}(GL_n(\mathcal{K}_x), GL_n(\mathcal{O}_x))$ is an associative algebra with a unit given by the characteristic function of $GL_n(\mathcal{O}_x) \subset GL_n(\mathcal{K}_x)$.

Let π be a (complex) representation of the group $GL_n(\mathcal{K}_x)$ (resp., $GL_n(\mathbb{A})$), and $v \in \pi$ a vector. Recall that v is said to be smooth if it is invariant with respect to an open compact subgroup. A representation π is called smooth if all its vectors are smooth. From now on, unless specified otherwise, all representations of $GL_n(\mathcal{K}_x)$ (resp., $GL_n(\mathbb{A})$) will be assumed to be smooth.

1.5. Proposition. *If π is a representation of $G(\mathcal{K}_x)$, then the subspace of $G(\mathcal{O}_x)$-fixed vectors $\pi^{G(\mathcal{O}_x)}$ is naturally a representation of $\mathcal{H}(GL_n(\mathcal{K}_x), GL_n(\mathcal{O}_x))$. Moreover, if π is irreducible, then $\pi^{G(\mathcal{O}_x)}$ is an irreducible $\mathcal{H}(GL_n(\mathcal{K}_x), GL_n(\mathcal{O}_x))$-module or 0.*

Proof. For $v \in \pi$ and any locally constant compactly supported function f on $G(\mathcal{K}_x)$ we can consider the action

$$f \cdot v = \int_{G(\mathcal{K}_x)} f(g) \cdot g(v)\, \mu(g).$$

When $v \in \pi^{G(\mathcal{O}_x)}$ and $f \in \mathcal{H}(GL_n(\mathcal{K}_x), GL_n(\mathcal{O}_x))$, we obtain that $f \cdot v \in \pi^{G(\mathcal{O}_x)}$ and thus we obtain an $\mathcal{H}(GL_n(\mathcal{K}_x), GL_n(\mathcal{O}_x))$-action on $> \pi^{G(\mathcal{O}_x)}$.

Suppose now that π is irreducible. We must show that every $v \in \pi^{G(\mathcal{O}_x)}$ generates the latter vector space under the action of $\mathcal{H}(GL_n(\mathcal{K}_x), GL_n(\mathcal{O}_x))$. By assumption, any $v' \in V$ can be obtained as $f' \cdot v$ for some locally constant compactly supported function f'. However, if $v' \in \pi^{G(\mathcal{O}_x)}$,

$$f' \cdot v = f \cdot v,$$

where f is obtained from f' by *averaging* with respect to $G(\mathcal{O}_x)$ on the left. □

1.6

Recall that an irreducible $G(\mathcal{K}_x)$-representation π is called spherical if $\pi^{G(\mathcal{O}_x)} \neq 0$. One can show (easily) that the above assignment $\pi \mapsto \pi^{G(\mathcal{O}_x)}$ establishes a bijection between the set of isomorphism classes of irreducible spherical representations of $G(\mathcal{K}_x)$ and irreducible $\mathcal{H}(GL_n(\mathcal{K}_x), GL_n(\mathcal{O}_x))$-modules.

Now, a fundamental theorem due to Gelfand combined with a result of Satake (both are fairly easy for GL_n) allows one to describe the algebra $\mathcal{H}(GL_n(\mathcal{K}_x), GL_n(\mathcal{O}_x))$ very explicitly.

First, it is known that $\mathcal{H}(GL_n(\mathcal{K}_x), GL_n(\mathcal{O}_x))$ is in fact commutative. Secondly, it is almost a polynomial algebra. More precisely,

$$\mathcal{H}(GL_n(\mathcal{K}_x), GL_n(\mathcal{O}_x)) = \mathbb{C}[T_x^1, \ldots, T_x^{n-1}, T_x^n, (T_x^n)^{-1}],$$

where $T_x^i \in \mathcal{H}(GL_n(\mathcal{K}_x), GL_n(\mathcal{O}_x))$ equals the characteristic function of the $GL_n(\mathcal{O}_x)$-double coset of the diagonal $n \times n$ matrix

$$(\underbrace{\varpi_x, \ldots, \varpi_x}_{i \text{ times}}, 1, \ldots, 1),$$

and $\varpi_x \in \mathcal{O}_x$ is the uniformizer.[1]

In particular, we see that every irreducible $\mathcal{H}(GL_n(\mathcal{K}_x), GL_n(\mathcal{O}_x))$-module is 1-dimensional, and hence for every spherical $G(\mathcal{K}_x)$-representation π, the subspace $\pi^{G(\mathcal{O}_x)}$ is 1-dimensional too.

Moreover, the set of isomorphism classes of irreducible $\mathcal{H}(GL_n(\mathcal{K}_x), GL_n(\mathcal{O}_x))$-modules can be identified with $\mathrm{Spec}(\mathbb{C}[T_x^1, \ldots, T_x^{n-1}, T_x^n, (T_x^n)^{-1}])$, which in turn can be recognized as the set of *semisimple conjugacy classes* in the group $GL_n(\mathbb{C})$.

[1] In fact, one has to normalize T_x^i by multiplying it by $(-|k_x|^{1/2})^{i(n-i)}$.

1.7

Now let us pass to the global Hecke algebra, $\mathcal{H}(GL_n(\mathbb{A}), GL_n(\mathbb{O}))$. By definition,

$$\mathcal{H}(GL_n(\mathbb{A}), GL_n(\mathbb{O})) = \bigotimes_{X\in|X|}' \mathcal{H}(GL_n(\mathcal{K}_x), GL_n(\mathcal{O}_x)),$$

where the restricted tensor product is taken in the sense that we consider finite linear combinations of elements $\underset{X\in|X|}{\otimes} f_x$, such that for almost all x, f_x is the unit element in the local Hecke algebra $\mathcal{H}(GL_n(\mathcal{K}_x), GL_n(\mathcal{O}_x))$.

Note that every irreducible $GL_n(\mathbb{A})$-representation π is the *restricted* tensor product

$$\pi \simeq \bigotimes_{X\in|X|}' \pi_x,$$

where almost all π_x are spherical. Again, we call an irreducible $GL_n(\mathbb{A})$-representation π spherical if $\pi^{GL_n(\mathbb{O})} \neq 0$, which is equivalent to the fact that in the above tensor product all the π_x are spherical.

Thus, given an irreducible spherical representation π of $GL_n(\mathbb{A})$, we can assign to it a collection $\{\gamma_x\}$, $\forall x \in |X|$ of semisimple conjugacy classes in $GL_n(\mathbb{C})$.

1.8

Now we return to the automorphic situation, i.e., to the quotient $GL_n(\mathcal{K})\backslash GL_n(\mathbb{A})$. Recall that a function f on the above quotient is called cuspidal if

$$\int_{U(\mathcal{K})\backslash U(\mathbb{A})} f(u \cdot g)\, \mu(u) = 0$$

for all $g \in GL_n(\mathbb{A})$ and all subgroups U, which are unipotent radicals of the *standard parabolic subgroups* of GL_n.

We will denote the subspace of cuspidal functions by

$$\mathrm{Funct}_{cusp}(GL_n(\mathcal{K})\backslash GL_n(\mathbb{A})).$$

By definition, it is invariant with respect to the $GL_n(\mathbb{A})$-action. One can show that every element in $\mathrm{Funct}_{cusp}(GL_n(\mathcal{K})\backslash GL_n(\mathbb{A}))$ is automatically compactly supported.

Recall that an irreducible representation π of $GL_n(\mathbb{A})$ is called cuspidal automorphic if

$$\mathrm{Hom}_{GL_n(\mathbb{A})}(\pi, \mathrm{Funct}_{cusp}(GL_n(\mathcal{K})\backslash GL_n(\mathbb{A}))) \neq 0.$$

An important result (the multiplicity 1 theorem) says that the above Hom is at most 1-dimensional. Moreover, the entire space $\mathrm{Funct}_{cusp}(GL_n(\mathcal{K})\backslash GL_n(\mathbb{A}))$ splits as a direct sum (and not a direct integral) of irreducible (and hence cuspidal automorphic) representations.

In what follows we shall study spherical cuspidal automorphic representations of $GL_n(\mathbb{A})$. From the above discussion it follows that our problem is equivalent to describing the decomposition of the space $\text{Funct}_{cusp}(GL_n(\mathcal{K})\backslash GL_n(\mathbb{A})/GL_n(\mathbb{O}))$ into a direct sum of common eigenspaces of the operators T_x^i for $i = 1, \ldots, n$ and all $x \in |X|$. We know already that such a decomposition indeed exists and has a simple spectrum, i.e., each set of eigenvalues corresponding to

$$\{\gamma_x \in GL_n(\mathbb{C})^{ss}/\text{Conj}\}, \ \forall x \in |X|$$

enters with multiplicity at most 1.

An additional important observation is that the harmonic analysis of the $GL_n(\mathbb{A})$-action on $\text{Funct}_{cusp}(GL_n(\mathcal{K})\backslash GL_n(\mathbb{A})$ is completely algebraic. This means that the topology of the field $\mathbb{C}$, which was where all our functions took values, is irrelevant. In particular, we can replace it by any other field of 0 characteristic, which we will take to be algebraically closed in order to have eigenspace decompositions. Our choice will be $\overline{\mathbb{Q}}_\ell$, where $(\ell, q) = 1$.

1.9

Let now σ be an n-dimensional irreducible $\overline{\mathbb{Q}}_\ell$-representation of the Galois group of K. We will assume that σ is everywhere unramified. That is, it can be regarded as a representation of the *étale fundamental group* $\pi_1(X) \to GL_n(\overline{\mathbb{Q}}_\ell)$. By definition, such a representation is the same as a *lisse* ℓ-adic sheaf (aka local system) on X that we will denote by E_σ.

For every point $x \in |X|$, we have a well-defined conjugacy class Fr_x (corresponding to the *Frobenius element*) in $\pi_1(X)$. We will denote by $\sigma_x \in GL_n(\overline{\mathbb{Q}}_\ell)^{ss}/\text{Conj}$ the *semisimple* part of $\sigma(\text{Fr}_x)$. (It is known that if σ is irreducible, the collection of $(\sigma_x, x \in |X|)$ determines it uniquely.) In terms of the local system E_σ, σ_x corresponds to the action of the Frobenius element in $\text{Gal}(\overline{k}_x/k_x)$ on the fiber of E_σ at x.

Now we come to the crucial definition: we say that an n-dimensional unramified irreducible Galois representation σ corresponds in the sense of Langlands to a cuspidal automorphic representation π of $GL_n(\mathbb{A})$ if $\forall x \in |X|$,

$$\sigma_x = \gamma_x,$$

where $\gamma_x \in GL_n(\overline{\mathbb{Q}}_\ell)^{ss}/\text{Conj}$ is the corresponding point in $\text{Spec}(\overline{\mathbb{Q}}_\ell[T_x^1, \ldots, T_x^{n-1}, T_x^n, (T_x^n)^{-1}])$ attached to π_x.

The Langlands conjecture, which is now a theorem due to Lafforgue, says that there is a bijection between the set of σ's and π's as above, determined by this correspondence relation.

2 Interpretation via vector bundles

2.1

A key observation for us is that the set $GL_n(\mathcal{K})\backslash GL_n(\mathbb{A})/GL_n(\mathbb{O})$ can be reinterpreted in geometric terms:

Let us denote by Bun_n the set of isomorphism classes of rank n vector bundles on our curve X. We claim that there is a natural bijection between Bun_n and $GL_n(\mathcal{K})\backslash GL_n(\mathbb{A})/GL_n(\mathbb{O})$.

Indeed, given a vector bundle $\mathcal{M}$, locally in the Zariski topology it is isomorphic to the trivial bundle $> \mathcal{O}_X^{\otimes n}$. Let us choose its trivialization $\mathcal{M} \underset{\mathcal{O}_X}{\otimes} \mathcal{K} \simeq \mathcal{K}^n$ at the generic point of X and also trivializations at all local places $\mathcal{M} \underset{\mathcal{O}_X}{\otimes} \mathcal{O}_x \simeq \mathcal{O}_x^n$.

By composing, for each $x \in |X|$ we obtain

$$\mathcal{K}_x^n \simeq \mathcal{O}_x^n \underset{\mathcal{O}_x}{\otimes} \mathcal{K}_x \simeq \mathcal{M} \underset{\mathcal{O}_X}{\otimes} \mathcal{O}_x \underset{\mathcal{O}_x}{\otimes} \mathcal{K}_x \simeq \mathcal{M} \underset{\mathcal{O}_X}{\otimes} \mathcal{K} \underset{\mathcal{K}}{\otimes} \mathcal{K}_x \simeq \mathcal{K}^n \underset{\mathcal{K}}{\otimes} \mathcal{K}_x \simeq \mathcal{K}_x^n,$$

that is an element of $GL_n(\mathbb{A})$. The effect of changing the above trivializations is the multiplication by an element of $GL_n(\mathcal{K})$ on the left and by an element of $\prod\limits_{x\in|X|} GL_n(\mathcal{O}_x) = GL_n(\mathbb{O})$ on the right.

2.2 *Example*

Note that for $n = 1$, $GL_1(\mathbb{A})/GL_1(\mathbb{O})$ is the set of divisors on X. Therefore, the double quotient $GL_1(\mathcal{K})\backslash GL_1(\mathbb{A})/GL_1(\mathbb{O})$ is the set of divisor classes (i.e., divisors modulo the principal ones). And the above bijection reestablishes the well-known fact that the group of divisor classes is the same as the group of (isomorphism classes of) line bundles.

2.3

Now let us explain how to view the Hecke operators in geometric terms. We fix a point $x \in X$ and an integer i, $1 \leq i \leq n$.

Let us denote by H_x^i the set of isomorphism classes of the following data: $(\mathcal{M}, \mathcal{M}', \beta)$, where $\mathcal{M}$ and $\mathcal{M}'$ are rank n vector bundles on X and β is an embedding of $\mathcal{M}$ into $\mathcal{M}'$, viewed as coherent sheaves on X, such that the quotient $\mathcal{M}'/\mathcal{M}$ has length i and is scheme-theoretically supported at x. (The latter condition is equivalent to the fact that $\mathcal{M}'/\mathcal{M}$ is noncanonically isomorphic to k_x^i, where k_x is the residue field at x.)

There are two natural maps, call them $h^{\leftarrow}$ and $h^{\rightarrow}$ from H_x^i to Bun_n, which send a triple as above to $\mathcal{M}$ and $\mathcal{M}'$, respectively.

2.4. Lemma. *For a given $\mathcal{M} \in \mathrm{Bun}_n$, the fibers of $h^{\leftarrow}$ over it can be canonically identified with the set of dimension i subspaces inside the vector space $\mathcal{M}_x$ over k_x. In other words, $(h^{\leftarrow})^{-1}(\mathcal{M})$ is the set of k_x-points of the Grassmannian $\mathrm{Gr}^i(\mathcal{M}_x)$.*

The same is true for $h^{\rightarrow}$, in which case we obtain that $(h^{\rightarrow})^{-1}(\mathcal{M}')$ is the space of k_x-points in $\mathrm{Gr}^{n-i}(\mathcal{M}'_x)$.

Proof. Let $(\mathcal{M}, \mathcal{M}', \beta)$ be a triple as above. Although β is an injection of coherent sheaves, when we pass to fibers at x, it is no longer injective and we obtain an exact sequence

$$0 \to \ker(\beta|_x) \to \mathcal{M}_x \to \mathcal{M}'_x \to (\mathcal{M}/\mathcal{M}')_x \to 0.$$

In particular, $\ker(\beta|_x)$ is a subspace of dimension i of $\mathcal{M}_x$.

Conversely, given a subspace $V \subset \mathcal{M}_x$, we define $\mathcal{M}'$ to be the subsheaf of $\mathcal{M}(x)$, whose local sections are of the form $t^{-1} \cdot m$, where t is the local parameter on X at x, and m is a local section of $\mathcal{M}$ whose value at x belongs to V. □

2.5

Note that since k_x is a finite field, the fibers of both $h^{\leftarrow}$ and $h^{\rightarrow}$ are finite sets.

Consider the operator from the space $\mathrm{Funct}(\mathrm{Bun}_n)$ to itself given by $f \mapsto (h^{\rightarrow})_!((h^{\leftarrow})^*(f))$. (Here the superscript $*$ means pull-back of a function with respect to the specified morphism, and the subscript ! means summation along the fibers; compare also with the notation introduced in the appendix.)

$$\mathrm{Bun}_n \xleftarrow{h^{\leftarrow}} \mathrm{H}^i_x \xrightarrow{h^{\rightarrow}} \mathrm{Bun}_n.$$

2.6. Proposition. *The above operator equals $T^i_x(f)$.*

2.7 *Digression*

Let us observe the following analogy with the case of modular forms. Let N be a natural number and p a prime with $(N, p) = 1$. We will study functions (or rather sections of certain bundles) on the modular curve $X(N)$ (it has nothing to do with our curve X).

By definition, $X(N)$ classifies pairs (E, α), where E is an elliptic curve (over $\mathbb{C}$) and α is an isomorphism between E^N and $(\mathbb{Z}/N \cdot \mathbb{Z})^2$ (here E^N is the group of points of order N on E).

Consider the Hecke correspondence at p. This is a curve, let us denote it by H_p, that classifies quintuples $(E, \alpha, E', \alpha', \beta)$, where (E, α) and (E', α') are two points of $X(N)$ and β is an isogeny $E \to E'$ such that $\ker(\beta)$ is of order p. Since N and p are co-prime, β induces an isomorphism $E^N \to E'^N$ and we require that it commutes with the identification of both groups with $(\mathbb{Z}/N \cdot \mathbb{Z})^2$.

As before, we have the natural maps

$$X(N) \xleftarrow{h^{\leftarrow}} \mathrm{H}_p \xrightarrow{h^{\rightarrow}} X(N)$$

that define an operator $f \mapsto (h^{\rightarrow})_!((h^{\leftarrow})^*(f))$, which coincides with *the Hecke operator* acting on modular forms.

2.8

Going back to the geometric situation, let us spell out the cuspidality condition in geometric terms. Let $n = n_1 + n_2$ be a partition. Let us consider the set $\mathrm{Fl}^n_{n_1,n_2}$ of isomorphism classes of short exact sequences

$$0 \to \mathcal{M}_1 \to \mathcal{M} \to \mathcal{M}_2 \to 0,$$

where $\mathcal{M}_1$ and $\mathcal{M}_2$ are vector bundles of ranks n_1 and n_2, respectively.

We have natural maps p and q

$$\mathrm{Bun}_n \xleftarrow{p} \mathrm{Fl}^n_{n_1,n_2} \xrightarrow{q} \mathrm{Bun}_{n_1} \times \mathrm{Bun}_{n_2},$$

which send a short exact sequence as above to $\mathcal{M}$ and $(\mathcal{M}_1, \mathcal{M}_2)$, respectively.

We define the constant term operator $r^n_{n_1,n_2} : \mathrm{Funct}(\mathrm{Bun}_n) \to \mathrm{Funct}(\mathrm{Bun}_{n_1} \times \mathrm{Bun}_{n_2})$ by

$$f \mapsto q_!(p^*(f)).$$

It is easy to see now that a function f is cuspidal if and only if $r^n_{n_1,n_2}(f) = 0$ for all $n_1 + n_2 = n$ with $n_1, n_2 > 0$.

3 From functions to sheaves

3.1

Our next step is to view Bun_n not just as a set, but as a set of rational points of an algebraic variety defined over $\mathbb{F}_q$, and consider ℓ-adic sheaves on it (cf. Section 4 for a brief review of ℓ-adic sheaves).

A slight complication is that the geometric object standing behind Bun_n is not a variety, but an algebraic stack (cf. [L-MB]), denoted $\mathcal{B}un_n$. The main difference between stacks and scheme is the following:

According to Yoneda, an object of a category (in our case, a scheme $\mathcal{Y}$) is completely determined by the functor Schemes $\Rightarrow$ Sets given by $S \mapsto \mathrm{Hom}(S, \mathcal{Y})$.

In the case of $\mathcal{B}un_n$, the above Hom is indeed defined, but it is no longer a set, but rather a category. This could be seen already when we talked about Bun_n: it is much more natural to consider the *category* of vector bundles on X than *the set of isomorphism classes* of vector bundles.

In general, for a scheme S we define $\mathrm{Hom}(S, \mathcal{B}un_n)$ to be the category, whose objects are rank n-bundles on $X \times S$ and morphisms are isomorphisms between vector bundles. It turns out that $\mathcal{B}un_n$ defined in this way is indeed an *algebraic stack.*

However, from the point of view of ℓ-adic sheaves, stacks behave as if they were algebraic varieties. In particular, if $\mathcal{Y}$ is a stack, the category $D^b(\mathcal{Y})$ makes sense (cf. Section 4). Moreover, if Y is *the set of isomorphism classes of objects* in the category $\mathrm{Hom}(\mathrm{Spec}(\mathbb{F}_q), \mathcal{Y})$, every object $\mathcal{F}$ of $D^b(\mathcal{Y})$ defines a function f on Y, and all the properties of the sheaf-function correspondence hold.

In particular, the category $D^b(\mathcal{B}un_n)$ makes sense, and every object $\mathcal{F} \in D^b(\mathcal{B}un_n)$ gives rise to a function f on Bun_n.

3.2

Let us now discuss what happens to the Hecke operators in our geometric context.

First, let $x \in |X|$ be fixed. We consider the Hecke stack $\mathcal{H}^i_x$, such that $\mathrm{Hom}(S, \mathcal{H}^i_x)$ is the category of triples $(\mathcal{M}, \mathcal{M}', \beta)$, where $\mathcal{M}$ and $\mathcal{M}'$ are rank n vector bundles on $X \times S$ and β is an embedding $\mathcal{M} \hookrightarrow \mathcal{M}'$ of coherent sheaves such that $\mathcal{M}'/\mathcal{M}$ is scheme-theoretically supported on $x \times S \subset X \times S$ and is a rank i vector bundle on this subscheme.

The set of isomorphism classes of $\mathbb{F}_q$-points of $\mathcal{H}^i_x$ is our old H^i_x. As before, we have the maps $h^{\leftarrow}, h^{\rightarrow} : \mathcal{H}^i_x \to \mathcal{B}un_n$. However, now for a given point of $\mathcal{B}un_n$, i.e., a vector bundle $\mathcal{M}$, the preimage $(h^{\leftarrow})^{-1} \subset \mathcal{H}^i_x$ is not a set, but a scheme isomorphic to $\mathrm{Gr}^i(\mathcal{M}_x)$.

We introduce the *Hecke functors* $T^i_x : D^b(\mathcal{B}un_n) \to D^b(\mathcal{B}un_n)$ by

$$\mathcal{F} \mapsto (h^{\rightarrow})_!((h^{\leftarrow})^*(\mathcal{F}))[i(n-i)],$$

where the expression in the parenthesis means cohomological shift.

3.3

However, we can do better by letting the point x move along the curve. We introduce the "big" Hecke stack $\mathcal{H}^i$ as follows:

$\mathrm{Hom}(S, \mathcal{H}^i)$ is the category of quadruples $(\phi, \mathcal{M}, \mathcal{M}', \beta)$, where $\mathcal{M}, \mathcal{M}'$ are as before, ϕ is a map $S \to X$ and β is once again an embedding of coherent sheaves $\mathcal{M} \hookrightarrow \mathcal{M}'$, such that the quotient is supported on the graph $\Gamma_\phi \subset X \times S$ and is a rank i vector bundle on it.

In addition to the projections $h^{\leftarrow}, h^{\rightarrow} : \mathcal{H}^i \to \mathcal{B}un_n$ as before, we have now a natural map $s : \mathcal{H}^i \to X$ which sends a quadruple $(\phi, \mathcal{M}, \mathcal{M}', \beta)$ as above to ϕ. For a fixed $x \in X$, we have $s^{-1}(x) \simeq \mathcal{H}^i_x$.

We define the Hecke functor $T^i : D^b(\mathcal{B}un_n) \to D^b(X \times \mathcal{B}un_n)$ by

$$\mathcal{F} \mapsto (s \times h^{\rightarrow})_!((h^{\leftarrow})^*(\mathcal{F}))[i(n-i)].$$

3.4

We are now ready to give the definition of Hecke eigen-sheaves. Our definition will be slightly crude (a better one, valid for an arbitrary group can be found in [BD]).

Let $\sigma : \pi_1(X) \to GL_n(\overline{\mathbb{Q}}_\ell)$ be a representation, and recall that by E_σ we denote the corresponding n-dimensional local system on X.

An object $\mathcal{F}_\sigma \in D^b(\mathcal{B}un_n)$ is called a Hecke eigen-sheaf with respect to σ if for every $i = 1, \ldots, n$ there exists an isomorphism

$$T^i(\mathcal{F}_\sigma) \simeq \Lambda^i(E_\sigma) \boxtimes \mathcal{F}_\sigma,$$

where Λ^i denotes the i-th exterior power.

If our base field $k = \mathbb{F}_q$, and $\mathcal{F}_\sigma$ is a Hecke eigen-sheaf with respect to σ, then the corresponding function on $GL_n(K)\backslash GL_n(\mathbb{A})/GL_n(\mathbb{O})$ is a common eigenfunction for Hecke operators with the eigenvalues given by σ, as in Section 1.9.

The Geometric Langlands conjecture (suggested by [L] following Drinfeld's original paper, which did completely the case of GL_2, [Dr]) predicts that for each σ as above, which is irreducible, there indeed exists a Hecke eigen-sheaf $\mathcal{F}_\sigma$, which is moreover cuspidal (in the sense that the functors $r^n_{n_1,n_2}$ defined as in Section 2.8 vanish on $\mathcal{F}_\sigma$).

3.5

To conclude, let us address the following question. Why did we limit our discussion to the spherical (=unramified) situation?

The reason is the following: as was explained earlier, to a *spherical* cuspidal automorphic representation, there corresponds a distinguished function f on $GL_n(K)\backslash GL_n(\mathbb{A})/GL_n(\mathbb{O})$, defined uniquely up to a scalar. Hence, geometrically, we wanted to construct a specific object in $D^b(\mathcal{B}un_n)$, which for $k = \mathbb{F}_q$ gives rise to this function.

For a general cuspidal automorphic representation π, it is difficult to find a canonical choice of a function lying inside $\pi \subset \mathrm{Funct}_{cusp}(GL_n(K)\backslash GL_n(\mathbb{A}))$.[2] Therefore, it is not clear what sheaves on the corresponding version of $\mathcal{B}un_n$ one would want to construct.

4 Appendix: the sheaf-function correspondence

4.1

Here we will recall several facts about Grothendieck's sheaf-functions correspondence.

First, if $\mathcal{Y}$ is a scheme (over an arbitrary base field), one can define the category $D^b(\mathcal{Y})$, called the derived category of ℓ-adic sheaves with constructible cohomologies.

If $\phi : \mathcal{Y}_1 \to \mathcal{Y}_2$ is a morphism between schemes, there are the pull-back and direct image functors

[2] Since our group is GL_n such a choice exists, namely, the so-called newforms, but this does not seem so natural from the geometric point of view.

$$\phi^* : D^b(\mathcal{Y}_2) \to D^b(\mathcal{Y}_1)$$
$$\phi_! : D^b(\mathcal{Y}_1) \to D^b(\mathcal{Y}_2)$$

If $\mathcal{F}_1$ and $\mathcal{F}_2$ are two objects of $D^b(\mathcal{Y})$, one can define their tensor product $\mathcal{F}_1 \otimes \mathcal{F}_2 \in D^b(\mathcal{Y})$.

4.2

Now let us suppose that $\mathcal{Y}$ is over the base field $\mathbb{F}_q$. We will denote by Y the corresponding set of $\mathbb{F}_q$-valued points of $\mathcal{Y}$ (which is a finite set) and by $\text{Funct}(Y)$ the vector space of $\overline{\mathbb{Q}}_\ell$-valued functions on it. Every object $\mathcal{F}$ of $D^b(\mathcal{Y})$ gives rise to an element $f \in \text{Funct}(Y)$:

First, if $\mathcal{Y}$ is a point-scheme $\text{Spec}(\mathbb{F}_q)$, an object of $D^b(\text{Spec}(\mathbb{F}_q))$ is the same as a complex of finite-dimensional $\overline{\mathbb{Q}}_\ell$-vector spaces, acted on by $\text{Gal}(\overline{\mathbb{F}}_q/\mathbb{F}_q)$. The corresponding function, which in this case amounts to just one element of $\overline{\mathbb{Q}}_\ell$, is the alternating sum of the traces of the Frobenius element in $\text{Gal}(\overline{\mathbb{F}}_q/\mathbb{F}_q)$. For general $\mathcal{Y}$, $\mathcal{F} \in D^b(\mathcal{Y})$ and $y \in Y$, the value of f at y equals the number assigned above to $\mathcal{F}_y$, where $\mathcal{F}_y \in D^b(\text{Spec}(\mathbb{F}_q))$ is the pull-back of $\mathcal{F}$ under the morphism $\text{Spec}(\mathbb{F}_q) \to \mathcal{Y}$ corresponding to the point y.

The basic properties of the assignment $\mathcal{F} \mapsto f$ are as follows: The tensor product $\mathcal{F}_1 \otimes \mathcal{F}_2$ goes over to $f_1 \cdot f_2$ (pointwise product). For a map $\phi : \mathcal{Y}_1 \to \mathcal{Y}_2$, the pull-back $\phi^*(\mathcal{F})$ goes to $\phi^*(f)$ and similarly for the push-forward. (As far as the proofs are concerned, the first two properties are immediate from the definitions, whereas the third one is known as the Grothendieck–Lefschetz trace formula).

4.3

It turns out that in many cases "interesting" functions defined on sets of the form Y for $\mathcal{Y}/\mathbb{F}_q$, come from ℓ-adic sheaves in the above way. In certain cases, one can go backwards. Here is an example.

Let $\mathcal{A}$ be a commutative connected algebraic group. An object $\mathcal{X} \in D^b(H)$ is called a character sheaf if it is a 1-*dimensional local system*, its fiber over the unit element on $\mathcal{A}$ is trivialized and $m^*(\mathcal{X}) \simeq \mathcal{X} \boxtimes \mathcal{X}$, where $m : \mathcal{A} \times \mathcal{A} \to \mathcal{A}$ is the multiplication map.

In this case, the corresponding function χ on the finite group $A = \mathcal{A}(\mathbb{F}_q)$ is a character.

Vice versa, given a character χ of A, one can reconstruct $\mathcal{X}$ as follows. Consider the Lang isogeny $L : \mathcal{A} \to \mathcal{A}$ given by

$$a \to \text{Fr}(a)/a,$$

where $\text{Fr} : \mathcal{A} \to \mathcal{A}$ is the Frobenius endomorphism. Then L is an *étale* Galois cover, with the group of covering transformations equal to $\ker(L) = A$.

The sought-for character sheaf $\mathcal{X}$ is defined as the ℓ-adic sheaf corresponding to the 1-dimensional representation of $\pi_1(\mathcal{A})$ given by

$$\pi_1(\mathcal{A}) \to \ker(L) = A \xrightarrow{\chi} \overline{\mathbb{Q}}_\ell^*.$$

By applying this construction to the case when $\mathcal{A}$ is the affine line $\mathbb{A}^1$, we obtain an ℓ-adic sheaf on $\mathbb{A}^1$ known as the Artin–Schreier sheaf. Similarly, when $\mathcal{A}$ is the multiplicative group $\mathbb{G}_m$ we obtain Kummer's sheaf on it.

Another example is $\mathcal{A} = \mathrm{Pic}^0(X)$, we recover the construction involved in Lang's proof of global class field theory over function fields; cf. [Se].

Bibliography

In addition to references to research papers (cf. below), here are some sources for the background material.

For algebraic geometry (schemes, coherent sheaves, etc.): Hartshorne, *Algebraic Geometry*; Kempf, *Algerbraic Varieties*; Mumford, *The Red Book of Varieties and Schemes.*

For local rings and fields, adeles, Frobenius elements and other notions of algebraic number theory: Cassels–Frohlich, *Proceedings of the Instructional Conference on Algebraic Number Theory*; Serre, *Corps locaux* and *Groupes algébriques et corps de classes.*

For representations of $\mathfrak{p}$-adic groups and introduction to the theory of automorphic functions: Gelfand–Graev–Piatetskii–Shapiro, *Representation Theory and Automorphic Functions* and Gelbart, *Automorphic Forms on Adèle Groups.* A (highly recommended) advanced course on representations of $\mathfrak{p}$-adic groups is [Be].

For étale sheaves: Milne, *Etale Cohomology.*

References

[Be] J. Bernstein, *Lectures on Representations of p-Adic Groups*, preprint, School of Mathematical Sciences, Tel Aviv University, Tel Aviv, Israel; available online from www.math.tau.ac.il/~bernstei.

[BD] A. Beilinson and V. Drinfeld, *Quantization of Hitchin's Integrable System and Hecke Eigen-Sheaves*, preprint; available online from www.math.uchicago.edu/~benzvi.

[De] P. Deligne, La conjecture de Weil II, *Publ. IHES*, **52**(1981), 313–428.

[Dr] V. Drinfeld, Two-dimensional ℓ-adic representations of the fundamental group of a curve over a finite field and automorphic forms on $GL(2)$, *Amer. J. Math.*, **105**(1983), 85–114.

[L] G. Laumon, Correspondance de Langlands géométrique pour les corps de fonctions, *Duke Math. J.*, **54**(1987), 309–359.

[L-MB] G. Laumon and L. Morret-Bailly, *Champs algébriques*, Springer-Verlag, New York, 2000.

[Mi] J. Milne, *Etale Cohomology*, Princeton Mathematical Series 33, Princeton University Press, Princeton, NJ, 1980.

[Se] J.-P. Serre, *Algebraic Groups and Class Fields*, Graduate Texts in Mathematics 117, Springer-Verlag, New York, 1998.